Bird Monitoring Methods

a manual of techniques for key UK species

Gillian Gilbert, David W Gibbons and Julianne Evans

Published by the RSPB in association with British Trust for Ornithology, The Wildfowl & Wetlands Trust, Joint Nature Conservation Committee, Institute of Terrestrial Ecology and The Seabird Group

This reprint edition published by Pelagic Publishing, 2011
www.pelagicpublishing.com

Copy-editing, text layout and graphics by Duncan Brooks

ISBN-13 978-1-907807-22-0

This book is a reprint edition of ISBN-10 1-901930-03-3

Cover photographs (top to bottom and left to right)
Surveyor in the field - *Andrew Hay (RSPB Images)*
Gannets - *Bill Paton (RSPB Images)*
Hawfinch - *Mark Hamblin (RSPB Images)*
Nightjar - *M W Richards (RSPB Images)*
Kestrel - *C H Gomersall (RSPB Images)*
Black-throated diver - *C H Gomersall (RSPB Images)*
Oystercatchers - *C H Gomersall (RSPB Images)*
Background image - *R Revels (RSPB Images)*

Contents

Generic survey methods

Acknowledgements

The experience and knowledge of numerous individual consultants has been used in compiling this manual and we have tried to acknowledge this help wherever possible. The sources of previously published methods are given in the reference lists within each account. Where an account has been entirely written by individuals other than the authors of this manual, they are acknowledged at the foot of the relevant text. A huge debt of gratitude is owed to the authors of the *Seabird Monitoring Handbook* (Walsh et al 1995), as all methods for seabird species marked [1] below, originate from this handbook. Draft accounts were circulated for comments to relevant experts, who are listed below. Despite this consultation process, full responsibility for the methods and any errors in the final text remains with the authors of the manual.

Single-species methods

Detailed comments on the texts for individual species accounts were provided as follows:

Species	Consultants
Red-throated diver	Ian Bainbridge, Peter Ellis (both RSPB)
Black-throated diver	Mark Hancock, Roger Broad, Digger Jackson (all RSPB)
Great northern diver	Peter Cranswick (WWT)
Slavonian grebe	Roy Dennis, Colin Crooke (RSPB)
Black-necked grebe	Ken Shaw (RSPB)
Manx shearwater [1]	Kate Thompson (JNCC), Emma Brindley (RSPB)
Storm petrel	Norman Ratcliffe (RSPB)
Leach's petrel	Norman Ratcliffe (RSPB), Kate Thompson (JNCC)
Gannet [1]	Kate Thompson (JNCC), Emma Brindley (RSPB)
Cormorant	Baz Hughes, Robin Sellers, Graham Ekins (all WWT)
Shag [1]	Kate Thompson (JNCC), Emma Brindley (RSPB)
Bittern	Gillian Gilbert, Glen Tyler (both RSPB)
Mute swan	Jeff Kirby (WWT)
Bewick's swan	Ian Francis (RSPB), John Bowler, Carl Mitchell (both WWT)
Whooper swan	Ian Francis (RSPB), John Bowler, Carl Mitchell (both WWT)
Bean goose	Ian Francis (RSPB), John Bowler, Carl Mitchell (both WWT)
Pink-footed goose	Ian Francis (RSPB), John Bowler, Carl Mitchell (both WWT)
White-fronted goose	Ian Francis (RSPB), John Bowler, Carl Mitchell (both WWT)
Greylag goose	Ian Francis (RSPB), John Bowler, Carl Mitchell (both WWT)
Barnacle goose	Ian Francis (RSPB), John Bowler, Carl Mitchell (both WWT)
Brent goose	Ian Francis (RSPB), John Bowler, Carl Mitchell (both WWT)
Shelduck	Ian Patterson, Baz Hughes (WWT)
Wigeon	Jeff Kirby (WWT), Dave Beaumont (RSPB)
Gadwall	Jeff Kirby (WWT), Dave Beaumont (RSPB)
Teal	Jeff Kirby (WWT), Dave Beaumont, Iolo Williams (both RSPB)
Pintail	Jeff Stenning, Lennox Campbell, Richard Evans (all RSPB)
Garganey	Jeff Kirby (WWT), Dave Beaumont (RSPB)
Shoveler	Jeff Kirby (WWT), Dave Beaumont (RSPB)
Pochard	Jeff Kirby (WWT), Dave Beaumont (RSPB)
Scaup	Jeff Stenning, Lennox Campbell, Richard Evans (all RSPB), Peter Cranswick (WWT)
Eider	Mark Pollitt, Peter Cranswick (both WWT), Ian Patterson
Long-tailed duck	Jeff Stenning, Lennox Campbell, Richard Evans (all RSPB), Peter Cranswick (WWT)
Common scoter	Mark Underhill, Baz Hughes, Peter Cranswick (all WWT), Mark Hancock, Dave Allen, Norman Holton (all RSPB)
Velvet scoter	Mark Underhill, Baz Hughes, Peter Cranswick (all WWT), Mark Hancock, Dave Allen, Norman Holton (all RSPB)
Goldeneye	Jeff Kirby (WWT), Dave Beaumont (RSPB)
Red-breasted merganser	Peter Cranswick (WWT)
Goosander	Mark Underhill, Peter Cranswick (both WWT)
Red kite	Ian Bainbridge (RSPB)
White-tailed eagle	Ian Bainbridge, Richard Evans, Gwen Evans, Roger Broad (all RSPB)
Marsh harrier	John Day (RSPB)
Hen harrier	Brian Etheridge (RSPB), Roger Clarke
Buzzard	Innes Sim (RSPB)
Golden eagle	Mike McGrady (RSPB)
Osprey	Roy Dennis
Kestrel	Andrew Village
Merlin	Graham Rebecca, Eric Meek (both RSPB)
Peregrine	Humphrey Crick (BTO)

Black grouse	Brian Etheridge (RSPB), David Baines (GCT)
Capercaillie	Ron Summers, Stewart Taylor, Bob Proctor (all RSPB)
Grey partridge	Karen Blake, Nicholas Aebischer, Dick Potts (all GCT)
Quail	Les Street (RSPB)
Water rail	John Wilson, Simon Busutill (both RSPB), Des Callaghan (WWT), Steve Moyes, David Robertson
Spotted crake	Des Callaghan (WWT), Simon Busutill, Norbert Schaeffer (both RSPB)
Corncrake	Rhys Green, Glen Tyler, Tim Stowe (all RSPB), Catherine Casey (IWC)
Oystercatcher	Mark O'Brien, Ken Smith (both RSPB)
Avocet	Hilary Welch
Stone-curlew	Rhys Green, Glen Tyler (RSPB)
Ringed plover	Tony Prater (RSPB)
Dotterel	Andy Amphlett, Pat Thompson (both RSPB), Phil Whitfield (SNH)
Golden plover	Mark Hancock, Lennox Campbell (both RSPB), Ray Parr (ITE), Andy Brown (EN)
Grey plover	Mark Rehfisch (BTO)
Lapwing	Mark O'Brien, Ken Smith (both RSPB), Stephen Browne (BTO)
Knot	Mark Rehfisch (BTO)
Purple sandpiper	Ron Summers (RSPB), Keith Brockie
Dunlin	Mark O'Brien, Ken Smith, Digger Jackson, Mark Hancock (all RSPB)
Ruff	Mark Rehfisch (BTO)
Jack snipe	Mark Rehfisch, Steve Holloway (both BTO)
Snipe	Mark O'Brien, Ken Smith (both RSPB)
Woodcock	Andrew Hoodless (GCT)
Black-tailed godwit	Mark O'Brien (RSPB), Jeff Kirby (WWT)
Bar-tailed godwit	Mark Rehfisch (BTO)
Whimbrel	Murray Grant (RSPB)
Curlew	Murray Grant, Mark O'Brien, Ken Smith (all RSPB)
Redshank	Mark O'Brien, Ken Smith (both RSPB)
Greenshank	Mark Hancock (RSPB)
Turnstone	Mark Rehfisch (BTO)
Red-necked phalarope	Mark O'Brien (RSPB)
Arctic skua [1]	Kate Thompson (JNCC), Emma Brindley (RSPB)
Great skua [1]	Kate Thompson (JNCC), Emma Brindley (RSPB)
Common gull [1]	Kate Thompson (JNCC), Emma Brindley (RSPB)
Lesser black-backed gull [1]	Kate Thompson (JNCC), Emma Brindley (RSPB)
Herring gull [1]	Kate Thompson (JNCC), Emma Brindley (RSPB)
Great black-backed gull [1]	Kate Thompson (JNCC), Emma Brindley (RSPB)
Kittiwake [1]	Kate Thompson (JNCC), Emma Brindley (RSPB)
Sandwich tern [1]	Kate Thompson (JNCC), Emma Brindley (RSPB)
Roseate tern [1]	Kate Thompson (JNCC), Emma Brindley (RSPB)
Arctic tern [1]	Kate Thompson (JNCC), Emma Brindley (RSPB)
Little tern [1]	Kate Thompson (JNCC), Emma Brindley (RSPB)
Guillemot [1]	Kate Thompson (JNCC), Emma Brindley (RSPB)
Razorbill [1]	Kate Thompson (JNCC), Emma Brindley (RSPB)
Black guillemot [1]	Kate Thompson (JNCC), Emma Brindley (RSPB)
Puffin [1]	Kate Thompson (JNCC), Emma Brindley (RSPB)
Barn owl	Mike Toms (BTO)
Short-eared owl	David Glue (BTO)
Nightjar	Andy Evans, Tony Morris, Dave Burges, Ian Robinson (all RSPB), Ron Hoblyn
Green woodpecker	David Glue (BTO), Ken Smith (RSPB)
Great spotted woodpecker	David Glue (BTO), Ken Smith (RSPB)
Lesser spotted woodpecker	David Glue (BTO), Ken Smith (RSPB)
Woodlark	Dave Burges, Neil Lambert, Chris Bowden, Simon Wotton (all RSPB)
Skylark	Paul Donald (RSPB)
Sand martin	Gareth Jones
Yellow wagtail	Iain Gibson, John Marchant (BTO)
Nightingale	Andrew Henderson, Robert Fuller (BTO)
Black redstart	David Glue (BTO)
Whinchat	John Callion
Stonechat	John Callion
Ring ouzel	Innes Sim (RSPB)
Redwing	David Gibbons (RSPB)
Cetti's warbler	Simon Wotton (RSPB)
Marsh warbler	Martin Kelsey, Chris Corrigan (RSPB)
Dartford warbler	Brian Pickess, Simon Wotton (both RSPB)
Firecrest	John Marchant (BTO)

Bearded tit	Lennox Campbell, John Wilson, Norman Sills (all RSPB), John Cayford
Crested tit	Ron Summers, Stewart Taylor, Bob Proctor (all RSPB)
Golden oriole	Peter Dalton
Chough	Sian Whitehead, Reg Thorpe, Iolo Williams, Ian Bullock, Alistair Moralee (all RSPB)
Tree sparrow	Jeremy Wilson (RSPB)
Linnet	Jeremy Wilson (RSPB)
Twite	Sean Reed, Henry McGhie (both RSPB), Andy Brown (EN), Jim Reid (JNCC)
Scottish crossbill	Ron Summers, Stewart Taylor, Bob Proctor (all RSPB)
Hawfinch	Jerry Lewis, Steve Roberts
Snow bunting	Andy Amphlett (RSPB), Rik Smith (SNH)
Cirl bunting	Andy Evans, Paul St Pierre (both RSPB)
Corn bunting	Julianne Evans (RSPB)

Generic survey methods

Detailed comments on the texts were provided as follows:

Generic breeding bird monitoring methods

Common Birds Census	John Marchant (BTO)
Breeding Bird Survey	Richard Gregory (BTO)
Waders: Brown and Shepherd (1993)	Mark Hancock, Lennox Campbell, Ray Parr (ITE), Andy Brown (EN)
Waders: O'Brien and Smith (1992)	Mark O'Brien, Ken Smith, Murray Grant (all RSPB)
Waders: Reed and Fuller (1983)	Mark Hancock, Digger Jackson (both RSPB)
Dabbling and diving ducks	John Wilson, Les Street (both RSPB), Jeff Kirby (WWT)
Gull populations [1]	Kate Thompson (JNCC), Emma Brindley (RSPB)
Gull productivity [1]	Kate Thompson (JNCC), Emma Brindley (RSPB)
Tern populations [1]	Kate Thompson (JNCC), Emma Brindley (RSPB)
Tern productivity [1]	Kate Thompson (JNCC), Emma Brindley (RSPB)

Generic wintering bird monitoring methods

Wetland Bird Survey Core Counts	Peter Cranswick (WWT), Mark Rehfisch (BTO)
National wintering swan census	Peter Cranswick, John Bowler (both WWT), Ian Francis (RSPB)
National censuses of geese	Peter Cranswick, Carl Mitchell (both WWT), Ian Francis (RSPB)
Low Tide Counts	Peter Cranswick (WWT), Mark Rehfisch, Andy Musgrove (both BTO)
Waterfowl on non-estuarine coastlines	Mark Rehfisch, Stephen Holloway (both BTO), Peter Cranswick (WWT)
Inshore marine waterfowl	Peter Cranswick (WWT), Jeff Stennings, Lennox Campbell, Richard Evans (all RSPB)
Waterfowl and seabirds at sea	Peter Cranswick (WWT)
Productivity of swans and geese	Peter Cranswick (WWT)
Pinewood bird survey	Ron Summers, Stewart Taylor, Bob Proctor (all RSPB)

Our greatest thanks go to all those individuals and organisations listed above who helped ensure that the methods presented are as reliable and contemporary as possible. We also thank: JNCC and the authors of the *Seabird Monitoring Handbook* for allowing us to reproduce their seabird methods; RSPB's Species Protection Department for commenting on drafts and contributing text for the *Introduction* and those authors who allowed us to use unpublished material. Mark Avery, Ken Smith, Graham Hirons and Mary Painter were all involved in various stages of the preparation of this manual and without them it would have remained an idea, rather than a publication. We thank Jayne Lillywhite, Justine Fieth, Duncan Brooks and Sylvia Sullivan of RSPB's Communications Services Department for their equanimity when the authors continually missed agreed deadlines, and for ensuring that the manual finally made its way into print. Line drawings in the text are the work of Dan Powell, John Busby, Philip Snow, Rob Hume, Mike Langman, Trevor Boyer and N Smith. Finally, we apologise to those we have forgotten and those who commented on texts which did not, in the end, find their way into the manual.

Introduction

Bird monitoring

Decisions often have to be made about which species of bird are most in need of conservation action. The basis for these decisions and the effectiveness of resulting conservation measures can only be judged by monitoring the numbers and distribution of the species in question. The task of monitoring the United Kingdom's birds and translating this monitoring information into conservation action is a huge one.

The term 'monitoring' is used quite loosely in everyday conversation and can appear to have a fairly broad definition, but it is important to be clear about what is meant by it from the outset. It is useful to distinguish between monitoring, surveillance and survey. A *survey* is a method of data collection (like mapping), which provides a framework for the systematic measurement of variables. *Surveillance* is the systematic measurement of variables over time, with the aim of establishing a series of time-related data. *Monitoring* refers to the measurement of variables over time in a systematic way with specific objectives in mind (Spellerberg 1991).

The distinction between a survey, surveillance and monitoring is that monitoring assumes a specific reason for the collection of data, such as ensuring targets are being met. For birds, and indeed for other taxa, the targets are to maintain or enhance the numbers and distribution of each species. One of the most important developments in UK conservation in recent years has been the coordinated setting of such national targets (Anon 1994, 1995, Wynne et al 1993, 1995).

An increasing number of organisations contribute to bird monitoring in the UK; some of the main players were involved in producing this manual, but others play a part, and all work in partnership. Between them, they monitor UK bird species through a variety of generic schemes (ie those covering many species) and single-species surveys.

The need for a manual

Effective long-term monitoring requires methods to be comparable from year to year and from site to site. Several recent publications – most notably Bibby et al 1992, Buckland et al 1993, Gibbons et al 1996 and Greenwood 1996 – have partly addressed this problem by discussing the various generic methods (territory mapping, point counts, line transects, etc) that can be used to census birds. These publications are extremely valuable in that they provide a sound theoretical understanding of such methods and general practical advice on how to undertake them. They do not, however, describe the detailed practical techniques that a fieldworker needs to monitor a given species or suite of species in the field. This manual attempts to fill this gap by providing methods to monitor a large number of UK species, mainly those of conservation concern. The *Seabird Monitoring Handbook* is an excellent example of such a practical guide, and here we have built on the approach used in that handbook and extended it to many more species. We hope that the methods proposed here will be adopted as UK-wide standards by all those involved in bird monitoring (reserve wardens, national survey organisers, bird clubs, environmental consultants, etc), at least until such time as new, better methods become available.

Reliability

We see this manual as an important starting point for standardising and improving bird monitoring methods, rather than being the final word. Most of the methods documented already existed, but in a wide variety of forms – some published, many not, some exceptionally detailed and frequently used, others less so. As a result, there is some discrepancy between methods in the level of detail provided. Most, if not all, of the methods have been used in the field. In very few cases, however, has anyone set out to test whether the proposed methods are actually the 'best' (broadly defined as yielding the most reliable result for the least effort). Many of the methods are based on common sense, a detailed knowledge of the species' natural history, and a general understanding of the sorts of methods that could be used. Although we consider that we have documented the best available methods, in many cases we are unable to judge their reliability with any certainty. The more we know of the ecology and behaviour of the species in question the more accurate the method will be. Future work should seek to calibrate what is measured by the method with the 'true' numbers present.

At some sites, a number of different species may have to be surveyed in a given season. Ideally each might be surveyed with a species-specific method, each with different visit dates, times, survey methods, etc. In practice, however, there may be insufficient resources for this work, and a generic method – one covering several similar species – may be more cost-effective. The second section of this manual lists a number of such generic methods. The species for which these methods are appropriate are documented.

Absolute population estimate or index?

One of the most fundamental decisions to make before undertaking a survey is to decide whether or not you need to obtain an absolute estimate of the numbers of a particular species on a site, or whether a relative index will suffice. Both can be used to measure between-year changes (which may be all that is required), but only an absolute estimate will tell you how many birds are on the site. A reliable index is, however, often preferable to a poor count. The BTO/JNCC/RSPB UK Breeding Bird Survey is an excellent example of a generic method which provides an accurate year-to-year index of breeding bird population trends.

In many cases, however, a year-to-year index is insufficient, and a reserve warden or a site manager needs to know how many pairs or how many territories there are of a particular species in a given season. Because of this, most of the methods presented here yield absolute estimates of population sizes.

Sampling

It is not always necessary to count all the birds present on a site to obtain an estimate of the total population size. In many cases it is more effective to survey representative samples of the population and to extrapolate the results to obtain an estimate of the entire population. Sampling is commonly used where the area to be surveyed is particularly large, or where counting all individuals is impractical as there are simply too many of them. A seabird colony may be an example where there are too many individuals to count, and many of the methods for seabirds presented here involve sampling. Sampling is an entire subject in its own right, so here we give a brief overview based on the comprehensive treatise of Greenwood (1996).

The samples must be representative of the whole. There must be more than one sample and the precision of the overall estimate increases with an increase in the number of replicate samples. It is simplest if all samples are the same size and if they are not so small as to cause edge-effects or so large that too few can be covered. Generally it is better to have many smaller samples than a few large ones. Greenwood (1996) outlines how to calculate the number of units to sample and the 'percentage relative precision' attainable for a given sample size.

Samples must be selected by non-subjective means. Those selected haphazardly or for convenience have serious drawbacks. Different parts of the same colony of birds, for example, will have different densities, and non-random sampling will bias plot selection so that the sample does not reflect the population as a whole. Samples should be randomly located within the survey area. Ideally all birds in the survey area or breeding sites in a colony should have an equal probability of inclusion within a sample. Particular attention should be paid to the scale of grids used when selecting random quadrats or transect locations such that all sites within the survey area or colony could potentially be included. The best way to ensure randomness is to give each potential sample unit a number (superimpose a grid onto a map or photograph of the site and number the individual grid squares or transects), and then choose which numbers to include by using a table of random numbers or the random number generator on a calculator or computer. It is always useful to select a few extra random plots in case one is found to be unusable. There are likely to be areas within the site which cannot be covered, perhaps because they are inaccessible. It is important to record what proportion of (a) the total and (b) the 'monitorable' population is present in the sample.

Stratified random sampling can be more efficient than simple random sampling and should be used wherever possible. There are usually systematic variations in population density across a site and it is valuable to divide the site into sub-areas differing in density and to sample randomly within each. Such sub-areas are termed strata. Choose strata such that the within-stratum variance is minimised and the between-stratum difference is maximised. It is difficult to do this without some prior knowledge of population density, but this is often available or can be gained from a rough preliminary survey. Stratification may lose its advantages if the number of strata is large relative to the total number of sampling units (so that each stratum contains just a few units). It is usually sufficient to use only 3–6 strata. Stratification is valuable also if the costs of sampling are different in different parts of the study area. If there is variation in both density and cost, both need to be considered, but variation in density is generally the more important (Greenwood 1996).

Remember that human minds are not capable of randomness, and even if your result does not 'look' random you should not adjust the results as long as the method is unbiased.

Statistics

Some of the methods require the calculation of means, standard errors, and/or standard deviations. A scientific calculator is very useful for basic calculations. The mean is used to describe the central tendency of a distribution of counts or measurements and is calculated by dividing the sum of the observations by the number of sampling units.

$$\text{mean} = \frac{\Sigma x}{n}$$

where x is each observation and n is the number of sampling units in a sample.

The standard deviation is a useful comparative measure of variance about a mean value of a sample. The standard deviation (s.d.) is calculated as:

$$\text{s.d.} = \sqrt{\frac{\Sigma (x - \text{mean})^2}{n - 1}}$$

where x is each observation and n is the number of sampling units in a sample.

The standard error of the mean can be more informative than the standard deviation because it indicates how close the population mean is likely to be to the sample mean. The standard error (s.e.) is calculated by dividing the sample standard deviation by the square root of the number of sampling units.

$$\text{s.e.} = \frac{\text{s.d.}}{\sqrt{n}}$$

where n is the number of sampling units in a sample.

We recommend that readers consult Fowler and Cohen (1986) if they need further statistical advice.

Species included

We have not included methods for all species that occur in the UK in this manual. Rather, the species covered are broadly those on the red and amber lists of Birds of Conservation Concern (BoCC: Gibbons et al 1996). Species can be included on these lists because of their status in the breeding or non-breeding season, and the methods presented here largely reflect this, eg we do not give a non-breeding season method for Dartford warbler *Sylvia undata* as it is red listed because of its status in the breeding season. Species-specific methods are not given for a number of breeding species on the red and amber lists as they are either sufficiently widespread to be considered well-covered by the BTO/JNCC/RSPB UK Breeding Bird Survey (eg song thrush *Turdus philomelos*) or are so rare (eg wryneck *Jynx torquilla*) in the UK as to warrant exclusion (see below). Several species not on the red or amber lists have been included if they are currently poorly monitored, difficult to monitor or germane to the interests of the organisations involved in compiling the manual.

The red and amber listed species for which no breeding season method is given because of effective BBS coverage are: stock dove *Columba oenas*, turtle dove *Streptopelia turtur*, swallow *Hirundo rustica*, dunnock *Prunella modularis*, redstart *Phoenicurus phoenicurus*, song thrush, blackbird *Turdus merula*, spotted flycatcher *Muscicapa striata*, marsh tit *Parus palustris*, starling *Sturnus vulgaris*, goldfinch *Carduelis carduelis*, bullfinch *Pyrrhula pyrrhula* and reed bunting *Emberiza schoeniclus*. An outline of the Breeding Bird Survey method is given in the generic survey methods section. If site population totals are required for these widespread species, we recommend use of the Common Birds Census (see generic methods section), although the number of visits can be reduced to tie in with a given species breeding season if it alone is being censused. If an example of a 'reduced visit CBC' survey method is required, please refer to the text for redwing *Turdus iliacus* (p 339).

The red and amber listed species for which no breeding season method is given because of rarity in the UK are: red-necked grebe *Podiceps grisegena*, pintail *Anas acuta*, scaup *Aythya marila*, honey buzzard *Pernis apivorus*, Montagu's harrier *Circus pygargus*, crane *Grus grus*, black-winged stilt *Himantopus himantopus*, Temminck's stint *Calidris temminckii*, ruff *Philomachus pugnax*, wood sandpiper *Tringa glareola*, Mediterranean gull *Larus melanocephalus*, little gull *Larus minutus*, wryneck, fieldfare *Turdus pilaris*, Savi's warbler *Locustella luscinioides*, icterine warbler *Hippolais icterina*, red-backed shrike *Lanius collurio*, brambling *Fringilla montifringilla*, serin *Serinus serinus*, parrot crossbill *Loxia pytyopsittacus* and scarlet rosefinch *Carpodacus erythrinus*. Aquatic warbler *Acrocephalus paludicola*, a globally threatened species which visits the UK on migration, is also missing from the manual; this is largely because no practical method has yet been devised to census this species effectively.

Format of the manual

The methods are split into two sections: single-species and generic. Within the single-species section, species are listed in taxonomic order. If a generic method is recommended for a given species, the reader is referred to the generic section of the manual. Within the generic section, methods are split into breeding season methods or wintering methods, and the suite of species to which the method refers should be evident from its title.

We anticipate that this manual will be dipped into when trying to find a method for a given species or group of species, rather than read comprehensively from cover to cover. Because of this, a lot of information is repeated between accounts. Although this is unfortunate in that it makes the manual lengthy, we hope it will make it of more practical use to the reader. Occasionally, example recording forms are provided with the method. These are meant to illustrate the manner in which data could be collected. In many cases, it may be necessary to customise recording forms for your own use. Fisher (1995) gives many practical tips on the design of recording forms.

A quick reference table showing where to find the method(s) for each species in the manual is given at the end of this *Introduction*.

Structure of single-species accounts

Each method in each species account follows the same format using standard headings, explained below (*italicised* text indicates abbreviations and headings that appear in the accounts).

Status

The conservation status of each species. Given below are the criteria (and their abbreviations) for inclusion on the red and amber lists in *Bird Species of Conservation Concern* (Gibbons et al 1996).

Red listed

BD Rapidly declining species: ≥50% decline in population or range in UK over the last 25 years.

HD Historical population decline in the UK between 1800 and 1995.

SPEC 1 Of global conservation concern.

Amber listed

BDM	Moderately declining species: declined by 25–49% in the UK in numbers or range in the last 25 years.
BR	Rare breeder: five-year mean of 0.2–300 breeding pairs in the UK.
BI	Internationally important breeding species: ⩾20% of European breeding population in the UK.
WI	Internationally important non-breeding species: ⩾20% of north-west European (wildfowl), East Atlantic Flyway (waders) or European (others) non-breeding populations in the UK.
BL	Localised breeders (⩾50% of the UK breeding population found in ten or fewer sites), but not BR.
WL	Localised non-breeders (⩾50% of the UK non-breeding population can be found in ten or fewer sites).
SPEC 2, SPEC 3	Species of unfavourable conservation status in Europe.

SPEC categories

SPEC categories (Species of European Conservation Concern, as defined by Tucker and Heath 1994) were used as one criterion for the revised red and amber national listings (see above). All European bird species have been allocated to one of five categories of conservation concern:

SPEC 1	Species of global conservation concern which regularly occur in Europe.
SPEC 2	Species whose global populations are concentrated in Europe and whose European populations have an unfavourable conservation status.
SPEC 3	Species whose global populations are not concentrated in Europe, but whose European populations have an unfavourable conservation status.
SPEC 4	Species whose global populations are concentrated in Europe and whose European populations have a favourable conservation status.
Non-SPEC	Favourable conservation status and not concentrated in Europe (ie all other species).

The European Threat Status (ETS) is given in brackets after the SPEC category, eg SPEC 1 (V). Whether or not a species is listed as SPEC 2–4, or as Non-SPEC, is determined partly by its ETS (Tucker and Heath 1994). The ETS can be:

E	Endangered
V	Vulnerable
R	Rare
D	Declining
L	Localised
Ins	Insufficiently Known
S	Secure
P	Provisional (provisionally any of the above)

W Indicates that the SPEC category or the ETS relates to winter populations.

Schedule 1 of WCA 1981

Schedule 1 species are rare breeding species which receive additional protection under the Wildlife and Countryside Act 1981 in England, Scotland and Wales. The equivalent legislation in Northern Ireland is the 1985 Northern Ireland Wildlife Order, and the list of Schedule 1 species is different to that in Britain.

Only Schedule 1 (or Schedule 1 – Part II) species are highlighted in this section; other Schedule classifications are not mentioned. It is an offence to intentionally disturb a Schedule 1 species while it is building a nest or is in, on or near a nest containing eggs or young. When monitoring such species during the breeding season you may need an appropriate government licence which permits disturbance for specific reasons. The necessity for such licences will depend on the type of monitoring you plan and the species concerned. In cases of doubt it is always advisable to obtain the appropriate licence. The licences are issued by the country agencies (EN, CCW, SNH) or via the BTO. In Northern Ireland licences are issued by the Environment and Heritage Service, Department of the Environment Northern Ireland. The licences will state what activities are permitted and should always be carried during fieldwork.

Although the process may appear something of an administrative exercise, it is important to appreciate the necessity of such licences and to adhere to any conditions imposed. If the law is to give additional protection to rare breeding birds from, for example, egg-collectors, falconry thieves and unlicensed photographers, it is essential that the scientific community keeps its licensing in order.

Annex I of EC Wild Birds Directive

This European community directive on bird conservation has become the principal measure protecting Europe's wild birds. It is wide-ranging and encompasses habitat protection and controls on trapping, killing and selling wild birds. Species are categorised into three regularly updated Annexes which are summarised below:

Annex I	'shall be the subject of special conservation measures concerning their habitat in order to ensure their survival and reproduction in their area of distribution'.
Annex II	'II/1 species may be hunted in the geographical sea and land area where this Directive applies'. 'II/2 species may be hunted only in the Member States in respect of which they are indicated'.
Annex III	'III/1 Member states shall not prohibit the sale, transport for sale or keeping for sale of any recognisable parts of these species provided the birds have been legally acquired'. 'III/2 Member states may restrict the activities above within their own territories, provided the birds have been legally acquired'.

For further information on this, see Pritchard et al (1992), McNiven (1994) and Tucker and Heath (1994).

National monitoring

Current, past and future significant national monitoring schemes and surveys are mentioned for each species.

CBC and BBS

Species monitored by the BTO/JNCC Common Birds Census (CBC) and/or the BTO/JNCC/RSPB Breeding Bird Survey (BBS). Inevitably, some species are monitored well by these schemes, others less so. We have followed the suggestions of Crick et al (1997) in deciding which species are monitored reliably by each of these schemes. Wherever they considered that the sample size of survey plots was low (<20 CBC plots or <100 BBS 1-km squares) we have assumed that they are not monitored effectively by these schemes, even though we recognise that for some of these species it may be the best monitoring information that exists.

Rare Breeding Birds Panel

Species monitored by the Rare Breeding Birds Panel which is sponsored and supported by RSPB, *British Birds*, JNCC and BTO. Please contribute records of these rare breeding species to the RBBP via your County Recorder.

Seabird Colony Register
Seabird Monitoring Programme

The UK Seabird Monitoring Programme (run by JNCC, in partnership with RSPB, SOTEAG and the Seabird Group), established in 1989, aims to ensure that sufficient data are collected to provide governmental and non-governmental organisations with sound advice relevant to the conservation needs of Britain's seabirds. The collation, analysis and dissemination of the data collected is an essential part of the programme. Colony data are stored on the Seabird Colony Register, a national database of seabird colony counts which is continually updated using data routinely supplied to the annual, sample Seabird Monitoring Programme.

It is important that records of seabirds are sent to the JNCC Seabird Colony Register. Three separate forms are used to collate data for the Seabird Colony Register: the 10-km-square summary form, a colony register form and a data sheet. Copies of these forms and instructions on how to complete them are given in Walsh et al (1995). Records of terns *Sterna* are collated by the RSPB, and all records should be sent to the nominated member of staff in the relevant RSPB regional or country office. The Seabird Monitoring Programme is heavily dependent on its volunteer fieldworkers, without whom it could not function. If you wish to contribute, or already hold data which could be submitted to the programme, please contact: Co-ordinator, Seabird Monitoring Programme, Seabirds and Cetaceans Team, JNCC, Dunnet House, 7 Thistle Place, Aberdeen AB10 1UZ.

Operation Seafarer (1969–1970)
Seabird Colony Register (1985–1987)
Seabird 2000 (1999–2001)

These three schemes are complete censuses of all British and Irish seabird colonies undertaken at approximately 15-year intervals, each covering several years. The 1985–87 census is the most recent comprehensive count, and the data are held within the Seabird Colony Register, parts of which are updated annually (see above). Seabird 2000 will be the next complete census, with fieldwork planned for 1999–2001.

WeBS

Non-breeding waterfowl species covered by the BTO/WWT/JNCC/RSPB Wetland Bird Survey (WeBS). Always use the appropriate methods and recording forms and ensure that you contact the local organiser before undertaking any WeBS counts. Details of WeBS are given in the generic survey methods section.

Population and distribution

A paragraph outlining the current distribution and population of the species and, where informative, recent trends.

Ecology

Key aspects of the species' ecology which may be relevant to the design and execution of the method, eg habitat requirements, timing of breeding season, behaviour, non-breeding season feeding requirements.

Breeding season survey

For all relevant species, methods are given for estimating breeding population sizes (or an index). In addition, methods for estimating breeding success and/or productivity are given where these are known (many seabirds) and practical. In this manual, methods to determine productivity are those which provide estimates of the total number of young fledged (enabling young fledged per pair to be calculated if the population size is known). All other methods which estimate some parameter of reproductive success (eg clutch size, hatching success, brood size, etc), but which fall short of an estimate of productivity, are referred to as 'breeding success' survey methods.

Methods for estimating breeding success/productivity are available for more species than are documented here but have been omitted if they are thought to be too disruptive to the species involved. Often such surveys are best carried out as part of a specific research project rather than as part of routine population monitoring.

Most of the methods documented in this manual do not require nests to be found. However, if you come across nests, please contribute the appropriate information to the BTO Nest Record Scheme. The BTO uses these data to monitor the breeding success of Britain's birds in different habitats and regions and through time. If you wish to contribute, please write to: Nest Record Scheme, BTO, National Centre for Ornithology, The Nunnery, Thetford, Norfolk IP24 2PU.

Winter survey

Winter survey methods are included where appropriate. Waterfowl are covered by the Wetland Bird Survey (WeBS), a full explanation of which is given in the generic survey methods section. WeBS will not always fulfil winter monitoring requirements on all sites (and does not claim to), so more detailed methods are given for some non-breeding waterfowl species. Winter survey methods follow the same format as the breeding survey methods.

The following sections provide the detailed, technical information that is needed to undertake the survey. The same format is followed for breeding (population, breeding success and productivity) and for winter survey methods.

> *Information required*
>
> This box explains succinctly the data that are required from the surveyor in order that either the breeding population, breeding success or productivity, or the winter population, can be assessed.

Number and timing of visits, Time of day

These should be adhered to. Alternative dates are given where possible for species with a wide latitudinal distribution, but such information is not always available.

Weather constraints

Information here is mostly common sense, although there are some species for which surveying in specific weather conditions is recommended.

Sites/areas to visit

This section provides information on the types of area in which the species is likely to be found, and details of any specific habitat requirements.

Equipment

Any equipment needed for a survey *in addition* to the standard requirements of binoculars, notebook, pencils, etc.

Safety reminders

The monitoring manual is *not* a key source of Health and Safety advice, nor is it a Health and Safety handbook. The organisations responsible for compiling this manual do not accept responsibility for accidents that occur while monitoring. All fieldworkers and their managers should refer to their organisations' Health and Safety advice and take adequate precautions to prevent an emergency and adequate provisions in case an emergency does happen. The safety reminders in this manual are largely commonsense suggestions for what can be difficult tasks undertaken in difficult conditions. General health and safety guidelines are given below.

Disturbance

This section is designed to clarify just how much, if any, disturbance can be justified while monitoring. All species will be affected to some degree by human disturbance (either directly or indirectly) and many are particularly vulnerable during the breeding season. 'As little disturbance as possible' is a good general guideline. More specifically, however, those species that are known to be particularly affected by disturbance are highlighted, and some acceptable limits to disturbance are given where these may otherwise be unclear. Be aware of, and sympathetic towards, any local or national animal welfare issues that may influence your decision to undertake the survey. See above for licensing requirements and below for further information on species protection.

Methods

The survey method points out any sampling required to survey widely dispersed or highly colonial species. The method is then outlined. Methods are generally based on line transects, territory mapping, point counts, nest counts, time-based surveys, etc. Having read the methods section, the surveyor should be confident about how to conduct the survey.

References

Lists all references cited in the text but is not an exhaustive list of references relevant to that method.

General field guidelines

Maps

Maps are essential for monitoring. You may need to photocopy Ordnance Survey maps for field use. There are strict copyright laws concerning the photocopying of Ordnance Survey maps; please find out whether or not you are covered to do this. Very large scale maps of about 1:2,500 (25 inches to the mile) are ideal for detailed territory mapping (CBC-type) surveys. Each sheet of these maps covers just a single kilometre square. They are available from the National Map Centre, 22–24 Caxton Street, London, SW1H 0QU and cost about £45 per sheet. Alternatively try to obtain maps from local landowners or libraries, or scale up a 6-inch sheet (1:10,000 or 1:10,560) then check it on the ground and add any useful extra features.

Access

In *ALL* circumstances obtain permission before gaining access to private land. Please remember the following:

- It is good practice to cultivate and maintain good relations with landowners/tenants and their staff.
- Do not carry out any survey work without prior access permission, preferably in writing.
- Respect the wishes of the landowner, eg give advance warning of the dates of your visits.
- Be aware of special circumstances for some sensitive species or habitats, eg night surveys and access to military land or shooting estates.

Species protection

Confidentiality

It is particularly important to keep information on rare breeding species confidential as such information is of great interest to many people involved in illegal activities. Do not pass on sensitive information, no matter how genuine the recipient may appear. In addition, consider the security of notebooks, marked maps, survey forms, computer records, etc. Ensure these are not left in circumstances which could compromise the information they contain.

Evidence of offences

For a number of species, particularly raptors, human persecution has, and continues to have, a significant effect on wild populations. Any evidence of such activities may be highly relevant to conservation objectives, and fieldworkers should always be alert to such human factors. Typical incidents might include birds being shot, poisoned or trapped, or nests being robbed or destroyed.

Currently, the only centralised record of offences against wild birds is maintained by the RSPB's Investigation Section. Please pass any information about possible offences or activities of suspicious individuals to this department at The Lodge, Sandy, Bedfordshire SG19 2DL. All information will be treated in confidence and in accordance with the provisions of the Data Protection Act. Where offences are actually witnessed being committed, the police should be alerted as soon as possible. When reporting such incidents, try and ensure that the information is brought to the attention of the police Wildlife Liaison Officer (WLO). These are designated officers who will normally have a better understanding of the legislation and issues involving wildlife crime.

In any instance, it is important to try and record all relevant information at the time of the event if possible. This could be recorded straight into a field notebook or whatever is available. If you are with another observer it is quite in order to make joint notes; just record the fact this has been done and ensure the notes are kept secure as they could be a court exhibit at a future date. In Scotland, it is particularly beneficial to have more than one observer, as corroboration is normally required. The following information, most of which is common sense, should be recorded:

- Time, date, place (grid reference where possible).

- Details of incident.
- Descriptions of any suspects, with attention to behaviour, clothing, items carried, any conversation, etc.
- Details of any vehicles.
- In potential disturbance cases, note any effect on the bird's behaviour.
- Optical aids used, distance of observation and visibility conditions.
- Whether photographs were taken.

It is important to consider your own personal safety. The police would normally advise you not to tackle people suspected of committing offences. There may be circumstances where it is felt necessary to confront people, for example where Schedule 1 birds are perhaps being inadvertently disturbed. Think carefully about such situations, your own behaviour and any risks involved.

Another potentially hazardous situation is finding suspected poisoning incidents. The placing of poison baits in the open is illegal, although this continues to be a national problem. Such baits are normally targeted at foxes, corvids or raptors, although any carrion-eating species is vulnerable. Mammals or birds lying dead next to an apparent bait should immediately be viewed with suspicion. Other signs of illegal poisoning include staked-out carcasses, dead birds with no apparent sign of injury or hens' eggs left in the open (sometimes with discoloration where poison has been injected). Do not handle these items unless you know what you are doing; many of the poisons used are extremely toxic and can even be absorbed through the skin. Try to record accurate details at the site of the incident with particular attention to the location; take photographs or sketch the scene if possible. Try and cover the animals and/or bait with vegetation or similar. This will help preserve the scene and prevent other wildlife being poisoned. Report the incident immediately; there is a Freephone hotline for poisoning incidents: 0800 321600. Try to contact the police and RSPB if appropriate.

Remember that many forms of 'pest' control are perfectly legal, usually to protect agricultural and livestock interests, and should not be interfered with. Typical methods include:

- Shooting of corvids, pigeons *Columba* and certain mammals.
- Free-running snares for foxes.
- Spring traps placed under cover for rats and mustelids (eg stoats). (NB spring traps set in the open, particularly if mounted on poles, should be reported immediately.)
- Larsen traps: portable cage traps with a spring-loaded door. These are used to take corvids and normally have a live magpie *Pica pica* or crow *Corvus corone* inside the trap as a decoy. Traps must be checked daily with adequate food, water and shelter for the decoy birds. Non-target birds must be released. (NB These traps have been used illegally to take raptors by using alternative decoys such as pigeons or gamebirds.)
- Crow cage traps. These are normally fixed structures, baited with eggs or carrion using a funnel or ladder system to allow corvids to enter the trap. These must be checked daily and when not in use they must be left so that birds cannot be accidentally caught.

If in any doubt about a particular incident try to seek the appropriate advice.

Section contributed by Guy Shorrock (RSPB Investigations)

Health and safety guidelines

The following are commonsense health and safety guidelines. Please consult a more thorough text if necessary (eg *RSPB Health and Safety Handbook*: RSPB 1994).

Think carefully about the risks involved before starting a survey. Consider the task, the environment, the weather, the time of day or other special circumstances; working alone may significantly increase the level of risk.

Be aware of the following occupational diseases:

- *Tetanus*. Preventable by regular inoculation (every five years).
- *Lyme disease*. A tick-borne spirochaetal infection transmitted to humans by the bite of an adult female tick. It is increasingly carried by seabird ticks. Try to prevent tick bites and regularly inspect skin for ticks. If the site of a tick bite remains unhealed, if a skin rash develops or if symptoms of ill health persist, then consult a doctor immediately.
- *Leptospirosis*. If you are working regularly near waterbodies, watercourses or farming operations, be aware of two varieties of leptospirosis. Cattle-associated leptospirosis (CAL) can be contracted by working in close contact with cattle. Human symptoms are a flu-like illness, severe headache and meningitis. Weil's disease or leptospiral jaundice is most commonly associated with rodents, particularly rats. The symptoms associated with this form of leptospirosis are jaundice, meningitis, conjunctivitis and renal failure.

Make sure your clothing is suitable for the environment and the time of year; carry extra clothing if working in areas of high ground, moorland or open water. When working in areas where the weather is likely to be poor or can change for the worse rapidly, eg mountains, offshore islands or high moorland areas, a plastic survival bag and food supplies (including high energy foods) should be carried. If an all-day trip is necessary, take appropriate food and drink. In remote areas, carry a map and compass and know how to take bearings from the map, and back-bearings from the ground. Carry a whistle on a string around your neck. Carry a watch, preferably waterproof. If the work is likely to extend into the hours of darkness, carry a torch and spare batteries. Always carry a first-aid kit.

Always leave a note of your whereabouts with a responsible person. This should include: date and time of departure, method of travel to and around the site, proposed itinerary, any potentially hazardous technique you will be using, expected time of leaving the site and return to base, and vehicle identification details. The person to whom these details are given should be told who to contact if you do not return.

Don't forget . . .

. . . to enjoy yourself!

References

Anon (1994) *Biodiversity: the UK Action Plan*. HMSO, London.

Anon (1995) *Biodiversity: the UK Biodiversity Steering Group Report, volume 2: Action Plans*. HMSO, London.

Bibby, C J, Burgess, N D and Hill, D A (1992) *Bird Census Techniques*. Academic Press, London.

Buckland, S T, Anderson, D R, Burnham, K P and Laake, J L (1993) *Distance Sampling: Estimating Abundance of Biological Populations.* Chapman & Hall, London.

Crick, H Q P, Baillie, S R, Balmer, D E, Bashford, R I, Dudley, C, Glue, D E, Gregory, R D, Marchant, J H, Peach, W J and Wilson, A M (1997) *Breeding Birds in the Wider Countryside: their Conservation Status (1971–95)*. BTO Research Report 187. BTO, Thetford.

Fisher, I J (1995) *RSPB Conservation Surveys: Form Design and Data Entry.* RSPB, Sandy.

Fowler, J and Cohen, L (1986) *Statistics for Ornithologists.* BTO Guide 22. BTO, Thetford.

Gibbons, D W, Avery, M I, Baillie, S R, Gregory, R D, Kirby, J, Porter, R, Tucker, G and Williams, G (1996) Bird species of conservation concern in the United Kingdom, Channel Islands and Isle of Man: revising the red data list. *RSPB Conservation Review* 10: 7–18.

Gibbons, D W, Hill, D A and Sutherland, W J (1996) Birds. In Sutherland, W J (ed) *Ecological Census Techniques: a Handbook.* Cambridge University Press.

Greenwood, J J D (1996) Basic techniques. In Sutherland, W J (ed) *Ecological Census Techniques: a Handbook.* Cambridge University Press.

McNiven, D (1994) *A Review of the Effectiveness of the European Council Directive 79/409/EEC on the Conservation of Wild Birds.* RSPB, Sandy.

Pritchard, D E, Housden, S D, Mudge, G P, Galbraith, C A and Pienkowski, M W (eds) (1992) *Important Bird Areas in the UK including the Channel Islands and the Isle of Man.* RSPB, Sandy.

RSPB (1994) *Health and Safety Handbook.* RSPB unpubl.

Spellerberg, I F (1991) *Monitoring Ecological Change.* Cambridge University Press.

Tucker, G M and Heath, M F (1994) *Birds in Europe: their Conservation Status.* BirdLife Conservation Series 3. BirdLife International, Cambridge.

Walsh, P M, Halley, D J, Harris, M P, del Nevo, A, Sim, I M W and Tasker, M L (1995) *Seabird Monitoring Handbook for Britain and Ireland.* JNCC, RSPB, ITE, Seabird Group.

Wynne, G, Avery, M, Campbell, L, Juniper, T, King, M, Smart, J, Steel, C, Stones, T, Stubbs, A, Taylor, J, Tydeman, C and Wynde, R (1993) *Biodiversity Challenge: an Agenda for Conservation in the UK* (first edition). RSPB, Sandy.

Wynne, G, Avery, M, Campbell, L, Gubbay, S, Hawkswell, S, Juniper, T, King, M, Newbery, P, Smart, J, Steel, C, Stones, T, Stubbs, A, Taylor, J, Tydeman, C and Wynde, R (1995) *Biodiversity Challenge: an Agenda for Conservation in the UK* (second edition). RSPB, Sandy.

Useful addresses

RSPB offices

UK Headquarters
The Lodge, Sandy,
Bedfordshire SG19 2DL

Northern Ireland Headquarters
Belvoir Park Forest,
Belfast BT8 4QT

Scottish Headquarters
Dunedin House,
25 Ravelstone Terrace,
Edinburgh EH4 3TP

Welsh Headquarters
18 High Street,
Newtown, Powys SY16 2NP

North of England Office
4 Benton Terrace,
Newcastle upon Tyne NE2 1QU

North-West England Office
Westleigh Mews,
Wakefield Road,
Denby Dale, Huddersfield,
West Yorkshire HD8 8QD

Central England Office
46 The Green,
South Bar, Banbury,
Oxfordshire OX16 9AB

East Anglia Office
Stalham House,
65 Thorpe Road, Norwich,
Norfolk NR1 1UD

South-East England Office
2nd Floor, Frederick House,
42 Frederick Place, Brighton,
East Sussex BN1 1AT

South-West England Office
Keble House, Southernhay
Gardens, Exeter,
Devon EX1 1NT

East Scotland Office
10 Albyn Terrace,
Aberdeen AB1 1YP

North Scotland Office
Etive House, Beechwood Park,
Inverness IV2 3BW

South and West Scotland Office
Unit 3.1, West of Scotland
Science Park, Kelvin Campus,
Glasgow G20 0SP

North Wales Office
Maes y Ffynnon,
Penrhosgarnedd,
Bangor, Gwynedd LL57 2DW

South Wales Office
Sophia House,
28 Cathedral Road,
Cardiff CF1 9LJ

Other bodies

British Trust for Ornithology
The Nunnery, Thetford,
Norfolk IP24 2PU

The Wildfowl & Wetlands Trust
Slimbridge,
Gloucestershire GL2 7BT

Joint Nature Conservation
Committee, Headquarters
Monkstone House, City Road,
Peterborough PE1 1JY

Joint Nature Conservation
Committee, Seabirds and Cetaceans
Dunnet House, 7 Thistle Place,
Aberdeen AB10 1UZ

Scottish Natural Heritage
Research and Advisory Services
2 Anderson Place,
Edinburgh EH6 5NP

English Nature, Headquarters
Northminster House,
Peterborough PE1 1UA

Countryside Council for Wales,
Headquarters
Plas Penrhos, Ffordd Penrhos,
Bangor, Gwynedd LL57 2LQ

Department of the Environment
Northern Ireland,
Environment and Heritage Service
10–18 Adelaide Street,
Belfast BT2 8GB

Institute of Terrestrial Ecology
Hill of Brathens,
Banchory Research Station,
Banchory,
Kincardineshire AB31 4BY

Seabird Group
c/o The Lodge,
Sandy, Bedfordshire SG19 2DL

Frequently cited references

A few publications are cited frequently in the main text and are referred to by abbreviated titles, as follows:

68–72 Atlas	Sharrock, J T R (1976) *The Atlas of Breeding Birds in Britain and Ireland*. Poyser, Berkhamsted.
88–91 Atlas	Gibbons, D W, Reid, J B and Chapman, R A (1993) *The New Atlas of Breeding Birds in Britain and Ireland 1988–1991*. Poyser, London.
Birds in Europe	Tucker, G M and Heath, M F (1994) *Birds in Europe: their Conservation Status*. BirdLife Conservation Series 3. BirdLife International, Cambridge.
BWP	Cramp, S et al (eds) (1977–93) *The Birds of the Western Palearctic*, 9 vols. Oxford University Press.
Census Techniques	Bibby, C J, Burgess, N D and Hill, D A (1992) *Bird Census Techniques*. Academic Press.
Population Estimates	Stone, B H, Sears, J, Cranswick, P A, Gregory, R D, Gibbons, D W, Rehfisch, M M, Aebischer, N J and Reid, J B (1997) Population estimates of birds in Britain and in the United Kingdom. *British Birds* 90: 1–22.
Red Data Birds	Batten, L A, Bibby, C J, Clement, P, Elliot, G D and Porter, R F (1990) *Red Data Birds in Britain*. Poyser, London.
Seabird Monitoring Handbook	Walsh, P M, Halley, D J, Harris, M P, del Nevo, A, Sim, I M W and Tasker, M L (1995) *Seabird Monitoring Handbook for Britain and Ireland*. JNCC/RSPB/ITE/Seabird Group.
WeBS 92–93	Waters, R J and Cranswick, P A (1993) *The Wetland Bird Survey 1992–93: Wildfowl and Wader Counts*. BTO/WWT/RSPB/JNCC, Slimbridge. (Other WeBS reports are referred to similarly.)
Winter Atlas	Lack, P (1986) *The Atlas of Wintering Birds in Britain and Ireland*. Poyser, Calton.

Abbreviations used in the text

AIA	Apparently Incubating Adult
AOB	Apparently Occupied Burrow
AON	Apparently Occupied Nest
AOS	Apparently Occupied Site
AOT	Apparently Occupied Territory
BBS	Breeding Bird Survey
BST	British Summer Time
BTO	British Trust for Ornithology
CBC	Common Birds Census
CCW	Countryside Council for Wales
DETR	Department of the Environment, Transport and the Regions
EC	European Community
EN	English Nature
EU	European Union
GCT	Game Conservancy Trust
GWGS	Greenland White-fronted Goose Study
ITE	Institute of Terrestrial Ecology
IWC	Irish Wildbird Conservancy (now Birdwatch Ireland)
IWC	International Waterfowl Census
I-WeBS	Irish Wetland Bird Survey
IWRB	International Waterfowl and Wetlands Research Bureau (now Wetlands International)
JNCC	Joint Nature Conservation Committee
NEWS	Non-estuarine coastal Waterfowl Survey
OS	Ordnance Survey
RBBP	Rare Breeding Birds Panel
RSPB	Royal Society for the Protection of Birds
RMP	RSPB's Reserves Monitoring Programme
SCR	Seabird Colony Register
SD	Standard Deviation
SE	Standard Error
SMP	Seabird Monitoring Programme
SMS	RSPB's Seabird Monitoring Strategy
SNH	Scottish Natural Heritage
SOC	Scottish Ornithologists' Club
SOTEAG	Shetland Oil Terminal Environmental Advisory Group
SPEC	Species of European Conservation Concern
SSDB	RSPB's Sites and Species Database
tetrad	2 × 2 km square of the British or Irish National Grids
WCA	Wildlife and Countryside Act
WeBS	Wetland Bird Survey
WSG	Wader Study Group
WWT	The Wildfowl & Wetlands Trust

Finding species methods in the manual
a quick reference

For each species this table gives the **page number(s)** of the method(s) included in the manual – whether they are single-species or generic, for the breeding season or for winter (non-breeding season), or for the monitoring of population (Pop.) or of productivity/breeding success (Prod./Succ.). Any one species may have several different methods.

	SINGLE-SPECIES METHODS			GENERIC METHODS		
	Breeding		Wintering	Breeding		Wintering
	Pop.	Prod./Succ.		Pop.	Prod./Succ.	
Red-throated diver *Gavia stellata*	32	36	—	—	—	430, 448, 451
Black-throated diver *Gavia arctica*	38	41	—	—	—	430, 448, 451
Great northern diver *Gavia immer*	—	—	—	—	—	430, 448, 451
Slavonian grebe *Podiceps auritus*	43	45	—	—	—	430, 448
Black-necked grebe *Podiceps nigricollis*	46	47	—	—	—	430, 448
Manx shearwater *Puffinus puffinus*	49	52	—	—	—	451
Storm petrel *Hydrobates pelagicus*	54	—	—	—	—	451
Leach's petrel *Oceanodroma leucorhoa*	60	—	—	—	—	451
Gannet *Morus bassanus*	64	68	—	—	—	451
Cormorant *Phalacrocorax carbo*	72	74	75	—	—	430
Shag *Phalacrocorax aristotelis*	78	80	—	—	—	451
Bittern *Botaurus stellaris*	84	86	86	—	—	—
Mute swan *Cygnus olor*	87	90	—	389	454	430, 436
Bewick's swan *Cygnus columbianus*	—	—	—	—	454	430, 436
Whooper swan *Cygnus cygnus*	—	—	—	—	454	430, 436
Bean goose *Anser fabalis*	—	—	—	—	454	430, 438
Pink-footed goose *Anser brachyrhynchus*	—	—	—	—	454	430, 438
White-fronted goose *Anser albifrons*	—	—	—	—	454	430, 438
Greylag goose *Anser anser*	—	—	—	—	454	430, 438
Barnacle goose *Branta leucopsis*	—	—	—	—	454	430, 438
Brent goose *Branta bernicla*	—	—	—	—	454	430, 438
Shelduck *Tadorna tadorna*	99	101	—	—	—	430
Wigeon *Anas penelope*	—	—	—	402	404	430
Gadwall *Anas strepera*	—	—	—	402	404	430
Teal *Anas crecca*	—	—	—	402	404	430
Pintail *Anas acuta*	—	—	—	402	404	430
Garganey *Anas querquedula*	—	—	—	402	404	430
Shoveler *Anas clypeata*	—	—	—	402	404	430
Pochard *Aythya ferina*	—	—	—	402	404	430
Scaup *Aythya marila*	—	—	110	—	—	430, 448, 451
Eider *Somateria mollissima*	111	113	—	—	—	430, 448, 451
Long-tailed duck *Clangula hyemalis*	—	—	114	—	—	430, 448, 451
Common scoter *Melanitta nigra*	115	118	118	—	—	430, 448, 451
Velvet scoter *Melanitta fusca*	—	—	120	—	—	430, 448, 451
Goldeneye *Bucephala clangula*	—	—	—	402	—	430, 448
Red-breasted merganser *Mergus serrator*	122	124	125	—	—	430, 448
Goosander *Mergus merganser*	128	130	131	—	—	430
Red kite *Milvus milvus*	133	136	—	—	—	—
White-tailed eagle *Haliaeetus albicilla*	138	140	—	—	—	—
Marsh harrier *Circus aeruginosus*	143	144	—	—	—	—
Hen harrier *Circus cyaneus*	146	—	148	—	—	—

cont.

	SINGLE-SPECIES METHODS			GENERIC METHODS		
	Breeding		Wintering	Breeding		Wintering
	Pop.	Prod./Succ.		Pop.	Prod./Succ.	
Buzzard *Buteo buteo*	151	—	—	389	—	—
Golden eagle *Aquila chrysaetos*	153	155	—	—	—	—
Osprey *Pandion haliaetus*	157	159	—	—	—	—
Kestrel *Falco tinnunculus*	160	—	—	386, 389	—	—
Merlin *Falco columbarius*	164	—	—	—	—	—
Peregrine *Falco peregrinus*	168	170	—	—	—	—
Black grouse *Tetrao tetrix*	172	—	—	—	—	—
Capercaillie *Tetrao urogallus*	176	176	177	—	—	457
Grey partridge *Perdix perdix*	179	180	—	386, 389	—	—
Quail *Coturnix coturnix*	182	—	—	—	—	—
Water rail *Rallus aquaticus*	184	—	—	—	—	430
Spotted crake *Porzana porzana*	187	—	—	—	—	430
Corncrake *Crex crex*	189	192	—	—	—	—
Oystercatcher *Haematopus ostralegus*	—	—	—	389, 394	—	430, 445
Avocet *Recurvirostra avosetta*	197	199	—	—	—	430
Stone-curlew *Burhinus oedicnemus* *	—	—	—	—	—	—
Ringed plover *Charadrius hiaticula*	203	—	—	—	—	430, 445
Dotterel *Charadrius morinellus*	206	—	—	—	—	—
Golden plover *Pluvialis apricaria*	—	—	—	394	—	430
Grey plover *Pluvialis squatarola*	—	—	—	—	—	430
Lapwing *Vanellus vanellus*	212	—	—	386, 389, 394	—	430
Knot *Calidris canutus*	—	—	—	—	—	430
Purple sandpiper *Calidris maritima*	216	218	—	—	—	430, 445
Dunlin *Calidris alpina*	—	—	—	394	—	430
Ruff *Philomachus pugnax*	—	—	—	—	—	430
Jack snipe *Lymnocryptes minimus*	—	—	221	—	—	430
Snipe *Gallinago gallinago*	—	—	—	389, 394	—	430
Woodcock *Scolopax rusticola*	225	—	226	—	—	430, 457
Black-tailed godwit *Limosa limosa*	229	—	—	—	—	430
Bar-tailed godwit *Limosa lapponica*	—	—	—	—	—	430, 445
Whimbrel *Numenius phaeopus*	232	—	—	—	—	430
Curlew *Numenius arquata*	—	—	—	389, 394	—	430, 445
Redshank *Tringa totanus*	—	—	—	394	—	430, 445
Greenshank *Tringa nebularia*	236	—	—	—	—	430, 445
Turnstone *Arenaria interpres*	—	—	—	—	—	430, 445
Red-necked phalarope *Phalaropus lobatus*	240	241	242	—	—	—
Arctic skua *Stercorarius parasiticus*	244	246	—	—	—	—
Great skua *Catharacta skua*	250	250	—	—	—	—
Common gull *Larus canus*	—	—	—	406	413	—
Lesser black-backed gull *Larus fuscus*	—	—	—	406	413	—
Herring gull *Larus argentatus*	—	—	—	406	413	—
Great black-backed gull *Larus marinus*	—	—	—	406	413	—
Kittiwake *Rissa tridactyla*	255	257	—	—	—	451
Sandwich tern *Sterna sandvicensis*	—	—	—	420	425	—
Roseate tern *Sterna dougallii*	—	—	—	420	425	—
Arctic tern *Sterna paradisaea*	—	—	—	420	425	—
Little tern *Sterna albifrons*	—	—	—	420	425	—
Guillemot *Uria aalge*	266	268	—	—	—	451
Razorbill *Alca torda*	272	275	—	—	—	451
Black guillemot *Cepphus grylle*	279	282	—	—	—	451
Puffin *Fratercula arctica*	286	290	—	—	—	451
Stock dove *Columba oenas*	—	—	—	386, 389	—	—
Turtle dove *Streptopelia turtur*	—	—	—	389	—	—
Barn owl *Tyto alba*	293	—	—	—	—	—

cont.

	SINGLE-SPECIES METHODS			GENERIC METHODS		
	Breeding		Wintering	Breeding		Wintering
	Pop.	Prod./Succ.		Pop.	Prod./Succ.	
Short-eared owl *Asio flammeus*	298	—	—	—	—	—
Nightjar *Caprimulgus europaeus*	301	—	—	—	—	—
Green woodpecker *Picus viridis*	304	—	—	386, 389	—	—
Great spotted woodpecker *Dendrocopos major*	307	310	—	386, 389	—	—
Lesser spotted woodpecker *Dendrocopos minor*	312	—	—	—	—	—
Woodlark *Lullula arborea*	315	—	—	—	—	—
Skylark *Alauda arvensis*	318	—	319	386, 389	—	—
Sand martin *Riparia riparia*	322	—	—	—	—	—
Swallow *Hirundo rustica*	—	—	—	389	—	—
Yellow wagtail *Motacilla flava*	325	—	—	389	—	—
Dunnock *Prunella modularis*	—	—	—	386, 389	—	—
Nightingale *Luscinia megarhynchos*	327	—	—	—	—	—
Black redstart *Phoenicurus ochruros*	329	—	—	—	—	—
Redstart *Phoenicurus phoenicurus*	—	—	—	389	—	—
Whinchat *Saxicola rubetra*	332	—	—	—		
Stonechat *Saxicola torquata*	334	—	—	—	—	—
Ring ouzel *Turdus torquatus*	336	—	—	—	—	—
Blackbird *Turdus merula*	—	—	—	386, 389	—	—
Song thrush *Turdus philomelos*	—	—	—	386, 389	—	—
Redwing *Turdus iliacus*	339	—	—	—	—	—
Cetti's warbler *Cettia cetti*	342	—	—	—	—	—
Marsh warbler *Acrocephalus palustris*	344	—	—	—	—	—
Dartford warbler *Sylvia undata*	347	—	—	—	—	—
Firecrest *Regulus ignicapillus*	349	—	—	—	—	—
Spotted flycatcher *Muscicapa striata*	—	—	—	386, 389	—	—
Bearded tit *Panurus biarmicus*	352	354	—	—	—	—
Marsh tit *Parus palustris*	—	—	—	386, 389	—	—
Crested tit *Parus cristatus*	356	—	356	—	—	457
Golden oriole *Oriolus oriolus*	357	—	—	—	—	—
Chough *Pyrrhocorax pyrrhocorax*	359	—	—	—	—	—
Starling *Sturnus vulgaris*	—	—	—	386, 389	—	—
Tree sparrow *Passer montanus*	362	—	—	389	—	—
Goldfinch *Carduelis carduelis*	—	—	—	386, 389	—	—
Linnet *Carduelis cannabina*	364	—	—	—	—	—
Twite *Carduelis flavirostris*	366	—	—	—	—	—
Scottish crossbill *Loxia scotica*	370	—	370	—	—	457
Bullfinch *Pyrrhula pyrrhula*	—	—	—	386, 389	—	—
Hawfinch *Coccothraustes coccothraustes*	371	—	—	—	—	—
Snow bunting *Plectrophenax nivalis*	374	—	—	—	—	—
Cirl bunting *Emberiza cirlus*	377	—	379	—	—	—
Reed bunting *Emberiza schoeniclus*	—	—	—	386, 389	—	—
Corn bunting *Miliaria calandra*	381	—	—	389	—	—

* Details of the methods for this species are available only on written request.

Single-species methods

Red-throated diver
Gavia stellata

Status

Amber listed: SPEC 3 (V)
Schedule 1 of WCA 1981
Annex I of EC Wild Birds Directive

National monitoring

Breeding surveys: Shetland 1983.
National surveys: 1994 (RSPB/SNH), 2004.
WeBS.

Population and distribution

Red-throated divers breed predominantly in Shetland, Orkney, the Outer Hebrides and the Caithness flows, although they do occur at lower densities in south-west Scotland (*88–91 Atlas*). Their breeding success varies considerably from year to year (Okill and Wanless 1990). In 1994, there were an estimated 935 breeding pairs in Scotland (Gibbons et al 1995).

The majority of the wintering population of around 4,850 birds (*Population Estimates*) is found on the British east coast. The numbers fluctuate considerably in response to weather conditions.

Ecology

Red-throated divers usually breed on small pools or lochans in open moorland but will also use pools in forested areas. Most pools and lochs hold only one pair, but where pools are abundant there may be several pairs within a few hundred metres, and densities over large areas may be high (about 1 pair per km^2) (*Red Data Birds*). The nest is usually near the water's edge, on the loch shore or on an island. Shallow pools with well-vegetated banks, promontories and islets are preferred. Nesting pools or lochans are seldom used for feeding. In coastal areas, the bulk of feeding is carried out at sea, while those breeding inland move to larger lochs close to the nesting loch to feed. Peak egg laying is from late May to early June, and clutch size is 2–3. There is one brood, although early losses are often replaced. The average incubation period lasts 27 days and the nidifugous young fledge after 34–48 days. The chicks tend to remain on the nesting loch until fledged but some pre-fledging movement to larger/other lochs nearby occurs. The chicks are not normally left alone until they are about three weeks old.

Breeding season survey – population

Survey methods are based on those developed by Gomersall et al (1984) and modified by Gibbons et al (1995).

Information required

- number of breeding pairs
- maximum number of non-breeding adults
- map showing boundary of the survey area.

Number and timing of visits

Two visits, one visit at the end of May or in June and one in July. There should be at least 14 days between visits.

Time of day

Any time of day.

Weather constraints

Avoid days with poor visibility, persistent rain or winds stronger than Beaufort force 4.

Sites/areas to visit

Small pools and lochans in open moorland and forested areas.

Equipment

- Schedule 1 licence
- 1:25,000 OS map of the area
- A4 photocopied map of the survey area for use in the field
- recording forms.

Safety reminders

Ensure someone knows where you are going and when you expect to return. Take a compass and always carry a survival bag, waterproofs, whistle, extra clothing, food and a first-aid kit in remote areas. Take extra care when working close to water and if any boat trips are necessary, wear life-jackets and make sure that at least two people are present.

Disturbance

No visits should be made to active nests. Observers do not need to search all shorelines on foot – every effort should be made to collect complete data without any disturbance to a sitting bird. Egg-collectors are a threat to this species.

Methods

You should have prepared field maps and recording forms (Figures 1–2). Mark the boundary of the survey area clearly on the field map. On each visit mark any area excluded from the survey, due to the ground or habitat being unsuitable, with an X. Circle all lochs and waterbodies visited, tick all other 1-km squares on the map that have been visited. Complete a new map for each visit, but use the same recording form for both visits.

Check all areas of standing water. OS maps may not show all suitable lochans, therefore visit all other flat, apparently dry areas. Exclude areas that are built-up, steeply sloping or have 100% plantation cover, as well as the sea and fast-flowing rivers.

Scan all areas of water and all shorelines. View from a distance to avoid disturbing incubating birds. Note that adults may lie flat on the water or on the nest in an attempt to be inconspicuous. It is not necessary to visit nests – the presence of an incubating bird is sufficient. If you are unsure whether the bird is incubating, brooding or resting, remain at a distance but observe for a longer period for signs of eggs being turned, a changeover or the presence of young.

If birds are only present on the water, or if the loch is sufficiently large to hold two or more territories, or if no birds are observed during

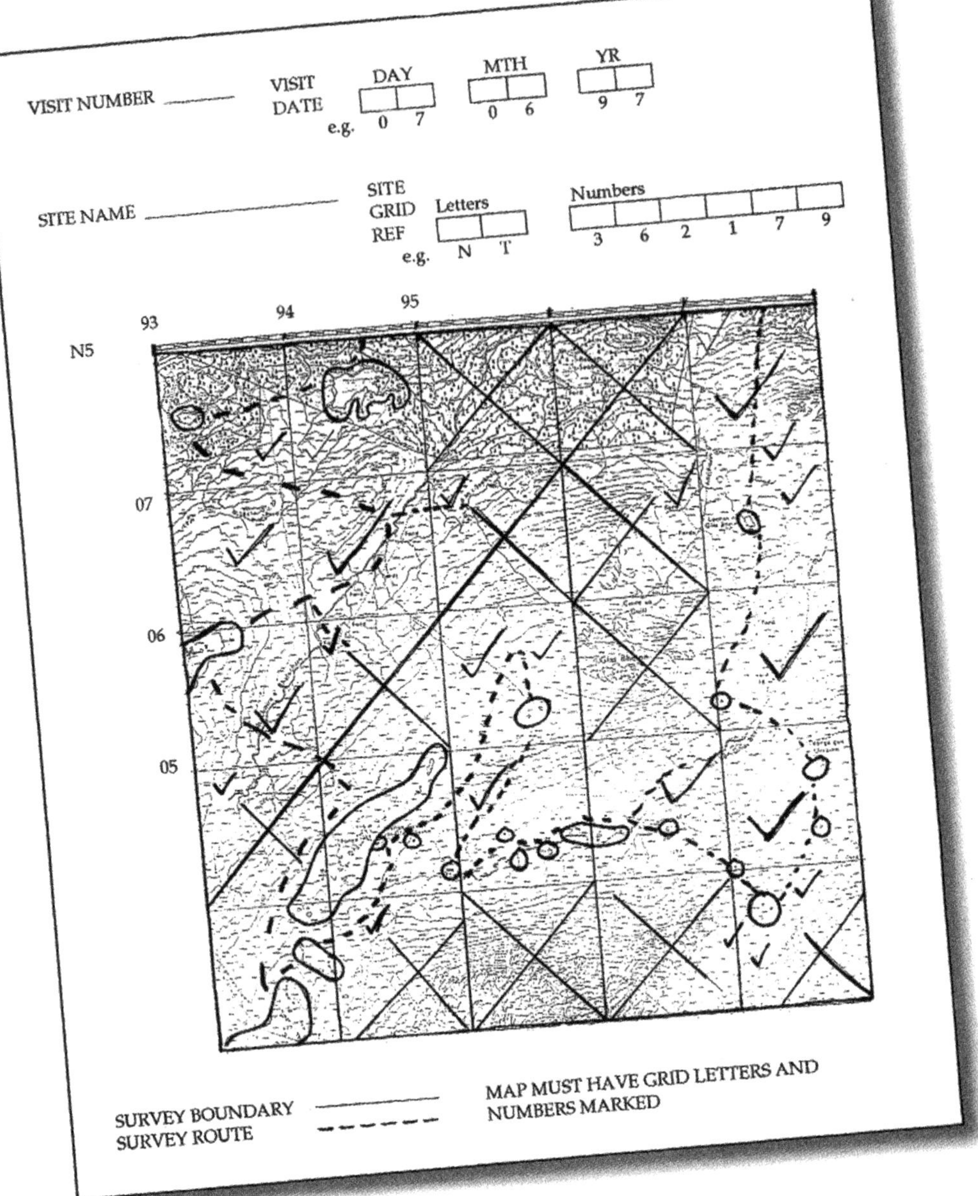

Figure 1
An example of a field map used in a red-throated diver survey. The areas crossed through are those deemed unsuitable habitat (too steep, built-up or dense plantation forest). The circled areas are waterbodies that have been visited, and ticked 1-km squares have also been searched for suitable waterbodies. The survey area should have its boundary clearly marked (in this case the survey area was a 5-km square).

scanning, walk the entire shoreline and carefully scan any islands for empty nest scrapes, incubating birds or signs that birds may have attempted to breed and failed (eg broken eggshells, dead chicks). At larger lochs, pay particular attention to sheltered bays and the shallower ends of the loch. At dubh (peaty) lochan pool systems, walk through the system to ensure that every pool and shoreline is checked. This should be possible without walking the perimeter of every pool.

Be careful that you do not record signs from the previous year. Eggshell fragments from the previous season will be bleached and almost colourless whereas fresh remains are olive-brown. There may also be the remains of scrapes from the previous year and water-worn hollows which can resemble scrapes. An active nest scrape is one that contains eggs that are still being incubated.

If no breeding attempt is noted after walking the shoreline but birds are present feeding or loafing on the water, observe their behaviour for up to one hour, taking particular note of birds pairing, birds attempting to go ashore or any other indications that the birds are on territory and may attempt to breed. Observe from a discreet distance so as not to disturb the birds and influence their behaviour. Note the direction in which birds fly if/when they leave the loch. Note that there may be 'assembly lochs' where there are large aggregations of divers which may be inexperienced and which have repeated clutch failures.

If clutch or chick loss is suspected, or evidence of loss is found, continue to search the shores on this and neighbouring lochs for replacement clutches and/or eggshell remains.

On the second visit, various factors can cause confusion, such as more than one pair breeding on the same loch, replacement clutches laid at different lochs by the same pair, unfledged chicks moving to a different loch, and chicks sheltering under banks or diving repeatedly. The water level may also have dropped considerably since the divers started nesting, thereby altering the shoreline.

Use a separate recording form for each breeding territory. Where lochs contain more than one territory (or potential territory) complete a form for each. Use a single recording form for lochs with no breeding territories. Ensure that you record all non-breeding birds on breeding lochs on a single form; generally this will be one of the breeding territory forms.

RED-THROATED DIVER SURVEY

Please write clearly in BLOCK LETTERS.
Please use 24 hr clock.

OBSERVER A. SPOTTER SITE NAME GLEN MHARI LOCH NAME MHARI

LOCH GRID REF Letters N H Numbers 1 2 3 4 5 6
e.g. N T 3 6 2 1 7 9

LOCH AREA (ha) 15ha Nº OF ISLANDS 1 NEST LOCATION: SHORE ____ ISLAND ✓ FLOATING VEG. ____ (tick one only)

SHORE TYPE AT NEST GRASS

NEST GRID REFERENCE N H 1 2 3 4 5 6

VISIT DATA "I" refers to an incubating adult, "A" refers to an active nest and "½" refers to the size of the young relative to an adult.

VISIT	DATE	START TIME	END TIME	NUMBER ADULTS	NUMBER NESTS	NUMBER EGGS	NUMBER EGG SHELLS	NUMBER OF YOUNG	SIZE OF YOUNG	COMMENTS
1	7/6/94	08:30	10:00	1I + 1	1A + 1	2	0	–	–	1 ADULT INCUBATING, THE OTHER FLEW IN, PREENED & LOAFED AROUND
2	11/7/94	09:05	10:30	1			–	2	½	1 ADULT WITH 2 CHICKS DIVING, YELLOW RING ON LEFT LEG OF ADULT
3	/ /	:	:							
4	/ /	:	:							

Figure 2
An example of a recording form used in a red-throated diver survey. This type of recording form helps to standardise the information collected. From the examples given 'I' refers to an incubating adult, 'A' refers to an active nest and ½ refers to the size of young relative to an adult.

On the survey form record:
- date
- start and finish time
- waterbody area (this can be estimated from the map)
- nest location (shore, island or floating vegetation)
- six-figure grid reference of active nest site (if present)
- number of incubating adults (I) plus the number of adults feeding, loafing, etc (eg 1I+1 = one incubating bird plus one other)
- number of nest scrapes (A) plus the number of empty scrapes (eg 1A + 1 = one active nest-site plus one empty scrape)
- number of eggs
- number of eggs estimated from broken shells
- number and size of any chicks seen (see below).

Report the following:
- number of breeding pairs (incubating birds seen, or young, eggshell fragments or dead chicks located)
- maximum number of non-breeding adults which is the sum of the following:
  - the maximum number of non-breeding adults on lochs where birds were recorded on one or both visits but there was no evidence of breeding; and
  - the maximum number of non-breeding adults on lochs containing breeding pairs (calculate this as the maximum number of adults across the two visits minus the number of breeding adults).

Breeding season survey – productivity

Information required
- number of viable young (ie about two-thirds grown) per pair.

Number and timing of visits

One visit (additional to the two to estimate the number of breeding pairs) after 23 July (best from 23 July to 15 August), checking those birds which laid earliest first.

Time of day

As for population survey (above). Allow sufficient time for the size of the loch.

Weather constraints, Sites/areas to visit

As for the population survey (above).

Disturbance

It is particularly important not to disturb any adults with young. It is possible to get all the information necessary by observing the area from a distance, eg from a small hill overlooking the water.

Methods

Check all areas of water within the survey where adults were present earlier in the season (whether proved breeding or not) for young. Scan the water and shoreline to locate any young present. If adults are located, scan the whole of the water surface and shoreline, even if there do not appear to be any young with them. The young are not always with their parents, often being close inshore and diving repeatedly, making them difficult to spot.

Record the number of adults present and the number and size of the chicks. Size chicks by comparing the water-line length of the chick with that of a nearby adult. They usually appear to be about 20–25% of adult length shortly after hatching. Record details of any sightings on the same recording form as used earlier in the season when locating adults and nests. Notes on the behaviour of the birds during the observation period can be made on the other side of the recording form.

Winter survey

WeBS.

Survey methods for *Inshore marine waterfowl* and *Waterfowl and seabirds at sea* are outlined in the generic survey methods section.

References

Gibbons, D W, Bainbridge, I and Mudge, G (1995) *The 1994 Red-throated Diver Gavia stellata Survey*. Unpublished RSPB report to SNH.

Gomersall, C H, Morton, J S and Wynde, R M (1984) Status of breeding red-throated divers in Shetland, 1983. *Bird Study* 31: 223–229.

Okill, J D and Wanless, S (1990) Breeding success and chick growth of red-throated divers *Gavia stellata* in Shetland 1979–88. *Ringing and Migration* 11: 65–72.

Black-throated diver *Gavia arctica*

Status

Amber listed: BR, SPEC 3 (V)
Schedule 1 of WCA 1981
Annex I of EC Wild Birds Directive

National monitoring

Annual survey of breeding black-throated divers at raft sites and paired natural sites (RSPB).
National surveys: 1985 (RSPB/SNH), 1994 (RSPB/SNH), 2004.
Annual counts of wintering birds in the Moray Firth (RSPB/BP).
WeBS.

Population and distribution

Black-throated divers breed almost exclusively on the larger freshwater lochs of mainland north-west Scotland and the Outer Hebrides, where there are estimated to be 150–160 breeding pairs. The population may have declined slightly in the last 20 years (*88–91 Atlas*), and breeding success has been shown to be relatively low, mainly due to predation and nest flooding (Campbell and Mudge 1989). Encouragingly, a proportion of the Scottish population has used artificial nesting rafts (Hancock 1994) with a resulting increase in breeding success. Approximately 1,300 black-throated divers winter around Britain's coast, tending to concentrate in the larger sandy bays and inlets, especially off the Scottish west coast.

Ecology

Black-throated divers breed on lochs in north and west Scotland, in a few cases as far south as Argyll. Breeding generally occurs on larger lochs or occasionally on smaller pools near larger lochs. Birds require areas of shallow water for feeding, and prefer vegetated islands without steeply sloping sides or sheltered and undisturbed parts of the mainland for nesting. More than one territory may occur on larger or more productive lochs. Adults return to breed in March/April and will remain until late August/September.

Breeding season survey – population

These methods are based on the RSPB black-throated diver survey instructions to fieldworkers (Jackson and Hancock 1994). Sites are defined as monitoring sites where birds are known to have bred (these are monitored on an annual basis); or other sites.

Information required

- number of summering territories
- number of breeding pairs.

Number and timing of visits

Two or more between 23 April and 23 July.

Time of day

Any time of day.

Weather constraints

Avoid adverse weather conditions as choppy water makes the birds more difficult to see.

Sites/areas to visit

Lochs in north and west Scotland.

Equipment

- Schedule 1 licence
- 1:25,000 OS map of the area
- recording form/map.

Safety reminders

Ensure someone knows where you are going and when you expect to return. Take a compass and always carry a survival bag, waterproofs, whistle, extra clothing, food and first-aid kit in remote areas. Take extra care when working close to water and if any boat trips are necessary make sure that at least two people are present and that you wear life-jackets.

Disturbance

Do not visit active nests. You can usually observe out of sight from the birds. Observers are not required to search all shorelines on foot and every effort should be made to collect complete data without any disturbance to a sitting bird. Egg-collectors are a serious threat to this species.

Methods

Visit monitoring sites between 7 and 31 May. If breeding is confirmed, you do not need to make further visits except to assess breeding success (see below). If you locate a pair without eggs or young, you should make a further visit before 15 June, to confirm occupancy. If you do not locate any birds, check all shores and all water and make at least two subsequent visits within two weeks to confirm occupancy.

Visit other sites twice, once in the period 23 April – 7 June, and once during 8 June – 23 July. If you confirm breeding on the first visit (ie adults with eggs or young), you do not need to make any further visits to the territory.

On each visit, spend at least two hours checking all areas of water and all shores for birds. When checking shorelines, look for a sitting, brooding or prospecting/nest-building bird. Where there are islands or a complex shoreline, you may need to carry out a partial perimeter walk. However, in most cases, you can achieve complete coverage by scanning from a number of viewpoints. At large territories, or where there are many islands, coverage by boat on a calm day is most efficient. Mark on the recording map the location, time and number of any adult black-throated divers observed. Record their behaviour on the other side of the recording map.

An example recording form is shown in Figure 1.

A summering territory is one where:

- a pair of adults is observed at the same site on two or more visits

- a pair of adults is observed on one visit and a single adult on one or more other visits to the same site
- a single adult is observed at the same site on two or more visits.

The number of breeding pairs is the number of summering territories where breeding is confirmed, ie an incubating adult, a nest with eggs, or chicks are seen on any of the visits.

If you see any young chicks, record their number and size. Size chicks by comparing the water-line length with that of a nearby adult. They usually look about 20–25% of adult length shortly after hatching.

In addition, record the state of the water surface (choppy, moderate or calm); the percentage of the water surface and the shoreline checked; visibility (poor, moderate or good) and predators or human disturbance (eg fishing boats).

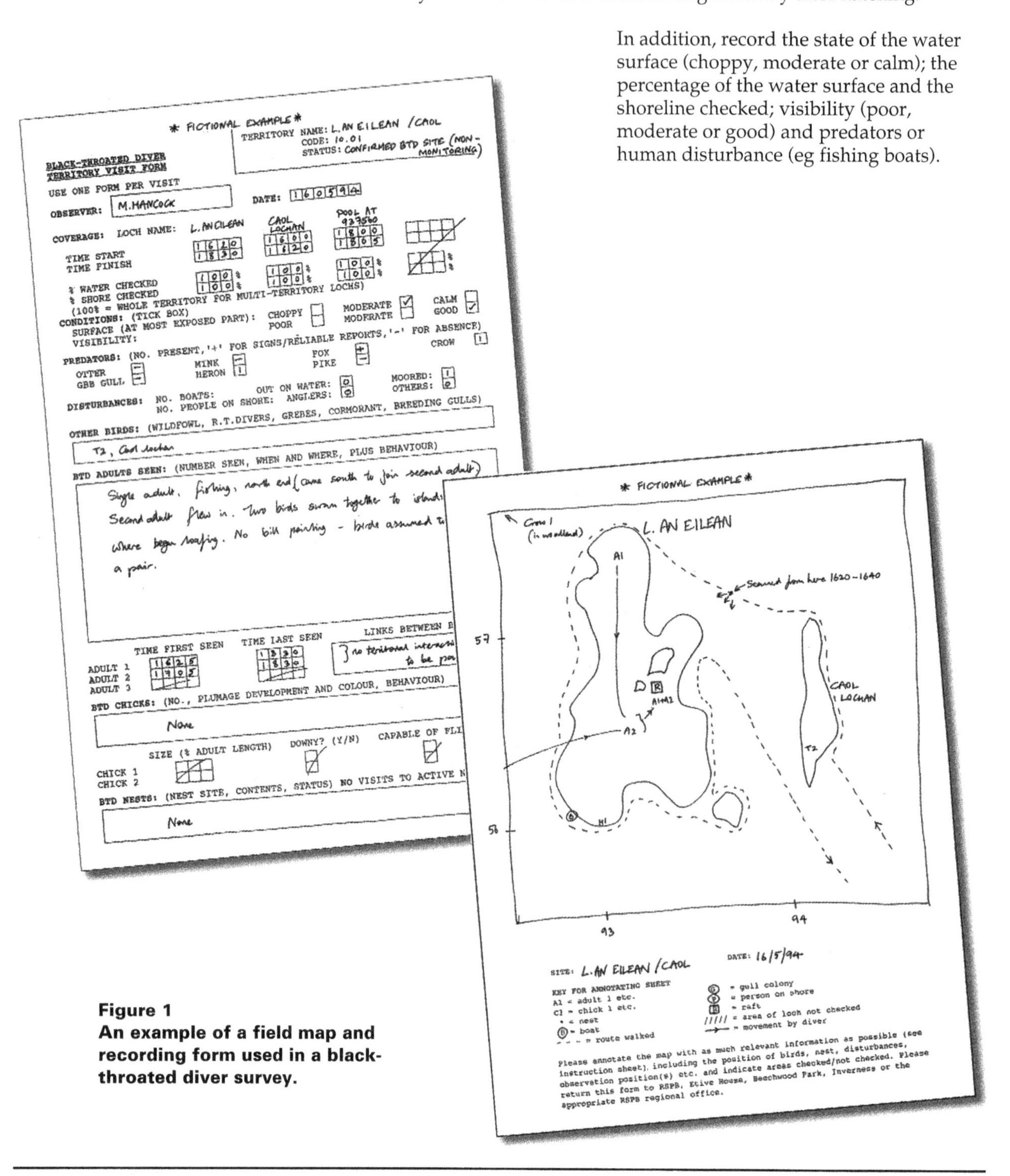

* FICTIONAL EXAMPLE *

BLACK-THROATED DIVER TERRITORY VISIT FORM

TERRITORY NAME: L. AN EILEAN / CAOL
CODE: 10.01
STATUS: CONFIRMED BTD SITE (NON-MONITORING)

USE ONE FORM PER VISIT

OBSERVER: M.HANCOCK

DATE: 160594

COVERAGE: LOCH NAME:	L. AN EILEAN	CAOL LOCHAN	POOL AT 937560	
TIME START	1620	1600	1800	
TIME FINISH	1830	1620	1805	
% WATER CHECKED	100 %	100 %	100 %	%
% SHORE CHECKED	100 %	100 %	100 %	%

(100% = WHOLE TERRITORY FOR MULTI-TERRITORY LOCHS)

CONDITIONS: (TICK BOX)
SURFACE (AT MOST EXPOSED PART): CHOPPY ☐ MODERATE ☑ CALM ☐
VISIBILITY: POOR ☐ MODERATE ☐ GOOD ☑

PREDATORS: (NO. PRESENT, '+' FOR SIGNS/RELIABLE REPORTS, '–' FOR ABSENCE)
OTTER – GBB GULL – MINK – HERON 1 FOX + PIKE – CROW 1

DISTURBANCES: NO. BOATS: OUT ON WATER: 0 MOORED: 1
NO. PEOPLE ON SHORE: ANGLERS: 0 OTHERS: 0

OTHER BIRDS: (WILDFOWL, R.T.DIVERS, GREBES, CORMORANT, BREEDING GULLS)
T2, Caol Lochan

BTD ADULTS SEEN: (NUMBER SEEN, WHEN AND WHERE, PLUS BEHAVIOUR)
Single adult, fishing, north end (came south to join second adult)
Second adult flew in. Two birds swam together to island
where began loafing. No bill pointing – birds assumed to
a pair.

	TIME FIRST SEEN	TIME LAST SEEN	LINKS BETWEEN B
ADULT 1	1625	1830	} no territorial interacti… to be pa…
ADULT 2	1705	1830	
ADULT 3			

BTD CHICKS: (NO., PLUMAGE DEVELOPMENT AND COLOUR, BEHAVIOUR)
None

	SIZE (% ADULT LENGTH)	DOWNY? (Y/N)	CAPABLE OF FLI…
CHICK 1			
CHICK 2			

BTD NESTS: (NEST SITE, CONTENTS, STATUS) NO VISITS TO ACTIVE N…
None

* FICTIONAL EXAMPLE *

SITE: L. AN EILEAN / CAOL

DATE: 16/5/94

KEY FOR ANNOTATING SHEET
A1 = adult 1 etc.
C1 = chick 1 etc.
• = nest
Ⓑ = boat
- - - = route walked
Ⓖ = gull colony
Ⓟ = person on shore
Ⓡ = raft
///// = area of loch not checked
→ = movement by diver

Please annotate the map with as much relevant information as possible (see instruction sheet), including the position of birds, nest, disturbances, observation position(s) etc. and indicate areas checked/not checked. Please return this form to RSPB, Etive House, Beechwood Park, Inverness or the appropriate RSPB regional office.

Figure 1
An example of a field map and recording form used in a black-throated diver survey.

Breeding season survey – productivity

Information required
- number of viable young per pair.

Number and timing of visits

Usually only one. Subsequent visit(s) may be necessary if small chick(s) are still present.

Time of day

Any time of day.

Weather constraints, Sites/areas to visit

As for the population survey (above).

Disturbance

It is especially important not to disturb adults with young. It should be possible to get all necessary information by observing from a distance.

Methods

Scan all water and shores (see above) to locate any young present. Chicks are often far from their parents and difficult to see, so even if you locate the adults and there are no young with them, you should still carefully scan the whole of the water surface and all shores. Make a further visit if young are less than two-thirds grown.

Mark on a recording map the number of adults and the number and size of the chicks present, the date and time when you saw them and where they were during the observation period. Size chicks by comparing the water-line length of the chick with that of a nearby adult. They usually appear to be about 20–25% of adult length shortly after hatching. Viable young are those over two-thirds grown (roughly one month old). If you are unsure whether any adults observed have young, record their behaviour on the reverse of the recording map. Calculate the number of viable young per pair by dividing the total number of viable young recorded by the total number of summering territories.

Record additional information as for the population survey (see above).

Winter survey

WeBS.

Survey methods for *Inshore marine waterfowl* and *Waterfowl and seabirds at sea* are outlined in the generic survey methods section.

References

Campbell, L H and Mudge, G P (1989) Conservation of black-throated divers in Scotland. *RSPB Conservation Review* 3: 72–74.

Hancock, M (1994) *Black-throated Diver Newsletter*. RSPB unpubl.

Jackson, D and Hancock, M (1994) *Black-throated Diver Survey Instructions*. RSPB unpubl.

Great northern diver *Gavia immer*

Status

Amber listed: WI
Non-SPEC
Schedule 1 of WCA 1981
Annex I of EC Wild Birds Directive

National monitoring

Rare Breeding Birds Panel.
WeBS.

Population and distribution

Great northern divers winter in greatest concentrations on the north and west coast of Scotland, the south-west of England and most of the Irish coast. An estimated 3,000 great northern divers winter around Great Britain (*Population Estimates*).

Ecology

They can occur off a variety of coastlines such as shallow sandy bays and rocky headlands, but characteristically winter in deep water further offshore than other diver species (*Winter Atlas*).

Winter survey

WeBS.

Survey methods for *Inshore marine waterfowl* and *Waterfowl and seabirds at sea* are outlined in the generic survey methods section. Extend the survey period to include April and May if passage flocks are to be surveyed.

Slavonian grebe *Podiceps auritus*

Status

Amber listed: BR
Non-SPEC
Schedule 1 of WCA 1981
Annex I of EC Wild Birds Directive

National monitoring

Annual breeding survey at known sites (RSPB).
National surveys: 1992, 2003.
Rare Breeding Birds Panel.
WeBS.

Population and distribution

Slavonian grebes started to colonise Scotland in 1908 (Crooke et al 1992). In the UK, their breeding stronghold is in the Scottish Highlands, Inverness-shire in particular (*88–91 Atlas*). At about 60 pairs, the population is small, and still regarded as vulnerable (*88–91 Atlas*).

Ecology

In Scotland, this species occurs on mesotrophic lochs where they nest in emergent vegetation, particularly bottle sedge *Carex rostrata* (Crooke et al 1992), although birds have been known to occupy more productive lochs and nest in loch-side bushes or overhanging branches (*88–91 Atlas*). More than one territory may occur on larger or more productive lochs. Eggs are laid mid-May to early August, mainly mid-May to June; clutches are generally of 4–5 eggs, and second broods are rare (*Red Data Birds*). The fledging period is about eight weeks (*BWP*).

Breeding season survey – population

This method is based on instructions for the national breeding surveys (Crooke et al 1992).

Information required

- number of pairs
- number of single birds.

Number and timing of visits

Monitoring sites (sites where Slavonian grebes have bred and are monitored on an annual basis) should be visited during 20–31 May. If breeding is confirmed (a pair with eggs or young), no further visits are required. If you locate a pair without eggs or young or a single bird, visit before 10 June to confirm occupancy. At all unconfirmed sites make at least two subsequent visits before July, to confirm or refute occupancy.

Other potential sites should be visited twice, once in late May and once in July. If you confirm breeding on the first visit (adults with eggs or young), you do not need to make further visits.

Time of day

Any time of the day.

Weather constraints

No specific constraints.

Sites/areas to visit

Mesotrophic lochs with areas of suitable nesting vegetation.

Equipment

- 1:25,000 OS map of the area
- A4 photocopied maps for use in the field
- Schedule 1 licence.

Safety reminders

Always inform someone where you are working and when you expect to return. Take a compass and always carry a survival bag, waterproofs, whistle, extra clothing, food and a first-aid kit in remote areas. Take extra care when working close to water, and if any boat trips are necessary make sure that at least two people are present and that you wear life-jackets.

Disturbance

Do not visit active nests. Observers are not required to search all shorelines on foot and you should make every effort to collect complete data without disturbance to sitting birds. Egg-collectors are a serious threat to this species.

Methods

Scan the loch from a suitable distance (>100 m) to avoid disturbance. Check all shores and all water thoroughly, especially sedge beds and areas of emergent vegetation. Watch from suitable vantage points, carrying out a complete circuit of the shoreline if appropriate. Note that grebes can be difficult to detect, particularly when concealed in emergent vegetation, and you should spend a minimum of two hours at each suspected territory.

Breeding is confirmed if:

- a bird with eggs or young is seen
- one or two birds are present on more than one occasion at sites with suitable nesting habitat.

If breeding is confirmed, you do not need to make further visits unless there is a need to assess breeding success (see below).

Record the numbers of pairs and single birds. Describe the birds' behaviour if there is any doubt about their breeding status. Mark their location and movements on the field map and cross-refer these locations to any description or notes made.

Record other additional information including: the state of the water surface (choppy, moderate or calm); the percentage of the water surface and the shoreline that has been checked; visibility (poor, moderate or good); and whether there were any obvious signs of predators or human disturbance.

Breeding season survey – productivity

Information required

- the number of viable young (ie about two-thirds grown) per pair.

Number and timing of visits

One visit between mid-July and mid-August, exact date depending on breeding stage observed during population survey (see above). Subsequent visit(s) may be necessary if small chick(s) are still present.

Time of day, Weather constraints, Sites/areas to visit, Equipment, Safety reminders

As for the population survey (see above).

Disturbance

It is particularly important not to disturb adults with young. It is possible to get all the necessary information by observing the area from a distance, sometimes from a small hill overlooking the loch.

Methods

Scan the whole water surface and shoreline to locate any young present, even if the adults are located and there are no young with them. Very small chicks are sometimes found hauled out by the nest-site on a bank or on an island, otherwise chicks will often be located very close to shore in the shallower calmer waters. This can make the chicks very difficult to spot, especially on complex shorelines with emergent vegetation.

Record the date, the number of adults present and the number and size of the chicks on a recording form and mark their location on a field map. Size the chicks by comparing their water-line length with that of a nearby adult. If the situation is unclear, then it may help to describe the behaviour of the birds during the observation period. Make a further visit if the young are less than two-thirds grown.

Record additional information as for the population survey (see above).

Reference

Crooke, C, Dennis, R, Harvey, M and Summers, R (1992) Population size and breeding success of Slavonian grebes in Scotland. In *Britain's Birds in 1990–91: the conservation and monitoring review*. BTO, Thetford.

Black-necked grebe *Podiceps nigricollis*

Status

Amber listed: BR
Non-SPEC
Schedule 1 of WCA 1981

National monitoring

Regular sites monitored annually (RSPB/EN).
Rare Breeding Birds Panel.
WeBS.

Population and distribution

First recorded nesting in the UK in 1904, the black-necked grebe has remained a rare breeding bird. Records are sparse and well-distributed in central Scotland and England, but there are only a few regular breeding sites. The European population fluctuates from year to year but since 1980 has tended to increase, at least in the northern and western edges of the species' range (*88–91 Atlas*). There are estimated to be 23–48 pairs breeding in the UK (*Population Estimates*), although in some years this figure is likely to be an underestimate.

Ecology

Black-necked grebes are found in lowland eutrophic meres, ponds, lochs and reservoirs with extensive emergent vegetation. Their nests are very well hidden in dense reeds or sedges. Clutches of 3–5 eggs are laid during late April to July; there are occasionally two broods and most young fledge by early August (*Red Data Birds*). It can be very difficult to tell a first brood from a second brood without frequent visits.

Breeding season survey – population

Information required
- number of breeding pairs
- maximum number of individual birds
- date of each visit.

Number and timing of visits

Twice weekly, from mid-May to the end of June.

Time of day

Any time.

Weather constraints

Avoid very wet and windy conditions.

Sites/areas to visit

Any potentially suitable site plus any sites which have had breeding black-necked grebes in the past. Check all potential sites each year. NB Potential sites can include areas with little open water.

Equipment

- telescope
- Schedule 1 licence.

Safety reminders

No specific reminders. See the general guidelines in the *Introduction*.

Disturbance

All breeding site information must be kept strictly confidential and minimal information should be taken on survey visits. This species is prone to disturbance, so keep out of sight when making observations.

Methods

Choose vantage points from which the waterbody can be observed without disturbing the birds. During each visit, record the total number of black-necked grebe seen and also the number of pairs. It will take several visits to establish how many pairs are present: most birds (including non-breeders) will participate in courtship displays initially, but birds which have formed a pair will move around together for most of the season.

Report the maximum number of adults seen on any one visit and the estimated number of breeding pairs. Also report the date of each visit.

Breeding season survey – productivity

Information required

- number of fledged young per pair.

Number and timing of visits

As for the population survey, together with frequent visits to follow the fate of chicks up to mid-August.

Time of day

Any time.

Weather constraints, Equipment, Safety reminders

As for the population survey (above).

Sites/areas to visit

All sites where adults have been seen.

Disturbance

It is particularly important not to disturb adults with young. It is possible to get all the information necessary by observing the area from a distance, sometimes from a vantage point overlooking the water.

Methods

On each visit, scan all of the water and shores to locate any young present. Even if the adults are located and there are no young with them, the whole of the water surface and shores should be scanned. The fledging period is about eight weeks (*BWP*); chicks near fledging are more than two-thirds adult size.

Chicks can be very difficult to spot, especially along complex shorelines with emergent vegetation. Adults can appear to have failed even when chicks are around, so it is worth making a visit as late as August to check whether or not young are present. Note the number of adults and the number and size of the chicks present. Chicks should be sized by comparing the water-line length of the chick with that of a nearby adult. Make frequent visits to record the presence and absence of well-grown chicks, and assume that those of adult size which disappear have fledged. Report the total number of fledged young and the number of fledged young per breeding pair.

Manx shearwater *Puffinus puffinus*

Status

Amber listed: BI, BL, SPEC 2 (L)
Annex I of EC Wild Birds Directive (Balearic subspecies *P. p. mauretanicus*)

National monitoring

Seabird Colony Register (SCR).
Seabird Monitoring Programme (SMP).
The colonies on Skomer and Skokholm will be censused by the Wildlife Trust, West Wales, by 2000; Bardsey was surveyed in 1996 and Canna in 1997. The Rum colony will require special methods including the use of aerial photographs in combination with ground surveys to assess colony boundaries.
Seabird 2000 (1999–2001).

Population and distribution

Over 90% of the population of this shearwater breeds in Britain and Ireland (*Birds in Europe*). The main breeding areas are on rat-free islands, particularly the island of Rum off the west coast of Scotland, the islands of Skokholm and Skomer off the south-west coast of Wales and islands off the south-west coast of Ireland (*88–91 Atlas*). There are estimated to be 220,000–250,000 pairs breeding in the UK (*Population Estimates*).

Ecology

Manx shearwaters are colonial. They breed on grassy islands and headlands, nesting within excavated burrows and in some instances in boulder fields. They lay one egg, from the end of April to late July. The early incubation to early nestling stage is early May to early July. Most of the young are fledged by the end of September, but a few not until mid-October. Chicks are left unattended by day once they are about a week old (*Red Data Birds*).

Breeding season survey – population

Breeding Manx shearwaters are difficult to count accurately, because of their nocturnal and burrow-nesting habits. Two methods are suggested here, both have been taken from the *Seabird Monitoring Handbook*.

Method 1 – Counting Apparently Occupied Sites/Burrows (AOSs) in randomly selected quadrats (or for the whole colony in the case of small colonies).
Method 2 – Using playback to count Apparently Occupied Sites/ Burrows in randomly selected quadrats (this is the preferred method of the two).

Information required

- estimated number of Apparently Occupied Sites (AOSs)
- map showing colony boundaries and location of sample quadrats.

Number and timing of visits

Method 1 – One visit to each quadrat, early May to early July.
Method 2 – One visit to each quadrat, late May to early June.

Time of day

Any time during the day.

Weather constraints

Avoid very wet and windy conditions.

Sites/areas to visit

Any known colony.

Equipment

- 1:10,000 map of the area
- compass for measuring angles and directions of quadrats
- enough stakes to mark each quadrat
- rope (2.52 m for 20 m^2 quadrats, or 2.96 m for 30 m^2 quadrats)
- *Method 2* – high-quality tape recording (15 seconds duration) of the male Manx shearwater call, with Walkman and portable speakers plus spare batteries.

Safety reminders

If working alone, always ensure someone knows where you are and when you intend to return. If cliffs or steep slopes are to be climbed, use the correct safety equipment and do not work alone. Beware of soft and eroding cliff edges and overhangs, wear footwear with plenty of grip and use ropes or safety harnesses on steep slippery slopes

Disturbance

Avoid walking in the colony for any longer than is absolutely necessary. Avoid any area of the colony containing burrows in danger of collapse.

Methods

Method 1: Sample quadrats (no playback)

This method is not suitable where shearwaters occur in mixed colonies with puffins or sizeable populations of rabbits because burrows within the sample quadrats may not be occupied by shearwaters. Nest-sites among boulders or scree, where entrances cannot easily be distinguished, also present major problems.

Map the boundary of the colony on a map and calculate its area. If the colony can be easily divided into areas of markedly different density associated with, for example, differences in vegetation or substrate, treat each area as a separate subcolony and sample within each (stratified random sampling); otherwise, sample the whole colony. Select a number of random locations for your quadrats within each sampling area (about 30 for a whole colony or 15 for each subdivision within a colony, eg Gibbons and Vaughan in press). Further information on sampling is provided in the *Introduction*. Some selected quadrats may not be usable for safety reasons or because there is a greater danger of collapsing burrows than elsewhere. If so, select an alternative random quadrat.

The best alternative to random positioning of quadrats is systematic positioning, whereby quadrats are placed at fixed intervals through the colony. Never use your own judgement to position plots.

Place quadrats at the selected points regardless of whether or not there are any burrows present, even if the whole area is bare rock. The location of the point should be fixed by measuring bearings using a compass and pacing out distances from fixed points, such as rock outcrops or other structures. Keep a record of the location of each point. These points may be permanently staked and used each year, or new points selected annually.

Circular quadrats of 20 m^2 or 30 m^2 are the most convenient shape and size of quadrat. Rotate a 2.52- or 2.96-m rope around a fixed stake at each randomly selected point and count the number of occupied sites/burrows (AOSs) within the area covered. Include all burrows for which more than half or the whole entrance is completely within the quadrat.

Signs that burrows are occupied include recent signs of digging (often the most obvious indication), fresh droppings, eggshell remains at the entrance, or one or more shearwater body feathers. Detecting shearwater burrows by smell is usually difficult.

To obtain an estimate of the total population, calculate the mean number of AOSs in each quadrat (sum of estimates for each quadrat divided by the number of quadrats taken). Divide this by the area of each quadrat (typically 20 m^2 or 30 m^2). Multiply the result (mean density) by the total area of the colony (in m^2) to give the final figure. Where subdivisions (different strata) have been sampled separately, this calculation should be performed for each subdivision separately and the results summed.

At colonies where shearwaters nest among scree or boulders, this method cannot be used. A very rough way of assessing the relative breeding densities in rocky versus turf habitats is to walk at a fixed pace through both habitats soon after sunset along random transects, recording the number of birds seen landing and departing per unit time and area. Ideally this should be done by the same observer on the same night, in all habitat types, within a short period of time. If this is not possible (eg if the colonies in the different habitats are widely separated), visit the different habitats, at the same time on different, randomly selected nights in similar weather conditions, within the same two-week period.

Method 2: Tape playback in sample quadrats

The count unit for assessing the population size in this method is still the apparently occupied site/burrow (AOS), but in this case the AOS figure is calculated from the number of burrows from which there is a response to playback.

Map the whole colony area and select the sampling areas and sampling points as above. It is very important that the same procedure is followed at all times. Use a recording of a male Manx shearwater call played at natural volume within 0.3 m of the burrow entrance for a maximum of 15 seconds.

Follow this procedure at all burrow entrances within the quadrat and record the number of entrances at which you get a response. Beware of counting responses from the same bird more than once if several entrances lead to the same nest-chamber.

A response will not be obtained from all occupied burrows. Multiplying the number of burrows responding by a correction factor of 1.98

(Brooke 1978) will provide an estimate of the actual number of occupied burrows (eg if responses are obtained from 50 burrows, the estimated number of occupied burrows will be 50 × 1.98 = 99 pairs). This assumes that each occupied burrow has only one entrance, and that each entrance leads to only one nest-chamber. The conversion factor may vary among colonies, but only this single estimate is available at present.

To obtain an estimate of the total population, first calculate the mean number of occupied burrows in each quadrat (sum of estimates for each quadrat divided by the number of quadrats taken). Divide this by the area of each quadrat (typically 20 m^2 or 30 m^2). Multiply the number obtained by the total area of the whole colony to obtain the final figure. Where different subdivisions have been sampled separately, this calculation should be performed for each subdivision separately and the results summed.

In colonies or subcolonies where shearwaters nest in boulders or scree, the tape playback method is less useful as there are so many potential nest-site entrances to check (many of which will be inaccessible). In such difficult habitats, adaptations of the tape playback method are probably more useful in arriving at crude whole-colony population estimates than for detailed monitoring of population change.

A minimum estimate of numbers of occupied burrows can be obtained by playing the call at a loud volume at several positions within each quadrat for 15-second periods, with 15-second gaps to listen for responses, over a period of several minutes. For more quantitative assessments, further work would be needed to determine the level of effort (loudness, duration and the number of positions for tape playback) required at a particular colony.

Alternatively, it might be possible to assess the relative breeding densities in rocky versus turf habitats by walking at a fixed pace through both habitats along random transects, playing male Manx shearwater calls at a loud volume a metre above the ground, and noting the number of birds audibly responding per unit time and area.

Breeding season survey – productivity

Information required

- number of large chicks located in a known number of burrows which contained eggs during the incubation period
- map of the extent of the colony showing the sample area.

Number and timing of visits

Two visits: one in early to mid-May, one in early to mid-August.

Time of day

Any time during the day.

Weather constraints, Sites/areas to visit, Safety reminders

As for the population survey (see above).

Equipment

- 1:10,000 map of the area
- series of *blunt* sticks (eg bamboo cane) of varying lengths, eg 15–60 cm long.

Disturbance

When feeling down Manx shearwater burrows for eggs, take extra care that you do so slowly and do not damage the egg. Try to determine the presence of a shearwater chick without actually removing it from its burrow; this can be done by feeling the bill to check that it is not a puffin chick.

Method

Avoid colonies where puffins are also known to breed – it is too difficult to tell whether an egg is Manx shearwater or puffin.

Check a series of burrows, dispersed through the colony, soon after the peak of laying in early to mid-May. Try to sample at least 100 burrows. Take the longest of your collection of sticks, lie on the ground, and push the stick and your arm down each burrow. Any incubating shearwater will move off the egg which can usually be felt with the stick on the floor of the nest-chamber. If the stick is too long to go around a bend in the burrow, try again with a shorter one. Do so slowly and carefully so as not to break the egg. Put a stake in the ground beside any burrow where an egg is felt, to enable you to relocate the burrow when any vegetation has grown tall. These checks are best made when the ground and burrow floor are dry.

Breeding success can vary markedly with burrow quality (Thompson and Furness 1991), so it is important not to be biased towards burrows that are easy to check (shallow). Very deep burrows may prove impossible to study using this technique. If such burrows are encountered, add a note on their relative numbers (compared with burrows where presence/absence of an egg could be confirmed) to highlight the possibility that results may not be representative of the colony.

Re-check the burrows in early to mid-August, when most chicks will be large (but before fledging begins in late August). It is usually easy to determine if the nest has been successful, either by feeling the chick or by searching for moulted down-feathers in the nest lining.

Productivity is reported as the number of chicks present, divided by the total number of burrows where the presence or absence of a chick was determined on the second visit.

References

Brooke, M de L (1978) Sexual differences in voice and individual recognition in the Manx shearwater (*Puffinus puffinus*). *Anim. Behav.* 26: 622–629.

Thompson, K R and Furness, R W (1991) The influence of rainfall and nest-site quality on the population dynamics of the Manx shearwater *Puffinus puffinus* on Rhum. *J. Zool. Lond.* 225: 427–437.

Gibbons, D W and Vaughan, D (1997) The population size of Manx shearwater *Puffinus puffinus* on 'the neck' of Skomer Island: a comparison of methods. *Seabird* 20: 3–11.

Storm petrel
Hydrobates pelagicus

Status

Amber listed: BL, SPEC 2 (L)
Annex I of EC Wild Birds Directive

National monitoring

Seabird Colony Register (SCR).
Seabird Monitoring Programme (SMP).
Seabird 2000 (1999–2001).

Population and distribution

Storm petrel breeding distribution in Britain and Ireland is restricted to the western coasts of remote uninhabited islands that are close to Atlantic water masses and the continental shelf edge. The strongholds for storm petrels are the islands on the west coast of Ireland (especially Co Kerry), but significant populations also occur at Annet (Scilly), Skokholm (Dyfed), North Rona (Western Isles) and Mousa (Shetland). Smaller colonies occur on many small islands along the north-west Scottish coast and the Hebrides (*BWP*, Lloyd et al 1991). These are extremely difficult birds to survey accurately but the current UK breeding population is estimated at 20,000–150,000 pairs (*Population Estimates*).

Ecology

Storm petrels breed in a variety of habitats: cavities among boulder beaches, scree or other loose rock; cavities in dry-stone walls and ruined buildings; cracks in rock (cliffs or level slabs) or peat banks; burrows in soil or peat, especially under or among partly buried rocks and boulders; disused burrows of other seabirds or rabbits and in side tunnels within active burrows; cracks in level peat under rank heather, even on the surface under a canopy of heather; and ledges in caves. Birds are colonial, although often rather loosely so. Eggs are laid mainly mid-June to early August. The peak of diurnal nest attendance is during the incubation period. There is a single clutch of one egg, and most of the young fledge by early October (*Red Data Birds*).

Breeding season survey – population

These methods are based on research work conducted on Mousa and Skokholm (Ratcliffe et al 1996, Ratcliffe et al 1998, Vaughan and Gibbons 1998) and experience gained from surveys conducted in north-west Scotland (Mainwood et al 1997, Gilbert et al 1998).

An assessment of storm petrel population status can take the form of a complete census or a sample survey. Both these approaches are outlined here, and much of the instruction is the same for both.

Information required

Census survey

- the probability of a response to tape playback
- estimated number of Apparently Occupied Sites for the whole census area
- a map showing the boundary of the census area and the location of colonies.

Sample survey

- the probability of a response to tape playback
- estimated densities within sample areas of all nesting habitats
- a map of the extent of all nesting habitats and an estimate of their area
- estimated number of Apparently Occupied Sites for the whole census area.

Number and timing of visits

Approximately seven visits to sample sites in all habitats to estimate detection probabilities, and single visits to other areas. About mid-July.

Time of day

Between 0800 and 2000 BST.

Weather constraints

Avoid very windy conditions as responses may be more difficult to hear.

Sites/areas to visit

Storm petrel nesting habitats are very diverse, and breeding can occur in several of the different habitat types outlined under *Ecology*, above.

Equipment

- 1:25,000 field map of the survey area, preferably enlarged by a photocopier
- a tape player with a good-quality recording of a male storm petrel purr call
- recording forms
- outdoor tape, permanent marker pens, tent pegs, compass, 50 m of string, 20-m tape-measure
- camping equipment if overnight stays are necessary (tent should have four-season specification).

Safety reminders

It is essential that you are accompanied by another person when working on remote uninhabited islands. Good communications with the rescue services by VHF radio or a cellphone are essential in case of an emergency. Nest-sites in cliff cracks will often be inaccessible and no attempt should be made to climb to them unless you have proper equipment and training. Be careful when conducting surveys on steep grass slopes as a slip could result in a dangerous fall.

Disturbance

Take care when walking around islands where storm petrels are present. Do not move boulders or disturb burrow entrances. Storm petrels are very sensitive to physical intrusion into the nest-site, so no attempt should be made to investigate nest contents. *No physical disturbance of the nest should occur.*

Methods

Birds in nest-chambers are usually well hidden and impossible to detect visually. In the cases of scree, boulder beach and wall habitats there are no discernible nest entrances that can be counted. As it is usually impossible to confirm breeding at a site, the count unit for both survey methods is the Apparently Occupied Site (AOS), estimated from responses to tape playback.

The boundary of the survey in the methods given below is assumed to be a single island, but the method could be applied to a close group of islands.

Census survey

Complete (census) surveys have the advantage of being more accurate than sample surveys. Storm petrel distribution is very patchy, even within areas of apparently suitable habitat, and this can result in large sampling errors, so it is recommended that complete surveys are undertaken wherever possible. This has been achieved on several relatively large islands off the west coast of Scotland (Mainwood et al 1997, Gilbert et al 1998).

These methods estimate the total number of AOSs by playback, but not all birds will respond, and estimating detection probabilities demands repeated playback surveys of study plots in all habitats present. Studies of response probability have demonstrated significant inter-annual and inter-colony variation. This means that the same correction factor cannot be applied to different years and colonies. In summary, the method is:

1. Locate each different habitat type occupied by storm petrels.
2. Select sample plots of each habitat type which are sufficient to provide 50+ responses for each habitat type; sample plots should be representative of the different densities of storm petrels.
3. Carry out repeated playback at the sample plots over at least seven days.
4. Meanwhile, carry out a full census using playback once in all areas of potential storm petrel habitat including the sample plot areas.
5. Calculate the probability of response to tape playback and thus a correction factor.
6. Multiply the number of AOSs detected during the complete census of the whole area by the correction factor to estimate the total number of AOSs present.

Expanding on this:

1. It is useful to survey the island for potential nesting habitats during the day and carefully map the locations and extent of all areas of boulder beach, walls, caves, scree, heather moorland and embedded boulders on a large-scale map. Visit these areas again at night and listen for calls to gain an indication of the distribution and relative densities of AOSs.

2. In each habitat choose sample plots which each contain a total of about 50 AOSs. Assume that you will detect at least 25–30% of AOSs on a single visit, so aim to find about 15 AOSs on the initial visit. If you find fewer than this, extend the area of the study plot. Mark the boundaries of the plots and split them into regular sections to make playback more systematic and responses easier to record. Different approaches to marking plots and tape playback are required for each habitat.
Dry-stone walls. Divide the walls into 2-m sections using tent pegs or

marked string and allocate numbers to the sections. Play the tape-recording of the purr call for 10 seconds in the middle of each 2-m section.

Boulder beaches and scree. Lay down 2-m-wide transects across the width of the boulder beach or scree. Play the tape for 10 seconds every 2 m along the centre of the transect and record the number of responses within each 2-m square. This method is applicable to sites where no burrow entrances are visible.

Burrows and cracks in peat or rock. Search for all burrows, holes among or under embedded rocks and cracks in peat banks and play the tape for 10 seconds at the entrance. Immediately after playback place your ear to the hole so that you can more easily hear any response.

Caves. Some birds may nest in full view inside dark caves. Make a direct count of incubating birds using a red-filtered torch. However, cracks *within* caves should be surveyed as described above.

3. Visit sample plots every one or two days. On the first visit, after every play of the tape, mark and number all the AOSs detected. Squares of light-coloured outdoor tape can be stuck to rocks, or tent pegs tipped with liquid paper can be used for burrows in soil. Write a unique number on these with a permanent marker pen. On the second visit mark any new sites and make a note of whether the ones recorded previously responded or not. This should be repeated until the number of new sites detected becomes consistently small, eg less than 5% of the total found on two previous consecutive visits (this could take seven or more visits, so a long stay will be necessary). This will indicate that almost all of the sites present have been found.

4. Visiting the sample plots will not take all day, so the census playback survey of the whole island can be carried out at the same time. Use the same techniques for each habitat as for the sample plots but without marking sections or AOSs. It is easier to remember where you have been by surveying as systematically as possible and carefully mapping where you have surveyed each day. Note the number of responses (detected AOSs) and the habitat type. The sample plot areas should be surveyed independently for the full census.

5. Calculate the detection probability, assuming that all sites within the sample plots were detected and that all AOSs located were present on all visits. This means that an AOS that is only detected on the last visit is assumed to have been present but not responding on the previous visits. Calculate the probability of detecting an AOS on a given visit by dividing the total number of responses on all repeat visits by the product of the total AOSs and total visits. Take the reciprocal of this figure to give a correction factor.

For example, the following table represents responses (1) and no responses (0) from 10 AOSs found over eight visits. The total number of responses (sum of all 1s) is 41; this is divided by 80 (total AOSs × total visits) to give 0.51 (proportion of AOSs that responded). The correction factor in this example is thus one divided by 0.51 = 1.95.

AOSs	Visits							
	1	2	3	4	5	6	7	8
1	1	1	0	1	0	0	0	0
2	0	0	1	0	1	1	1	1
3	0	0	0	0	0	0	0	1
4	0	0	0	1	0	1	1	1
5	1	1	1	1	0	1	1	0
6	1	1	1	1	1	1	0	0
7	0	1	1	0	1	0	1	1
8	1	0	0	0	1	0	1	1
9	1	1	1	0	0	0	0	0
10	0	0	0	1	0	1	1	1

6. Multiply the correction factor by the total number of AOSs detected in each habitat during the full census playback survey (the sample plot areas should be included in this) to give an estimate of the total number of AOSs in each habitat. Sum the totals in each habitat to yield an all-island(s) estimate.

Sample survey

Sample surveys are less accurate than complete censuses, but can be used where the area to be covered is particularly large, or where there are insufficient resources to do a complete census. Even the sample survey can be time consuming as it also requires repeated visits to some areas over a period of a week or so.

The sample survey is very similar to the complete census, except that only a sample of each habitat type (about 20%) is surveyed. Total estimates in each habitat are determined by extrapolation based on the sampling intensity. In summary, the method is:

1. Locate each different habitat type occupied by storm petrels and estimate its total area.
2. Randomly select sample plots which cover 20% or more of each habitat type.
3. Carry out tape playback in all sample plots and record the number of AOSs detected in each.
4. By extrapolation calculate the number of AOSs which would have been detected had the total area of each habitat type been surveyed.
5. Select sample plots of each habitat type which are representative of different densities of storm petrels and which are each sufficient to provide 50+ responses (these can include the random sample plots).
6. Carry out repeat visit playback at the sample plots over a period of about seven days.
7. Calculate a response probability and thus a correction factor from these repeat visits.
8. Multiply the number of AOSs 'detected' in the total area of each habitat by the correction factor to obtain an estimate of population size for that habitat. Sum the individual habitat totals to give an all-island(s) population estimate.

Expanding on this:

1. Mapping of habitat types is the same as for the census survey (see above).

2. Sampling should be carried out using randomly allocated plots which are stratified within habitat types and within discrete areas of similar habitat. The sampling strategy varies according to habitat type:

Walls. Measure the entire length of wall on the island. Divide the walls into sections 20 m long and allocate a number to each section. Use random numbers to select which 20-m sections to survey.

Boulder beach and scree. Map the extent of boulder beach to scale on graph paper and calculate the total area. Lay a string marked at 2-m intervals on the long axis of the area of habitat and allocate a number to each mark. Select a number of these marks using random numbers. For each selected mark run a 2-m-wide transect over the habitat at right angles to the long-axis string.

Heather and burrows. Map the extent of the habitat to scale on graph paper, calculate the area and divide it into 5 × 5 m quadrats. Allocate numbers to each quadrat and select a sample at random. In areas where densities of AOSs are very low, such as heather, use 10-m quadrats.

Aim to sample at least 20% of the total area of each habitat present on the island. In areas of low AOS density (eg heather), this percentage should be increased.

3. Use the same playback census technique as for the census survey (above).

4. For each habitat, calculate the mean density of detected AOSs across sample plots. Multiply the mean density in each habitat by the total area of each to give an estimate of the total number of AOSs which would have been detected in each habitat had a complete census been undertaken.

5–8. Calculate the detection probability and correction factors in the same way as for the complete census. Estimate the total number of AOSs in each habitat by multiplying the total number of AOSs 'detected' in the total area of that habitat by the correction factor. Sum the individual habitat population estimates to obtain an all-island(s) estimate. Calculate 95% confidence intervals around this estimate by bootstrapping (if you do not know how to do this, seek statistical advice).

References

Gilbert, G, Hemsley, D and Shepherd, M (1998) A survey of storm petrels in the Treshnish Isles 1996. *Scottish Birds* 19: 145–154.

Lloyd, C, Tasker, M L and Partridge, K (1991) *The Status of Seabirds in Britain and Ireland.* Poyser, London.

Mainwood, A R, Ratcliffe, N, Murray, S and Mudge, G P (1997) Population status of storm petrels *Hydrobates pelagicus* on islands off north west Scotland. *Seabird* 19: 22–30.

Ratcliffe, N, Suddaby, D and Betts, M (1996) *An Examination of the Census Methods Used for Storm Petrels Hydrobates pelagicus on Mousa and Skokholm.* RSPB unpubl.

Ratcliffe, N, Vaughan, D, Whyte, C and Shepherd, M (1998) The status of storm petrels on Mousa, Shetland. *Scottish Birds* 19: 155–163.

Vaughan, D and Gibbons, D W (1998) The status of breeding storm petrels *Hydrobates pelagicus* on Skokholm Island in 1995. *Seabird* 20: 12–21.

Leach's petrel *Oceanodroma leucorhoa*

Status

Amber listed: BL, SPEC 3 (L)
Schedule 1 of WCA 1981
Annex I of EC Wild Birds Directive

National monitoring

Seabird Colony Register (SCR).
Seabird Monitoring Programme (SMP).
Seabird 2000 (1999–2001).

Population and distribution

In the UK, Leach's petrels breed on a few, very remote, islands off the west coasts of Shetland, Orkney and the Hebrides. They are rarely seen, and nest in burrows, under boulders, in peat cracks and in dry-stone walls. They are extremely difficult birds to survey accurately but the current UK breeding population is estimated at 10,000–100,000 pairs (*Population Estimates*).

Ecology

Leach's petrels breed on undisturbed offshore islands, down burrows in open turf, or sometimes under a boulder or in scree. Burrows are often hidden in long vegetation, and entrances have to be located by hand as well as visually. They are loosely colonial and all activity onshore is nocturnal. One egg is laid from the end of May to mid-August. The peak of diurnal nest attendance is during incubation, and for most British and Irish colonies this is probably mid-June. There is one brood and the young fledge by mid-October (*Red Data Birds*).

Breeding season survey – population

This method is adapted from the method developed for storm petrel (Ratcliffe et al 1996, Mainwood et al 1997, Gilbert et al 1998, Ratcliffe et al 1998, Vaughan and Gibbons 1998) and advice on the problems of surveying Leach's petrel given in Ellis et al (in press). It requires repeated visits or an overnight stay for a week or more. Considering the remoteness of the locations concerned, considerable logistical planning would be required to undertake such a survey.

Information required

- probability of a response to tape playback
- estimated number of Apparently Occupied Sites for the whole census area
- a map showing the boundary of the census area and the location of colonies.

Number and timing of visits

Approximately seven visits to sample sites in all habitats to estimate detection probabilities, and single visits to other areas. Mid- to end of June; after this the probability of adults being present in burrows or of getting a response to playback declines (Ellis et al in press).

Time of day

Between 0800 and 2000 BST.

Weather constraints

Avoid very windy conditions as responses are difficult to hear.

Sites/areas to visit

All suitable nesting habitat (see *Ecology*, above).

Equipment

- 1:25,000 field map of the survey area, preferably enlarged by a photocopier
- tape player with a good-quality recording of male and female Leach's petrel chatter calls
- recording forms
- permanent marker pens, tent pegs tipped with liquid paper or coloured bamboo canes, compass, 50 m of string and a 20-m tape-measure
- camping equipment if overnight stays are necessary (tent should have four-season specification)
- Schedule 1 licence.

Safety reminders

It is essential that you are accompanied by another person when working on remote uninhabited islands. Good communications with the rescue services by VHF radio or a cellphone are essential in case of an emergency. Nest-sites in cliff edges will often be inaccessible and no attempt should be made to reach them unless you have proper equipment and training. Be careful when conducting surveys on steep grass slopes as a slip could result in a dangerous fall.

Disturbance

Take care when walking around islands where petrels are present: do not move boulders or disturb burrow entrances. Leach's petrels are very sensitive to physical intrusion into the nest-site so no attempt should be made to investigate nest contents. No physical disturbance of the nest should occur.

Methods

Without use of an endoscope it is usually impossible to confirm breeding, therefore the counting unit for this survey method is the Apparently Occupied Site (AOS), the presence of which is inferred from a response to tape playback.

The boundary of the survey area is assumed to be the coastline of a single island, but the method could be applied to a close group of islands.

A complete census is much more accurate than a sample survey because Leach's petrel distribution is very patchy, even within areas of apparently suitable habitat. Sampling could result in large errors.

This method estimates the total number of AOSs detected by playback. Not all birds will respond, therefore estimating detection probabilities demands repeated playback surveys of study plots in all habitats. From this, a correction factor can be calculated. In summary, the method is:

1. Locate each different habitat type occupied by Leach's petrels.
2. Select sample plots in each habitat type which are representative of different densities of Leach's petrels and which are sufficient to provide 20+ responses in each (total numbers may be low, so find the best sample plots you can).
3. Carry out playback at the sample plots over at least seven days.
4. Meanwhile, undertake a full census using playback once in all areas containing potential Leach's petrel habitat, including the sample plot areas.
5. Calculate the probability of response to playback and from this a correction factor.
6. Multiply the number of AOSs detected during the full census by the correction factor to estimate the total population size; this could be broken down by habitat.

Expanding on this:

1. It is useful to survey the island for areas of suitable habitat during the day and map them carefully. Mark all burrows found with coloured bamboo canes and plot their locations on a large-scale map. Visit these areas again at night and listen for calls to gain an indication of the distribution and relative densities of AOSs.

2. Choose sample plots in each habitat which contain as many AOSs as possible (ideally 20+). You will detect at least 75% of AOSs on a given visit, so aim to find about 15 AOSs on the initial visit. If you find less than this, extend the area of the study plot.

3. Visit each sample plot every one or two days and use tape playback (see below). On the first visit, mark and number all of the AOSs detected. Tent pegs tipped with liquid paper or coloured bamboo cane can be used to mark burrows in soil. Write a unique number on these markers with a permanent marker pen. On the second visit, mark any new sites and make a note of whether or not birds responded from previously recorded sites. Repeat until the number of new sites detected is less than 5% of the total on two previous consecutive visits (this could take seven or more visits, so a long stay will be necessary). This will indicate that almost all of the sites present have been located. Constructing a table of visits and responses like the one shown below will help to simplify the analysis of the results.

4. Visiting the sample plots will not take all day so carry out the full playback census at the same time. Use the same basic playback technique (see below) as for the sample plots. Note the number of responses and the habitat type. The sample plot areas should be re-surveyed as part of the full census.

For 3 and 4 use a playback of the male and female Leach's petrel chatter call to elicit vocal responses from birds in their nest-sites during the day. Different survey techniques are required for different habitats:

Boulder scree. Lay down 2-m-wide transects across the width of the boulder beach or scree. Play the tape for 30 seconds every 2 m along the centre of the transect and record the number of responses within each 2-m square. This method is applicable to sites where no burrow entrances are visible.

Burrows. Search for all burrows, if necessary by hand, eg in long vegetation. Play the tape down each burrow for 30 seconds. Immediately after playback place your ear to the hole so that any responses can be heard more readily. Note the number of burrows found and the number of responses detected.

5. Calculate the response probability, assuming that all sites within sample plots were detected and that all AOSs located were present on all visits. This means that an AOS that is only detected on the last visit is assumed to have been present but not responding on all earlier visits. Calculate the probability of detecting an AOS by dividing the total number of AOSs that responded on all repeat visits by the product of the total AOSs and total visits. Take the reciprocal of this as the correction factor.

For example, the table below represents responses (1) and no responses (0) from 10 AOSs found over eight visits. The total number of responses (sum of all 1s) is 41, this is divided by 80 (total AOSs × total visits), to give 0.51 (proportion of AOSs that responded on a given visit). The correction factor in this example is thus one divided by 0.51 = 1.95.

AOSs	Visits							
	1	2	3	4	5	6	7	8
1	1	1	0	1	0	0	0	0
2	0	0	1	0	1	1	1	1
3	0	0	0	0	0	0	0	1
4	0	0	0	1	0	1	1	1
5	1	1	1	1	0	1	1	0
6	1	1	1	1	1	1	0	0
7	0	1	1	0	1	0	1	1
8	1	0	0	0	1	0	1	1
9	1	1	1	0	0	0	0	0
10	0	0	0	1	0	1	1	1

6. Multiply the total number of AOSs detected in each habitat during the full census by the correction factor to give an estimate of the total number of AOSs in each habitat. Sum the individual habitat estimates to yield the all-island(s) population estimate.

References

Ellis, P, Ratcliffe, N and Suddaby, D (in press) Seasonal variation in diurnal attendance and response to playback by Leach's petrels *Oceanodroma leucorhoa* on Gruney, Shetland. *Bird Study.*

Gilbert, G, Hemsley, D and Shepherd, M (1998) A survey of storm petrels in the Treshnish Isles 1996. *Scottish Birds* 19: 145–154.

Mainwood, A R, Ratcliffe, N, Murray, S and Mudge, G P (1997) Population status of storm petrels *Hydrobates pelagicus* on islands off north west Scotland. *Seabird* 19: 22–30.

Ratcliffe, N, Suddaby, D and Betts, M (1996) *An Examination of the Census Methods Used for Storm Petrels Hydrobates pelagicus on Mousa and Skokholm.* RSPB unpubl.

Ratcliffe, N, Vaughan, D, Whyte, C and Shepherd, M (1998) The status of storm petrels on Mousa, Shetland. *Scottish Birds* 19: 155–163.

Vaughan, D and Gibbons, D W (1998) The status of breeding storm petrels *Hydrobates pelagicus* on Skokholm Island in 1995. *Seabird* 20: 12–21.

Gannet
Morus bassanus

Status

Amber listed: BI, BL, SPEC 2 (L)

National monitoring

Seabird Colony Register (SCR).
Seabird Monitoring Programme (SMP).
1994–95 North Atlantic gannet survey (Murray and Wanless 1997).
Seabird 2000 (1999–2001) or next North Atlantic gannet survey (2004–2005).

Population and distribution

Gannets breed at about 24 colonies around the coasts of Britain and Ireland. All gannetries are situated close to good fishing waters, usually on islands with cliffs. Colonies exist off Wales, England, Ireland and Scotland. However, the right combination of nesting and feeding habitat occurs mainly in the north and west of Scotland (*88–91 Atlas*). Numbers of gannets have been gradually increasing in the 20th century as they recover from heavy persecution in the 19th century (*88–91 Atlas*). There are an estimated 201,000 nests in the UK (*Population Estimates*). This represents about 70% of the world population (*Birds in Europe*).

Ecology

Gannets breed on cliff ledges or cliff-top slopes, on stacks, headlands and islands. They are highly colonial. Egg-laying is from March to June; there is a single egg and a single brood. Fledging occurs mainly between late August and early September, but a few young do not fledge until November (*Red Data Birds*).

Breeding season survey – population

The methods given here are taken from the *Seabird Monitoring Handbook*. *Method 1* involves field counts. *Method 2* involves counts from photographs.

Information required

- *Method 1* – number of Apparently Occupied Nests (AONs) or Apparently Occupied Sites (AOSs)
- *Method 2* – number of AOSs
- map of the extent of the colony (for both methods).

Number and timing of visits

Method 1 – two visits at least three weeks apart, one in June and one in July.
Method 2 – one visit in June or July.

For both methods, counts between mid-May and mid-August are better than none at all.

Time of day

0900–1600 BST. Visits made or photographs taken outside this range are better than none at all.

Weather constraints

The more benign the weather conditions, the easier and more accurate the counts will be.

Sites/areas to visit

Any colony not already covered (all British colonies were censused in 1994–95; see above).

Equipment

- 1:10,000 map of the area to be visited
- camera
- black and white film
- transparent overlays (overhead projection sheets) and marker pens
- a gannet data sheet for whole colony counts (see Figure 1)
- for many colonies, an aircraft!

Safety reminders

Someone should know where you are and when you are expected to return. All cliffs and crags are hazardous. Do not attempt to rock climb unless specifically required to do so, and never when alone or without training and equipment.

Disturbance

Aerial (helicopter or plane) surveys can cause massive disruption at colonies, and colonies are particularly vulnerable to disturbance when chicks are small.

Methods

There are two types of recording unit.

- An *Apparently Occupied Site* (AOS) is any site occupied by one or two adult gannets irrespective of whether or not any nesting material is present, so long as the site appears suitable for breeding. Sites with unattended chicks are included. Care is needed to ensure that single gannets on adjacent AOSs are not confused with a pair of gannets on a single AOS. This is the recommended count unit, particularly when counting gannets from aerial photographs of colonies.
- An *Apparently Occupied Nest* (AON) is a site at which one or two adults with nest material (however flimsy), or a large chick (which may not be on an obvious nest), are present. Only recommended for colonies which can be counted directly from land or sea, especially where large numbers of non-breeding birds are present and observers find assessment of AOSs confusing at close range.

Field counts

It is important to count the whole colony, as changes in size are more likely to occur through expansion or contraction rather than there being changes in density. It is also desirable to replicate counts, preferably using two counters.

Counting should be undertaken from suitable vantage points. Use the natural features of the site (rocks, crevices, etc) to divide the colony into clearly defined count sections. Make a note of any unoccupied sections. Mark the boundary of each count section and the best vantage point for counting it on a map. Mark the boundaries of each count section on photographs (or on accurate sketch maps) of the colony by covering the photograph with a transparent overlay. For many gannetries, such subdivisions have been established during earlier studies and will be on file with JNCC, RSPB, SNH or other bodies.

If nest material (even of small or partly constructed nests, no matter how flimsy) can be safely distinguished in all parts of the colony, count all AONs, otherwise count AOSs. If a comparison needs to be made with a previous count of a different census unit, try to count both AONs and AOSs at the same time. In subsequent years, AONs (or AOSs if AONs are not possible) should be the priority.

Mark on a map the parts of the colony not visible from land. Estimate and record on the map the minimum and maximum number of nests which might be hidden, based on the number of nests found in the visible sections. If possible, check and count the hidden sections from a boat on a calm day especially if the hidden sections are likely to contain more than 10% of the population. When reporting these estimates, clearly state the difficulties involved in order that the degree of comparability with other counts can be gauged.

Report the number of AOS/AONs counted on each visit and the mean of the two visits, for each count section, and for the whole colony. If a single count was made, indicate this clearly. Record how each section was counted (eg from land, from sea, or a combination of the two). If both the early and late counts are considered accurate, use the larger of the two (to allow for real seasonal variation) as the best estimate of the population. However, if the count accuracy is considered to be low, or is uncertain, the average count (with range) is more appropriate. If time allows, counts of individual adults and immatures in non-breeding 'club' areas are also useful.

Counts from photographs

Photograph the colony, or parts of it, at as close range as possible, provided birds in the colony are not disturbed (especially by aircraft) and provided you can distinguish any overlap between the photos from each part of the colony. Where the colony cannot be divided using physical features, a single sharp photograph is preferable to several larger-scale photographs. If possible, visit the colony on foot – or viewing it from a boat it will make identification of non-breeding ('club') areas from photographs easier. (NB Many non-breeders may leave the colony on the approach of the aircraft.)

Good-quality, large prints are essential; black-and-white photographs are ideal. Alternatively, project a transparency onto white paper fixed securely to a solid surface such as a wall.

Count all the AOSs. Where a number of photographs are used (particularly if taken from different vantage points), avoid double-counting parts of the colony by marking subgroup boundaries on a transparent overlay or print projection. Cross off AOSs with fine, coloured markers, in contiguous groups of 200–300 at a time. Use varied colours for each group in case you have to start again, and note the numbers in each group as you do them. Try to identify, and count separately, the number of individuals of any age in 'club' (non-breeding) areas of the colony.

Ideally, one or two people should do a repeat count of each photograph of each subgroup. If this is not possible, make at least one repeat count yourself. Report all counts made for each subgroup of the colony and the whole colony, and also the mean counts (± standard deviation). Report the numbers of individuals in clubs if counted. Mark on maps or photographs, the extent of the colony and of non-breeding areas for future reference.

The results from either method should be recorded on a gannet data recording sheet (see Figure 1).

Gannet colony count data sheet

Observer name:

Address:

Gannetry:

Date of count:

Time....................

*Census unit: (tick as appropriate) Apparently occupied site / Nest / Other (specify)..................

Count method: (tick as appropriate) Field count from land / Field count from sea / From photo from land / From photo from air / Other (specify)...............

Weather and sea conditions:

*Problems:

*Comments:

Subgroup (if used; refer to map)	Method	Accurate count	Estimate
Total			

**Counting unit:*
1. Apparently occupied site (AOS): one or two adult gannets, or a chick, on a site suitable for breeding, irrespective of whether any nest material is present.
2. Apparently occupied nest (AON): one or two adult gannets occupying a site suitable for breeding, with nest material, no matter how flimsy, present, plus any large chicks without an obvious nest late in season.

N.B. Please state if you used any other counting unit and define it precisely. Please do not correct or adjust counts in any way.

**Problems/Comments:* Add notes on any parts of the colony that were uncountable (and try to estimate proportion), any reservations about the count, etc.

Figure 1
An example of the gannet recording sheet which should be filled in and handed in with recorded results. This datasheet was recommended for use by the *Seabird Monitoring Handbook*.

Breeding season survey - productivity

The following two methods are taken from the *Seabird Monitoring Handbook*. Measuring productivity in gannets is relatively simple, and should ideally be based on repeated checks of individual nests within plots dispersed throughout the colony (Method 1). However, many gannetries are not visited sufficiently often to permit such an approach; in such cases, Method 2 should be applied. *Method 1* is intensive and based on repeated checks on individually mapped nests. *Method 2* involves comparisons of nest and chick counts and requires fewer visits.

Method 1 - mapped nests

Information required

- total number of fledged young divided by the number of nests where birds appeared to be incubating at some stage in the season.

Number and timing of visits

One visit every 7–10 days during late April/mid-May to late August.

Time of day

Not a constraint for this method.

Weather constraints

Visits should preferably be made in dry conditions.

Sites/areas to visit, Equipment, Safety reminders

As for the population survey (above).

Disturbance

Under no circumstances should you flush sitting birds. Be particularly careful not to cause disturbance to chicks before fledging.

Methods (*Method 1: Mapped nests*)

For small colonies, check all visible nests. For large colonies, choose sample areas (plots) containing 50–100 nests. In each colony, select at least five plots at random. Base the number of plots chosen on the time available to carry out the work. To select the plots, divide the whole colony into groups of 50–100 nests, excluding areas which cannot be viewed safely, and number them. Randomly select the study plots from these. Information on sampling is given in the *Introduction*. It is ideal, but not necessary, to use the same plots each season; if you do, ensure the plot boundaries are clearly marked.

Photograph the selected plots, preferably when birds are at their nests, and make large (A4) black-and-white prints. Tape a transparent overlay over the photographs, mark on the positions of nests and number them.

Visit the area every 7–10 days after your first visit and record the state of each nest (eg a few sticks, complete platform), contents (if visible – do not flush sitting birds), and whether an adult appears to be incubating or brooding. Chicks can fledge from 12 weeks old, so count disappeared birds which could have reached 12 weeks old during the 7–10 days before being revisited as having fledged (see Table 1 for a guide to ageing chicks).

When recording the results, also record the reasons for any losses if known (eg predation, egg did not hatch, chick died at nest).

Report the number of fledged young divided by the number of nests (where birds appeared to be incubating at some stage in the season), for each plot separately, and the mean (± standard error) of the results for individual plots. The latter gives an estimate of productivity for the whole colony. Report the numbers of fledged young and nests for each plot separately – do not add them together.

Method 2 - Nest and chick counts

Information required

- number of chicks that potentially fledged divided by the number of nests where an egg or apparent incubation was recorded.

Number and timing of visits

Two or three visits. The first visit should be in the period mid- to late May; the second visit should be during 10–15 August and, if possible, the third visit should be during 24–29 August.

Time of day

0900–1600 BST.

Table 1
Guide to ageing gannet chicks (cited in the *Seabird Monitoring Handbook*).

Age	Description
Newly hatched	Black and naked, egg tooth obvious.
Week 1	Fairly black, sparse hair-like down, very wobbly (normally brooded constantly by adult).
Week 2	Partly covered with down; larger than parent's feet; head and neck bare; movements well co-ordinated.
Week 3	Body and wings covered in white down, but lacks luxuriantly fluffy look of a four-week-old; cannot be covered by parent.
Week 4	Down long and fluffy; two-thirds adult size, takes up most of nest.
Week 5	Still fluffy; approaches adult size; pin primary and tail feathers show black through the down.
Week 6	Fluffy, but scapulars, wings and tail feathers clear of down; looks bigger than parent.
Week 7	Mantle and back a mixture of white down and black feathers; breast, underparts, head and neck covered in long white down.
Week 8	Mainly black above; down disappearing from forehead, mantle/back and tail.
Week 9	Down starts to go from ventral surface, but still thick on flanks, belly and parts of neck; looks scruffy.
Week 10	Some down on nape, flanks and back.
Week 11	Only wisps of down remain, on nape and flanks.
Week 12	Complete juvenile plumage.

Weather constraints

Visits should preferably be made in dry conditions.

Sites/areas to visit, Equipment, Safety reminders

As for the population survey (above).

Disturbance

Do not flush sitting birds under any circumstances. Be especially careful not to cause disturbance to chicks before fledging.

Methods (*Method 2: Nest and chick counts*)

This is the best method to use if time is limited. Select study plots as for Method 1. Try to cover as many nests as is safe and practicable; in small colonies this may be all nests.

Map or photograph the nests within each plot. On the first visit (mid- to late May) count all nests with an adult apparently incubating or an unattended chick; these count as breeding attempts. Keep a separate count of other occupied, well-built nests where the egg may not have been laid or where it may already have failed.

On the second visit (around the time when the chicks fledge; 10–15 August) search for any additional breeding attempts. Count all chicks present and (if possible) record their age using the guidelines in Table 1.

Productivity is expressed as the number of chicks that potentially fledged, divided by the number of nests where an egg or apparent incubation was recorded. Bear in mind that a figure based on only one count of chicks may be a substantial overestimate, and may not be directly comparable to figures from other colonies or years.

Reference

Murray, S and Wanless, S (1997) The status of the gannet in Scotland in 1994–95. *Scottish Birds* 19: 10–27.

Cormorant *Phalacrocorax carbo*

Status

Non-SPEC

National monitoring

Breeding colony survey (Sellers 1997).
Seabird Colony Register (SCR).
Seabird Monitoring Programme (SMP).
Seabird 2000 (1999–2001).
WeBS.
Christmas week survey (WWT).
National winter roost survey.

Population and distribution

Both the North Atlantic subspecies *Phalacrocorax carbo carbo* and the continental subspecies *P. c. sinensis* are now known to breed in the UK, with *sinensis* predominating at inland breeding colonies. Cormorants have attempted to breed at over 50 different inland sites in Britain, and colonies have been established at nine. The inland breeding population increased by 28% per year between 1991 and 1995 and by 17% between 1995 and 1996 to reach some 1,400 pairs in 1996 (Sellers and Hughes 1997). The UK population of cormorants has been estimated at 7,600 breeding pairs or 14,700 wintering individuals (*Population Estimates*), although both estimates are now out of date by five to ten years. Taking into account increases in both breeding and wintering numbers, UK populations now probably number in excess of 9,000 breeding pairs and 25,000 wintering birds.

Ecology

Coastal cormorant colonies form on cliffs, stacks or offshore islands; inland colonies are located in trees, usually on islands. There is a large spread in nesting dates, both within and between colonies, and a marked difference between coastal and inland colonies (Newson 1997). At inland colonies the first nests are established in early February, while the majority have begun by mid-March. Few first nesting attempts are initiated after mid-May. Following a 30-day incubation period, the first young hatch in mid-March with most hatching during April. Fledging takes place at 8–9 weeks of age, so the first young fledge in late May and most of the rest during late June and early July.

At inland colonies, the breeding season extends over about three months. At coastal colonies, birds generally breed about a month later than at inland colonies but in a somewhat more synchronised manner, with nest initiation usually being spread over less than two months (this is, however, still much more protracted than other seabirds such as gulls or terns). At coastal colonies, the first young cormorants hatch in mid-April, but most hatch in mid-May and fledge in mid- or late July. There is also inter-colony variation in the timing of breeding for both inland and coastal populations. For example, birds at Blockhouse Stack in Pembrokeshire typically breed two weeks earlier than a colony less than 30 km away at Green Scar. There is similar variation at inland colonies: cormorants at Besthorpe in Leicestershire breed nearly two months earlier than those nearby at Rutland Water. Cold weather delays the onset of breeding at both inland and coastal colonies.

Breeding season survey - population

Cormorant nests are generally large and easily counted. However, the ease of counting differs between coastal, ground-nesting birds which are generally easy to survey, and inland, tree-nesting birds which are much less straightforward, especially once there are leaves on nest trees. Furthermore, the timing of breeding differs between inland and coastal breeders with inland birds breeding earlier and having a more protracted breeding season than coastal birds.

To identify all nesting attempts, detailed weekly nest-mapping would be required. The method given here, which requires several visits, is likely to underestimate the population size.

Information required

- maximum number of Apparently Occupied Nests from any one count
- a map of the survey area, annotated with any subdivisions used and any parts of the colony not visible from land.

Number and timing of visits

Coastal sites

At least one visit, four if possible, between early May and late June (one visit must be made in mid-May). Counts should be made at the time when the maximum number of nests is occupied; local knowledge may be needed to obtain the optimum dates.

Inland sites

Before the breeding season make preliminary visits to mark each nest tree with an individual code. During the breeding season make at least one visit, up to four if possible, between early April and mid-May. If you are unable to enter the colony to mark nesting trees, ensure that at least one visit is made just before leaf-break in early or mid-April; if entry is possible, however, then make at least one visit towards the end of April.

Time of day

Any time.

Weather constraints

Good light and good weather conditions are recommended.

Sites/areas to visit

All breeding colonies are listed in the Cormorant Breeding Colony Register (Sellers 1997). For comprehensive regional or national surveys, an aerial survey may be required to check that known colonies still exist and to search for any which are newly established.

Equipment

- telescope
- boat (for colonies which cannot be observed adequately from land)
- aeroplane, camera and film for any aerial photography, also slide projector and marker pens
- paint or other means of marking large numbers on nesting trees at inland sites.

Safety reminders

Take all necessary precautions against the hazards which coastal areas present. Anyone surveying colonies by boat should wear a life-jacket and possess the necessary qualifications for boat-handling on the open sea. Inland colonies may also present unique hazards, especially if observers enter colonies for survey purposes. Fieldworkers require experience of boat-handling, tree-climbing and the use of ladders.

Disturbance

Do not enter cormorant breeding colonies unless you have gained experience under supervision of doing so; this cannot be overstressed, especially during the brood-rearing period, when chicks may jump from their nests up to five weeks before they can fly. Visit length should be kept to an absolute minimum, preferably less than 30 minutes. Cormorants are highly susceptible to human disturbance: birds readily leave their nests when humans approach, leaving unattended nests vulnerable to predation by gulls at coastal colonies or by corvids at inland colonies. Such predation risk may be reduced by covering nests or young chicks with vegetation. Eggs may also chill or overheat.

Methods

Nest counts are expressed as numbers of Apparently Occupied Nests (AONs). This includes birds that appear to be incubating, unattended broods of young, and other attended, well-built nests including empty ones apparently capable of holding eggs.

Coastal sites

The best means of surveying coastal cormorant colonies depends on the visibility of the colony from land. If the whole colony can be observed from land then land-based counts usually provide the most accurate survey method, while for colonies on sheer cliff faces or on offshore islands boat or aerial surveys or a combination of the two may be more accurate.

Aerial photography

Photograph the colony from the air during the peak of the breeding cycle, when most sites are occupied. Make large prints of the photographs, or project slides onto a large sheet of paper attached to a solid surface. Count or mark all nests with adults sitting, with chicks, or which appear to be substantially constructed. Aerial surveys, even those using aerial photography, can underestimate the numbers of nests present, especially in larger colonies. It is a good idea to take a number of different photographs of each colony from different angles and at different magnifications. However, beware of double-counting sites from more than one photograph. It is recommended that more than one observer counts the same photograph to allow for inter-observer variability and that the mean count is used as the population figure.

Field counts

By far the best means of counting cormorant nests is mapping the positions of nests by entering breeding colonies. However, do not do this unless you have previous experience of doing so. Instead count AONs from suitable vantage points. Whenever possible, counts should include an objective assessment of the general stage of the breeding cycle (eg ratio of trace to well-built nests, proportion of well-built nests that are empty). A subjective assessment of whether there seemed to be fewer nests than one would expect from the numbers of adults present may provide some indication of an unusually late or poor breeding season. Keep a note of and map any parts of a colony that might not be visible from land. Try to estimate the number of AONs likely to be hidden, based on the number of visible sections. When reporting these estimates be very clear about the method used and its reliability.

Inland sites

Mapping nests by entering colonies is especially useful at tree-nesting sites which can be difficult to map accurately from a distance (see *Disturbance*, above). Regardless of whether or not tree-nesting colonies are to be surveyed from within or from outside the colony, a preliminary visit should be made before birds begin breeding in order to mark each nest tree with an individual code (making sure these can be seen from shore-based observation points). If you are unable to enter the colony to mark nesting trees, ensure you make at least one visit just before leaf-break in early to mid-April. If entry is possible, make at least one visit towards the end of April. During subsequent monitoring visits, the positions of nests can then be mapped in each marked tree. In tree-nesting colonies where individual trees cannot be marked, the positions of nests should be mapped in relation to obvious landmarks or natural features as well as in relation to one another. Failed nests quickly disappear as nest material is stolen by other nesting pairs. Photographs provide an ideal means of double-checking these nest maps.

For all methods, report the maximum count of AONs made during any one visit.

Breeding season survey – breeding success and productivity

Information required

- number of Apparently Occupied Nests per colony
- proportion of nests successfully rearing any young
- clutch size and brood size at hatching and fledging in successful nests.

Number and timing of visits

An absolute minimum of three visits is required for each colony: one during late incubation, one immediately following hatching and one when chicks are 4–5 weeks old.

Time of day

Either at dawn or in mid-afternoon.

Weather constraints, Sites/areas to visit, Equipment

As for the population survey (above).

Safety reminders

See the population survey. You should always be accompanied by someone with previous experience of this work.

Disturbance

See the population survey. If nestlings are disturbed after they have been fed they will regurgitate their food, therefore it is important that the recommended timing of visits is strictly observed. Young cormorants over about four weeks old (five weeks before fledging) are easily displaced from their nests. Tree-nesting birds are especially at risk, and young birds will readily jump from their nests due to the presence of humans nearby. It is highly likely that most of these birds leaving the nest prematurely will die, therefore it is essential that only experienced or supervised observers enter cormorant breeding colonies. This point cannot be overstressed.

Methods

Breeding success data can be collected at the same time as and using similar nest-mapping methods to those used for population monitoring. The total number of pairs attempting to nest or re-nesting after failed attempts should already have been established using one of the population survey methods (above).

Make at least three visits, one during late incubation to record the number of nests and clutch size, one immediately after hatching to record brood size at hatching, and one when the chicks are four to five weeks old to record fledging success. In practice, because breeding is spread over such a protracted period, weekly visits are usually necessary in order to record complete productivity data for all nests and to keep track of second attempts by pairs which have failed.

If regular visits are not possible it is most important (and most straightforward) to monitor the number of pairs attempting to breed, and the brood size at fledging in successful nests, because both of these can be established without entering the colony. During monitoring visits, it is important to record the stage of the breeding season (eg in terms of the size and level of development of chicks present), as breeding success is known to decline as the season progresses. Chick mortality is highest during the first three weeks after hatching, so fledging success should be monitored when chicks are four weeks of age or older.

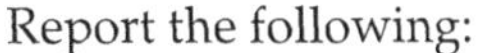

Report the following:

- The total number of active nests (eggs or apparent incubation seen) which failed or which fledged one, two, three or four chicks. Record any further details of losses, eg: predation; eggs did not hatch; chick dead in nest, possibly starved.
- The total number of young fledged divided by the number of successful nests (where birds were definitely or probably incubating).

Winter survey

WeBS

The time of day for WeBS counts of cormorants is important. In general, peak winter counts are obtained when birds are not feeding, around three hours after dawn when birds have just completed their early-morning feeding bout and have hauled out to loaf and dry their wings. However the timing of this differs with habitat, season and site, so, if possible, counters should conduct preliminary visits to sites to identify the best times to count. On estuaries, feeding rhythms may be determined by tidal cycles (eg birds feeding mainly at high tide) and peak counts may occur some time before or after high tide. Details for undertaking WeBS counts are given in the generic survey methods section.

Christmas week survey

Nearly 300 cormorant roost-sites have now been identified by the WWT/JNCC cormorant roost-site inventory, most situated on lakes in southern and central England, with notable concentrations along the Thames and Severn valleys, and in the Midlands gravel pits and reservoirs. Many birds are also thought to roost at sites that support breeding colonies in the summer. Inland roosts are mainly in trees on islands on relatively undisturbed sites while coastal roosts are on cliffs, islands, navigation posts or jetties.

Co-ordinated roost counts yield higher total population figures than do daytime counts (Ulenaers et al in press), and they may therefore be a more suitable method for population estimation. Counts at the top 52 inland roost sites in England located 18% more cormorants than were recorded during the top 52 WeBS daytime counts (Sellers and Hughes 1997).

Information required
- number of birds.

Number and timing of visits

One visit on a single date during the last week of December or the first week of January. For future surveys, a preferred date within this period will be set. From 1998, this may be extended to three dates during the year: once during peak autumn migration in October, once in mid-winter (the Christmas week count), and once during spring migration in March.

Time of day

Dusk. Observers should be present at roosts about one hour before dark to ensure that all birds entering roosts are recorded.

Sites/areas to visit

All known roost-sites.

Equipment

- telescope.

Safety reminders

See above. Particular attention should be paid to coastal cormorant roosts, access to which may be dangerous in winter, especially in freezing temperatures and high winds.

Disturbance

None necessary.

Methods

Count all the birds present at each roost as they arrive at the roost; light conditions will begin to deteriorate with the onset of dusk. If possible, the number of birds with white bellies (first-year *P. carbo carbo*) should also be recorded. Notes should also be made of any colour-ringed birds that are observed, including whether the ring is on the left or the right leg, the colour of the ring and the direction in which the inscription reads (up or down the leg). Even incomplete details are useful and are usually sufficient to identify the colony of origin and/or the year of ringing.

If the survey method is extended to included a further two visits, the Christmas week mid-winter count will remain the highest priority. If inclement weather conditions are experienced during a roost count, the count should be repeated at the earliest opportunity.

Compiled by Baz Hughes, Robin Sellers and Graham Ekins, with additional information from the *Seabird Monitoring Handbook*.

References

Newson, S (1997) *Reproductive Performance and Breeding Phenology of Great Cormorants at Coastal and Inland Colonies in Britain*. First Year PhD Report, August 1997. University of Bristol.

Sellers, R M (1997) *Cormorant Colony Register*. WWT report CBCS-R-004, issue 3.

Sellers, R M and Hughes, B (1997) *Inventory of Inland Cormorant Roosts and Breeding Sites in Great Britain*. Report to JNCC.

Ulenaers, P, Devos, K and Jacob, J-P (in press) Population development of wintering and breeding cormorants (*Phalacrocorax carbo sinensis*) in Belgium. *Suppl. Ric. Biol. Selvaggina* 26.

Shag
Phalacrocorax aristotelis

Status

Amber listed: BI
SPEC category 4 (S)

National monitoring

Seabird Colony Register (SCR).
Seabird Monitoring Programme (SMP).
Seabird 2000 (1999–2001).

Population and distribution

Around 55% of the European population of shags breeds in Britain and Ireland, with almost 50% in Britain (*Birds in Europe*). Shags breed around rocky coastal cliffs, but are absent from the coasts of south and east England (*88–91 Atlas*). Between 1970 and 1987 there was an overall increase of 40% in the UK breeding numbers, though no increase took place either in north-west Scotland or in the Northern Isles. There are estimated to be 37,500 pairs of shags breeding in the UK (*Population Estimates*).

Ecology

Shags breed in colonies on cliff ledges from just above the high-water mark to over 100 m, normally in sheltered locations and often just inside small caves. They will also nest on boulder beaches and in fissures under very large boulders. There are 1–6 eggs, laid from late March to June, with hatching from mid-April. Shags may have two broods in some cases, and second clutches may be laid later in the season. Incubation takes 30–31 days and the fledging period is about 53 days (*Red Data Birds*).

Breeding season survey – population

This method was taken from the *Seabird Monitoring Manual*. It is a straightforward procedure, following Potts (1969) and Harris and Forbes (1987).

Information required

- the maximum number of Apparently Occupied Nests from any one count
- a map of the survey area, annotated with any subdivisions used and any parts of the colony not visible from land.

Number and timing of visits

If possible, make at least four visits to the site between early May and late June. Counts should be made at the time when the maximum number of nests is occupied. For north and east Britain, this is normally in early June (the late incubation/early nestling period), but counts in mid-June are acceptable. Further south, counts in late May are likely to be best, but local knowledge may be needed to obtain the optimum dates.

Time of day

Any time of day. In the evening 'extra' immatures and sub-adults may come to roost, but this should not affect counts of occupied nests.

Weather constraints

Avoid very wet and windy conditions.

Sites/areas to visit

Any site not already covered.

Equipment

- 1:10,000 map
- boat (optional) and life-jackets.

Safety reminders

Cliff edges may not be as solid as they look and may become more unsafe in wet and windy conditions. If working alone always ensure someone knows where you have gone and when you intend to return. If working with boats, never work alone. The boat should be operated by an experienced, trained boat handler and life-jackets should be worn at all times. Take equipment necessary to deal with emergencies. Do not attempt to rock-climb unless specifically required to do so, and never when alone or without training and equipment.

Disturbance

Avoid disturbing nesting birds, especially those on cliffs, eg by appearing on the skyline.

Methods

Define clearly the boundaries of the census area on a map, whether a length of coastline or a colony. For counting purposes and future reference, it is useful to subdivide this area further using easily recognisable natural features definable on a map, and then annotate the map with nest counts. Keep the census area consistent between years.

The recommended count unit is the Apparently Occupied Nest (AON), ie active nests (birds sitting tight whether or not eggs or young were seen, or an unattended brood of young) and other attended, well-built nests (apparently capable of holding eggs). Record nests which do not fall into this category separately, as they are often abandoned, or destroyed by other pairs stealing nest material (Harris and Forbes 1987).

Make at least four counts of AONs between early May and late June. Report all counts, but the highest reliable count of the whole census area on a single occasion should be entered as the final population figure. Do not combine peak counts of individual subcolonies from different dates.

Mark on the map any parts of a colony that are difficult or impossible to see from land. Estimate (minimum and maximum) the number of AONs likely to be hidden, based on numbers on visible sections (although these may not necessarily show similar densities to hidden sections) or on previous sea-based v. land-based counts. However, in reporting these estimates be very clear that they are of unknown reliability and may not be directly comparable with other counts. If at all possible, check and count these sections from a boat on a calm day (especially if you estimate that hidden sections are likely to total more than about 10% of the population). If a proportion of the birds in these sections can be seen from either land or sea, keep an additional note of the number of adults visible. Counts of adults should not be included in any detailed assessment of population changes for a colony or coastline, but may be required if a whole-colony estimate would otherwise be incomplete.

If possible, make an objective assessment of the general stage of the breeding cycle by recording the ratio of trace to well-built nests and the proportion of well-built nests that are empty. A subjective assessment of whether there seem to be fewer nests than expected from the numbers of adults present may provide an indication of an unusually late breeding season, or a season where a large proportion of adults have not attempted to breed. Counts of loafing adults, including those well away from any nests, can be useful, but avoid counts in the evening, when 'extra' immatures and sub-adults may come to roost.

Breeding season survey – breeding success and productivity

These methods are taken from the *Seabird Monitoring Handbook*. Monitoring of shag productivity is straightforward, but an accurate result necessitates regular visits (*Method 1*). However, a few visits can produce a useful estimate for comparing success in different years (*Method 2*).

Method 1 – regular visits

Information required

- total number of nests where eggs or apparent incubation were recorded
- total number of nests which failed
- total number of nests which fledged one, two, three or four chicks
- further notes on losses such as predation; eggs did not hatch; chick dead in nest, possibly starved
- if plots are used, express the colony productivity as the mean of the plot means (± standard error) and present data for each plot
- annotated map of the survey area showing any subsections, sample plots or parts of the colony not visible.

Number and timing of visits

Every 7–10 days from mid-April to July.

Time of day

Any time.

Weather constraints

Avoid very wet and windy conditions.

Sites/areas to visit

Any site not already covered.

Equipment

- 1:10,000 map
- camera and black-and-white film
- transparent overlays
- marker pens
- recording forms and clipboard.

Safety reminders, Disturbance

As for the population survey (above).

Methods (*Method 1: Regular visits*)

The method as described in the *Seabird Monitoring Handbook* is taken largely from Harris (1989) with amendments, and is used for pairs nesting on cliffs, rocks and accessible boulder sites. It involves visits to the colony to check the progress of breeding at numbered nest-sites every 7–10 days from when birds start laying until the young are fully feathered. If the colony is small, try to check all the visible nests. Where it is large, however, you may need to sample. The higher the proportion of the population that can be checked the better.

When sampling in a large colony, choose plots containing 10–30 nests. Check at least three study plots but preferably five or more. Two methods for choosing the location of these plots have been used:

a) Identify all potentially suitable study plots and select randomly from these.
b) Divide the colony into (say) four or five approximately equal parts (either by area or number of nests) and pick the same number of plots in each area. This method is not as good as (a), but has been used where the number of possible plots is small. Further information on sampling is given in the *Introduction*.

Whatever method you use, document exactly how you made your choice. If you are constrained to check only specific plots for some reason (eg safety, time, places which do not disturb birds or the public), record this. It is not necessary to use the same plots each season, unless they are also being used for population monitoring.

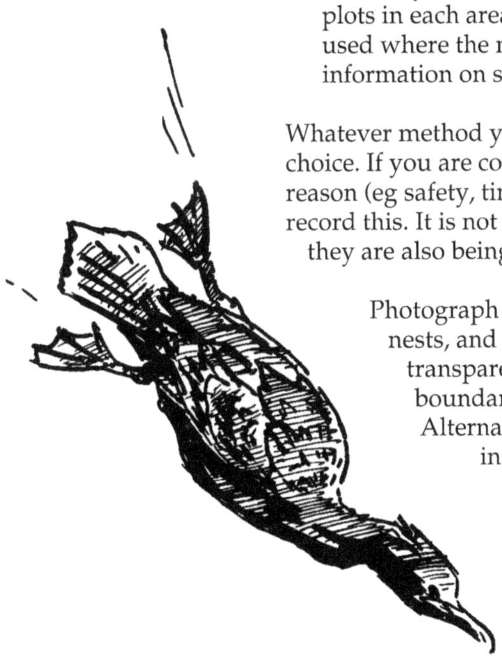

Photograph the selected plots, preferably when birds are at their nests, and make large (A4) black-and-white prints. Tape a transparent overlay over the photograph. Mark the plot boundaries and the positions of nests; number the nests. Alternatively, sketch the plot boundaries and nest positions in a notebook, although this is more likely to result in confusion during later checks.

Visit the area every 7–10 days from mid-April onwards and for each nest record the state of the nest (eg few sticks, complete platform), nest contents (if visible – do not flush sitting birds) and whether a bird appears to be incubating or brooding. Pay particular attention to large young on open ledges, as large young sometimes move away from the nests. You will have to assume that well-feathered young (with little or no down remaining on mantle and upperwings) which appear healthy will fledge. Figure 1 gives an example of a form that can be used to collect data on individual nests.

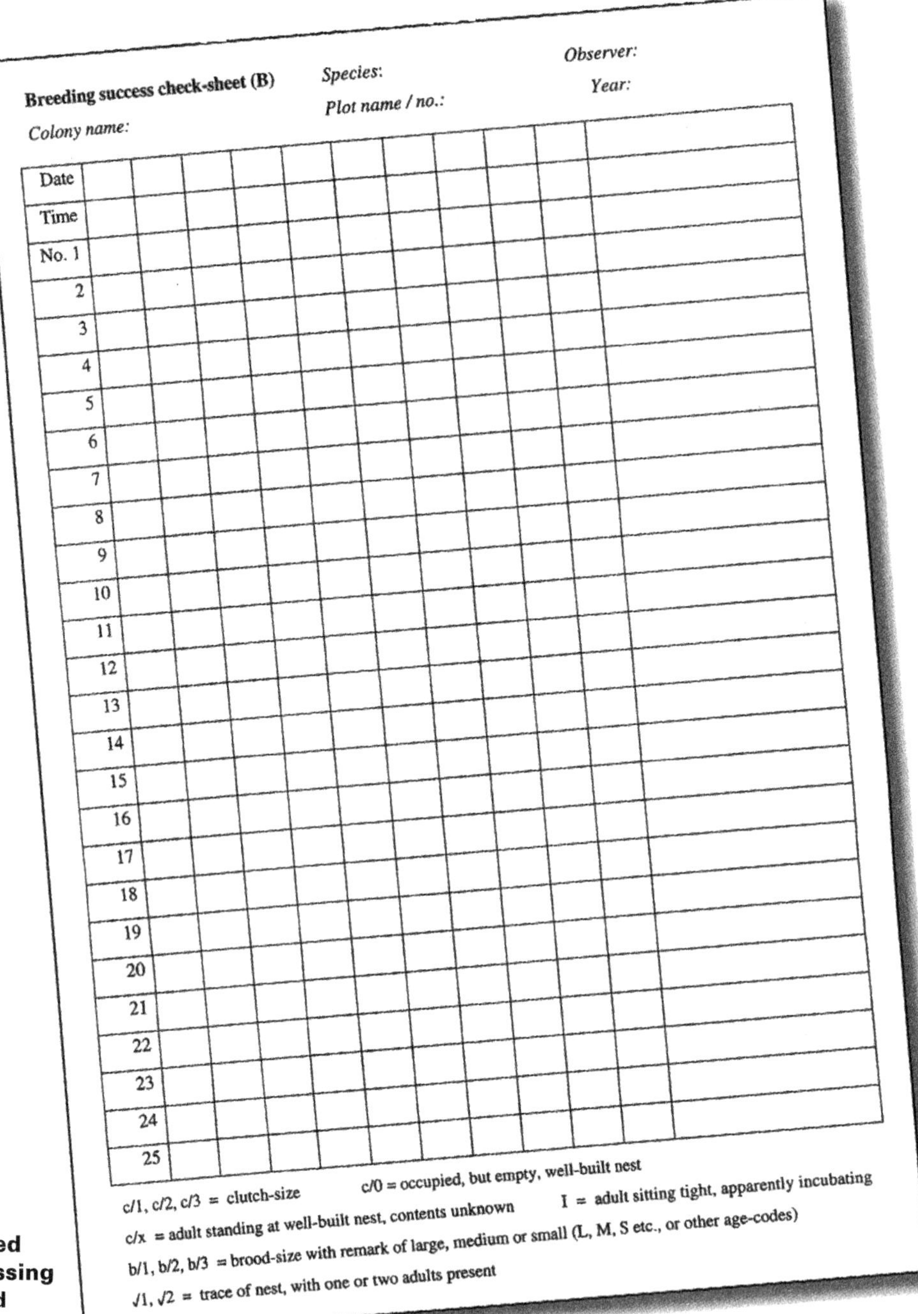

Breeding success check-sheet (B) *Species:* *Observer:*

Colony name: *Plot name / no.:* *Year:*

Date											
Time											
No. 1											
2											
3											
4											
5											
6											
7											
8											
9											
10											
11											
12											
13											
14											
15											
16											
17											
18											
19											
20											
21											
22											
23											
24											
25											

c/1, c/2, c/3 = clutch-size c/0 = occupied, but empty, well-built nest

c/x = adult standing at well-built nest, contents unknown I = adult sitting tight, apparently incubating

b/1, b/2, b/3 = brood-size with remark of large, medium or small (L, M, S etc., or other age-codes)

√1, √2 = trace of nest, with one or two adults present

Figure 1
The *Seabird Monitoring Handbook*'s recommended recording sheet for assessing productivity in shags and other nest-building seabirds.

Report the total number of young fledged divided by the number of nests where birds were definitely or probably incubating. If sample plots are used, give figures for each plot. Calculate colony productivity as the mean of plot means (± standard error). Do not pool results from plots; there may be marked differences between plots, and the mean productivity for the colony is best calculated as the mean of the plot figures (± standard error). Report the total number that failed, if possible with notes on causes (eg predation; eggs did not hatch; chick dead in nest, possibly starved), and the total number of nests that fledged one, two, three or four chicks.

Method 2 – three visits

Information required

- the number of chicks divided by the number of occupied, well-built nests
- productivity reported as the mean of the plot means (± standard error), and data presented for each plot
- an annotated map of the survey area showing any subsections or parts of the colony not visible.

Number and timing of visits

At least three visits, one during incubation, one when first chicks are about to fledge, and, if possible, one follow-up visit to check on smaller chicks from early May to July.

Time of day, Weather constraints, Sites/areas to visit, Equipment, Safety reminders, Disturbance

As for Method 1 (above).

Method (*Method 2: Three visits*)

This method is for use when limited time is available. Select study-plots as in *Method 1*. Again, try to cover as much of the population as is practical; in smaller colonies this may be all nests.

Check nests at least twice, once during incubation and once around the time when the first chicks are likely to fledge, when a search should also be made for additional well-built nests. Note the numbers of chicks in each nest and, if possible, their approximate size/age. If necessary, make a follow-up visit to check on smaller chicks from early May to July. Report the number of chicks divided by the number of occupied, well-built nests, for each plot separately and the mean of the results from each plot (± standard error). The latter gives an estimate of productivity for the whole colony.

This method will overestimate production, as it assumes all chicks survive to fledge. The overestimate can be lessened by a follow-up visit, or by making visits to check on chicks that were still small during the second visit.

References

Harris, M P and Forbes, R (1987) The effect of date on counts of nests of shags *Phalacrocorax aristotelis*. *Bird Study* 34: 187–190.

Potts, G R (1969) The influence of eruptive movements, age, population size and other factors on the survival of the shag, *Phalacrocorax aristotelis*, on the Farne Islands, Northumberland. *J. Animal Ecology* 49: 465–484.

Bittern
Botaurus stellaris

Status

Red listed: BD, HD, BR, SPEC 3 (V)
Schedule 1 of WCA 1981
Annex I of EC Wild Birds Directive

National monitoring

Annual breeding surveys of known sites (RSPB).
Rare Breeding Birds Panel.

Population and distribution

The bittern is extremely rare in Britain, and by 1886 had ceased to breed due to persecution. Recolonisation took place in Norfolk in the early 1900s, followed by a slow increase and spread up to the mid-1950s. After this there began a decline that still continues (*Red Data Birds*). This is mainly due to loss of habitat and gradual unsuitability of existing habitat. Less than 20 males have been recorded booming in recent years (RSPB 1998) and the RSPB is monitoring all recently occupied sites. Winter sightings are scattered throughout the UK, with higher numbers in south-east England. Between 50 and 150 birds winter in the UK (*Winter Atlas*), with higher numbers during more severe winters.

Ecology

Bitterns usually occur in large reedbeds (>20 ha). Some males start booming in February, but the main period is mid-March to mid-June, though this varies between sites. The booming vocalisation of the male is very variable in the frequency with which it is given and the distance over which it can be heard. Eggs are laid from late March to May. Chicks leave the nest and disperse into the reeds at about 12 days old, fledging between June and August. In winter, bitterns will be found in a slightly wider range of habitats, ranging from wet ditches with cover, to the more usual large stands of wet reedbed.

Breeding season survey – population

Information required

- maximum number of booming males heard at any one time on a single visit
- summary map showing location of all booming males heard over three visits.

Number and timing of visits

At least three visits: first week of April, end of April or beginning of May, and mid-May. If any booming is heard, several further visits a few days apart will be necessary to confirm that the bird remains present for at least a week.

Time of day

Preferably in the two hours before dawn, otherwise in the two hours after dusk.

Weather constraints

Calm conditions are best.

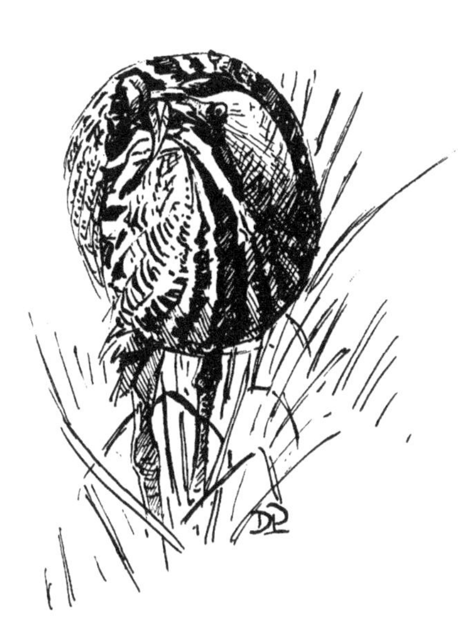

Sites/areas to visit

Reedbeds where booming has been reported or where wintering birds have been recorded in recent years.

Equipment

- Schedule 1 licence
- 1:10,000 map of the area to be visited
- A4 photocopied map of the survey area for use in the field
- compass
- clipboard.

Safety reminders

Ensure someone knows where you are working and what time you expect to return. If entry into the reedbed is unavoidable, try to work in pairs and use a map that shows safe and unsafe areas. Do not enter reedbeds in the dark, and if you are in one in the evening make sure you get out before it becomes dark.

Disturbance

Avoid trampling damage – newly created paths can invite predators. Be aware of the presence of other rare breeding birds such as bearded tits and marsh harriers which you may affect by disturbance.

Methods

If you hear a booming male at a site that is not already being covered by the RSPB, contact the Conservation Science Department at The Lodge (address in the *Introduction*). They will be able to obtain good-quality sound recordings of the booms for analysis to identify individuals and give a more accurate population estimate (Gilbert et al 1994).

It is important to confirm the presence of a booming male at a site for more than one week. On hearing one it will therefore be necessary to make several further visits, a few days apart. If more than one booming male is suspected within a site, listen at a position from which they can be heard together or co-ordinate a team of people to listen for, time and map booming males. Listen for unusual booming or characteristic booming patterns as this can help to distinguish between individuals. Assume that the number of males present at each site is equal to the maximum number of birds heard booming at any one time on a single visit, unless you have evidence to the contrary (eg very distinctive booming patterns). The more accurately the booming periods are known at each site the easier it is to produce a national population figure for the species.

Map the location of any booming males. This can be done by triangulation. Pick three vantage points around the bird and mark them on the field map. Take a bearing to the booming male from each. Mark the bearings on the map with a pen and ruler – the bittern is where all three lines intersect. Do this on at least three occasions, at least three

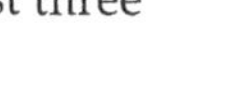

days apart. Record any sightings on the same map.

Breeding season survey - breeding success

The best way to obtain an indication of breeding success is by observing feeding flights and juveniles later in the season. Juveniles have brighter, sandier plumage and bright yellow legs. Report any sightings of summering birds away from breeding sites as above.

Winter survey

Record any sightings of bitterns between August and February.

References

Gilbert, G, McGregor, P K and Tyler, G (1994) Vocal individuality as a census tool: practical considerations illustrated by a study of two rare species. *J. Field Ornithology* 65: 335–348.

RSPB (1998) *The Bittern Monitoring Report 1998*. RSPB unpubl (confidential).

Mute swan
Cygnus olor

Status

Non-SPEC
Annex II/2 of EC Wild Birds Directive

National monitoring

National breeding surveys in 1955–56, 1978, 1983 and 1990, partial survey in 1961 (Rawcliffe 1958, Campbell 1960, Eltringham 1963, Ogilvie 1981, Brown and Brown 1985, Ogilvie 1986, Delany et al 1992).
BBS.
WeBS.

Population and distribution

Mute swans are widespread throughout most of Britain and Ireland but are most abundant in lowland areas. They rarely occur above 300 m (*88–91 Atlas*). Although absent from many northerly and westerly areas, where high ground predominates, they flourish on the southern Outer Hebrides and Orkney. Being generally resident or only partially migratory (Birkhead and Perrins 1986), the winter distribution is similar to that of summer (*Winter Atlas*). Many breeding pairs remain on their territories throughout the year, while immature and non-breeding birds join flocks at highly traditional sites.

The British population remained relatively stable, at about 20,000 birds, from the mid-1950s to the mid-1980s, although there was considerable regional variation, with dramatic declines in some areas (Ogilvie 1981). However, the population increased sharply with the introduction of legislation against the use of lead fishing weights (Kirby et al 1994, 1995). This is known to have considerably reduced the incidence of lead poisoning (Sears and Hunt 1991). Other important factors are likely to include a number of mild winters during the late 1980s, expansion in the acreage of winter cereals in parts of the range and an increase in the number of swan hospitals and rehabilitation programmes (Delany et al 1992, Kirby et al 1994). All of these are certain to have improved annual survival rates of mute swans. The UK population is now estimated to have reached 28,000–30,000 individuals (*Population Estimates*).

Ecology

Despite their wide distribution, mute swans are to some extent restricted by their preference in most areas for breeding on eutrophic waters. Breeding waters include lakes, reservoirs, canals, slow-flowing rivers, ponds and small streams. Nests are usually within 100 m of water and normally far apart. Early eggs are laid in mid-March, incubation is about 36 days and there is a single brood. The young are precocial with a fledging period of 120–150 days (*BWP*).

Breeding season survey – population

Survey methods are based on those developed for national breeding censuses, as described by Delany et al (1992).

Information required

- number of territorial birds broken down into three categories: at a nest, with cygnets, or failed breeders
- number of non-breeding birds
- positions of all birds or nests found.

Number and timing of visits

One visit in April. Additional visits may be necessary to determine breeding status.

Time of day

Any time of day, preferably in good light conditions.

Weather constraints

Visit in calm, dry weather.

Sites/areas to visit

All suitable habitat within the recording area. Check even the smallest and most public waterbodies.

Equipment

- 1:25,000 map of survey area
- recording forms.

Safety reminders

Breeding swans can be highly aggressive. Avoid close encounters; two people per survey team is recommended.

Disturbance

Some disturbance may be necessary to determine birds' breeding status.

Methods

Aim to cover systematically all suitable habitat for breeding and non-breeding birds in the recording unit, preferably a 10 × 10 km square. Record the date and the 10-km square designation. Use separate recording forms for breeders and non-breeders. The reverse of the form should contain a 10 × 10 numbered grid, to represent the 10 × 10 km squares, on which to mark the locations of the birds observed. An example of a suitable recording form is given in Figure 1.

Since non-breeders may move about, try to avoid counting the same birds twice by covering the site in as short a time as possible. Breeding and non-breeding swans can be censused efficiently by aerial survey, and microlight aircraft are useful in areas with complex ditch systems.

For breeding birds, assign a code to each pair, nest or brood located, describe its location and record a six-figure OS grid reference. Classify the habitat as pond or lake, reservoir, gravel (or other) pit, river, stream, canal, ditch, estuary or seashore (or specify). Provide status information by using the following codes: pair on territory, but without nest (T), pair

MUTE SWAN CENSUS 1990: BREEDING PAIRS: OBSERVATIONS

Site code	Location	Grid ref.	Habitat	Dates and observations (Dates as 00/0 please)
A	LITTLETON BRICKPITS, NR. WICK	716092	GRAV. PIT	03/4T, 15/4N, 01/5N
B	ALVESTON RES.	747007	RESERVOIR	03/4N, 01/5N, 27/5T
C	HILLSIDE LAKE THORNHAM	701065	LAKE	03/4T
D	LITTLE AVON NR. THORNBURY	784021	STREAM	15/4N, 07/5D
E	R. SEVERN, FROME	733062	RIVER	01/5N

BTO/WWT/SOC MUTE SWAN CENSUS 1990: BREEDING PAIRS

County (England & Wales) or District & Region (Scotland) AVON

Please mark on the grid the positions of all pairs and nests found, using the following symbols to represent the state you recorded on your last visit:

Territorial pair X
Pair with nest .... O
Pair with brood .... ●

Against each symbol write the letter used for the site code on the other side of this form.

Please shade any parts of the 10 km square that you were unable to cover.*

What is your best estimate of the number of pairs in the shaded part? What is your reason for this belief?*

O
NO SUITABLE HABITAT

Please write your name, address, and phone no. here:

J. SMITH 654321
1 FIELD ROAD
BRISTOL
AVON. BS1 1AA

As soon as possible after 31 May, please return to your local organizer.

Your local organizer is:
SIMON DELANY
WWT, SLIMBRIDGE, GLOS.
GL2 7BT (0453-890333)

*Note: estimates are much less useful than proper coverage.

Figure 1
Example recording form as used in the BTO/WWT/SOC Mute Swan Census in 1990. This form (front and back) is for breeding pairs; another form (identical but for its title) should be used for non-breeding birds.

with nest (N), pair with cygnets (B), pair known to have nested but failed (D). On the reverse of the form, mark the positions of the birds or nests found using the following symbols: territorial pair, no sign of breeding (X), pair with nest (O), pair with brood (●). A pair that nested and then failed should be marked with the symbol for its last known state before failing. Against each symbol write the code used for the site on the other side of the form.

For non-breeders, the site code, location, grid reference and habitat can be recorded in the same way as for breeding birds but on a different form (make clear that the form refers to *non-breeding birds*). Record the number of birds found in each place, and on the grid on the reverse side of the form mark positions of birds and flocks with an X.

Note that many mute swans are ringed with metal or plastic (darvic) leg rings which can be read in the field. Try to record the ring colour, code and which leg the ring is on. If the bird is paired, also try to note the partner's ring details (or specify 'unringed') and the number and ring details of any cygnets. Contact WWT for further details.

Breeding season survey – breeding success and productivity

The breeding and winter surveys provide sufficient information on breeding success and productivity. More detailed information need only be collected when carrying out specific research projects.

Winter survey

WeBS.

See *National wintering swan census* and *Productivity of swans and geese* in the generic survey methods section.

Contributed by Jeff Kirby

References

Birkhead, M E and Perrins, C M (1986) *The Mute Swan*. Helm, London.
Brown, A W and Brown, L M (1985) The Scottish mute swan census 1983. *Scottish Birds* 13: 140–148.
Campbell, B (1960) The mute swan census in England and Wales, 1955–56. *Bird Study* 7: 208–223.
Delany, S, Greenwood, J J D and Kirby, J (1992) *National Mute Swan Survey 1990*. WWT report to JNCC, Slimbridge.
Eltringham, S K (1963) The British population of mute swans in 1961. *Bird Study* 10: 10–28.
Kirby, J S, Delany, S and Quinn, J (1994) Mute swans in Great Britain: a review, current status and long-term trends. *Hydrobiologia* 279/280: 467–482.
Kirby, J S, Salmon, D G, Atkinson-Willes, G L and Cranswick, P A (1995) Index numbers for waterbird populations. III. Long-term trends in the abundance of wintering wildfowl in Great Britain, 1966/67–1991/92. *J. Applied Ecology* 32: 536–551.
Ogilvie, M A (1981) The mute swan in Britain, 1978. *Bird Study* 28: 87–106.
Ogilvie, M A (1986) The mute swan *Cygnus olor* in Britain 1983. *Bird Study* 33: 121–137.
Rawcliffe, C P (1958) The Scottish mute swan census 1955–56. *Bird Study* 5: 45–55.
Sears, J and Hunt, A (1991) Lead poisoning in mute swans, *Cygnus olor*, in England. Pp 383–388 in Sears, J and Bacon, P J (eds) *Proc. Third IWRB International Swan Symposium, Oxford 1989. Wildfowl Suppl.* 1.

Bewick's swan
Cygnus columbianus

Status

Amber listed: WI, WL, SPEC 3^{W} $(L)^{W}$
Schedule 1 of WCA 1981
Annex I of EC Wild Birds Directive

National monitoring

WeBS.

Population and distribution

Within the UK, the majority of Bewick's swans winter in England and Northern Ireland, with smaller numbers in Scotland and Wales. The largest flocks are found on the Ouse and Nene Washes, at Martin Mere and on the Ribble estuary in Lancashire, along river valleys in Gloucestershire and Hampshire, on the Norfolk, Somerset and Kent levels, and at Loughs Foyle, Neagh and Beg in Northern Ireland. The UK wintering population fluctuates considerably depending on the severity of the weather on the continent, varying between about 5,000 and 10,000 in the period between 1984 and 1995 (Beekman in press). Additional birds arrive from the continent during periods of freezing conditions and prolonged snow cover.

Ecology

In winter, Bewick's swans are found inland close to water on permanent pasture, winter cereals, root crops or flooded meadows and, at coastal sites, on brackish lagoons (*Winter Atlas*).

Winter survey

WeBS.

See *National wintering swan census* and *Productivity of swans and geese* in the generic survey methods section.

Reference

Beekman, J H (in press) International censuses of the NW-European Bewick's swan population, January 1990 and 1995. *Swan Specialist Group Newsletter* 6.

Whooper swan *Cygnus cygnus*

Status

Amber listed: BR, WI, WL
SPEC 4^{w} (S)
Schedule 1 of WCA 1981
Annex I of EC Wild Birds Directive

National monitoring

Rare Breeding Birds Panel.
WeBS.

Population and distribution

One or two pairs of whooper swan breed in the UK each year. In winter, whooper swans have a predominantly northerly distribution. During the 1990–91 winter, the largest flocks of whooper swans were associated with arable land, although the majority of birds (68%) occurred on permanent inland waters (Rees et al 1997). The UK wintering population was recorded as 8,700 in January 1991 and 7,800 in January 1995 (Kirby et al 1992, Cranswick et al 1997).

Ecology

Whooper swans will roost and occasionally feed on water, including quite small lochs and ponds. They will feed on emergent and submergent freshwater plants, grass, farmland (cereals and stubble fields) and intertidal plants, eg eel-grass (*Winter Atlas*, Rees et al 1997).

Winter survey

WeBS.

See *National wintering swan census* and *Productivity of swans and geese* in the generic survey methods section.

References

Cranswick, P A, Bowler, J M, Delany, S N, Einarsson, O, Gardarsson, A, McElwaine, J G, Merne, O J, Rees, E C and Wells, J H (1997) Numbers of whooper swans *Cygnus cygnus* in Iceland, Ireland and Britain in January 1995: results of the international whooper swan census. *Wildfowl* 47: 17–30.

Kirby, J S, Rees, E C, Merne, O J and Gardarsson, A (1992) International census of whooper swans *Cygnus cygnus* in Britain, Ireland and Iceland: January 1991. *Wildfowl* 43: 20–26.

Rees, E C, Kirby, J S and Gilburn, A (1997) Site selection by swans wintering in Britain and Ireland; the importance of geographical location and habitat. *Ibis* 139: 337–352.

Bean goose
Anser fabalis

Status

Amber listed: WL
Non-SPEC
Annex II/1 of EC Wild Birds Directive

National monitoring

WeBS.

Population and distribution

A small population of bean geese winters in the UK. These are mainly in two regular flocks: one in central Scotland and one in the Yare valley in England (both are *fabilis*). Small flocks also turn up from time to time at other east-coast locations (mainly *rossicus*), especially after cold spells on the continent. Around 450 bean geese winter in the UK (*Population Estimates*).

Ecology

Wintering bean geese are found on coastal grazing marshes, rough pasture, stubble and other agricultural fields, and roost on lakes, rivers and moorland (*BWP*).

Winter survey

WeBS.

See *National censuses of geese* and *Productivity of swans and geese* in the generic survey methods section.

Pink-footed goose
Anser brachyrhynchus

Status

Amber listed: WI, WL
SPEC 4 (S)
Annex II/2 of EC Wild Birds Directive

National monitoring

WeBS.

Population and distribution

The wintering population of pink-footed geese is found in east and south Scotland and in some parts of England, such as the north Norfolk coast and Lancashire. There are about 225,000 pink-footed geese wintering in the UK (C Mitchell pers comm).

Ecology

Pink-footed geese are associated with lowland farmland where barley stubble, potato fields, winter-sown cereals and pasture provide winter food. They roost on the mudflats and sandbanks of estuaries and on inland lochs and reservoirs (*Winter Atlas*).

Winter survey

WeBS.

See *National censuses of geese* and *Productivity of swans and geese* in the generic survey methods section.

White-fronted goose *Anser albifrons*

Status

flavirostris	Amber listed: WL Non-SPEC Annex I of EC Wild Birds Directive	*albifrons*	Amber listed: WL Non-SPEC Annex II/2 of EC Wild Birds Directive

National monitoring

WeBS.

Winter counts of *flavirostris* undertaken annually since 1980 (Greenland White-fronted Goose Study; Fox and Francis 1996).

Population and distribution

Two subspecies of white-fronted goose winter in Britain and Ireland: Greenland white-fronted goose *Anser albifrons flavirostris* and European white-fronted goose *Anser albifrons albifrons*. Their distribution does not overlap. Greenland white-fronted geese winter in Ireland, on the north and west coasts of Scotland, and at one site in Wales. There are about 19,000 Greenland white-fronted geese wintering in the UK (*WeBS 1994–95*). European white-fronted geese winter in the south of England and in south Wales. There are about 6,100 European white-fronted geese wintering in Britain (*Population Estimates*).

Ecology

Greenland white-fronted geese feed on a variety of habitats, including unimproved grassland, arable farmland, stubbles, potatoes, mires, callows and machair. European white-fronted geese prefer low-lying wet pasture, either bordering coastal marshes or along river valleys (*Winter Atlas*).

Winter survey

WeBS.

See *National censuses of geese* and *Productivity of swans and geese* in the generic survey methods section.

Reference

Fox, A D and Francis, I (1996) *Report to the 1995/96 National Census of Greenland White-fronted Geese in Britain*. GWGS, Kalø.

Greylag goose *Anser anser*

Status

Amber listed: WI, WL
Non-SPEC
Schedule 1/II (in Outer Hebrides, Caithness, Sutherland and Wester Ross only) of WCA 1981
Annex III/2 of EC Wild Birds Directive

National monitoring

WeBS.
North Scottish population survey, 1999.

Population and distribution

Feral breeding greylag geese are widespread, with an estimated 13,800 individual adults in the UK. There are also 500–700 pairs of native greylag breeding in the UK (*Population Estimates*), mainly resident in South Uist and Sutherland. The greylag geese which winter in Britain and Ireland originate from three separate populations: the native Scottish population, the Icelandic population and the feral population. Most of the greylags which breed in Iceland winter in Scotland and the west coast of Ireland. The largest wintering non-feral group is in north-west Scotland. Most other flocks which winter south of a line from the Isle of Man to Teesmouth are feral birds. There are about 90,000 Icelandic greylags (*WeBS 1994–95*), about 9,000 Hebridean (C Mitchell pers comm) and 14,000 feral birds wintering in Britain (*WeBS 1994–95*).

Ecology

Greylags feed on estuaries, on farmland with cereals or root crops and on grass. They roost on estuaries, lakes, rivers or reservoirs (*Winter Atlas*).

Winter survey

WeBS.

See *National censuses of geese* and *Productivity of swans and geese* in the generic survey methods section.

Barnacle goose
Branta leucopsis

Status

Amber listed: WI, WL, SPEC 4/2 (L)[W]
Annex I of EC Wild Birds Directive

National monitoring

WeBS.
Greenland barnacle geese monitored annually on Islay, Coll, Tiree and Hoy (SNH).
National survey of Greenland barnacle geese every five years (most recent 1998) (WWT).
Svalbard barnacle geese monitored annually (WWT).

Population and distribution

There are two separate populations of barnacle goose wintering in Britain and Ireland: birds from Greenland winter on the coasts of the north and west of Scotland and Ireland, and birds from Spitsbergen winter on the Solway Firth. Other records of barnacle geese in England and Wales arise from escaped captive individuals. Wintering in the UK are an estimated 22,000 birds from Svalbard (C Mitchell pers comm), 26,950 from Greenland and 800 feral birds (*Population Estimates*).

Ecology

Barnacle goose flocks roost at night on estuaries, on sandbanks and on sea lochs 1–2 km offshore. They feed on a wide range of grasses, as well as saltmarsh plants and stubble fields (*Winter Atlas*).

Winter survey

WeBS.

See *National censuses of geese* and *Productivity of swans and geese* in the generic survey methods section.

Brent goose
Branta bernicla

Status

Amber listed: WI, WL, SPEC 3 (V)
Annex II/2 of EC Wild Birds Directive

National monitoring

WeBS.
Annual monitoring of light-bellied brent geese (Svalbard, Greenland) (Sunderland University, EN).
Annual monitoring of light-bellied brent geese (Canada) throughout Ireland (I-WeBS/WWT).

Population and distribution

In Britain, brent geese winter on the east and south coast of England, with increasing numbers of birds inland. In Ireland, the birds winter on the west and east coasts. There are two subspecies of brent goose: the dark-bellied brent goose *Branta bernicla bernicla* winters in south-east England from the Wash to the Exe estuary; Irish records are of light-bellied brent geese *Branta bernicla hrota* which also occur on the Northumberland coast. There are an estimated 103,300 dark-bellied brent geese and 17,000 light-bellied brent geese wintering in the UK. The majority of the light-bellied birds are in Northern Ireland (*Population Estimates*).

Ecology

Brent geese prefer large estuaries and areas of intertidal mudflat, feeding preferentially on eel-grass and saltmarsh vegetation. Birds inland will feed in grassland and in cultivated crops such as barley, wheat and oilseed rape (*Winter Atlas*).

Winter survey

WeBS.

See *National censuses of geese* and *Productivity of swans and geese* in the generic survey methods section.

Shelduck
Tadorna tadorna

Status

Amber listed: BI, WI, WL
Non-SPEC

National monitoring

National breeding survey 1992–93 (WWT).
WeBS.

Population and distribution

Shelducks breed around most of Britain's coast and at some inland sites. The 1992 survey of breeding shelducks in Britain reported a total of 44,700 adult birds (*WeBS 1992–93*). Great Britain has an estimated 73,500 wintering shelducks (UK has 76,400) (*Population Estimates*). The wintering population of shelducks is expanding in Europe, but the reasons for this are not known. It is likely that better protection throughout their range has been at least partly responsible (Kirby et al 1995).

Ecology

Shelducks may breed in colonies or solitarily. They need foraging areas of fairly high biological productivity, especially sands and mudflats over which periods of inundation with shallow water alternate with periods of drying out brought about by the tidal cycle or evaporation, eg estuaries and shallow coasts. For nesting, birds require suitable holes or patches of vegetation providing shelter or concealment. Nest-holes may be in sand dunes and other light soils, or in ground covered by dense prickly or thorny shrubs. A clutch of 8–10 eggs is laid during May–June. Shelducks are single-brooded. The young fledge by August and often gather in crèches of up to 100. The winter feeding habitats in Britain and north-west Europe are almost exclusively shallow coastal and estuarine waters with extensive low-tide areas of sand and mudflats, and inland shallow waters near the coast (*Red Data Birds*).

Breeding season survey – population

This survey method is based on that used in the 1992–93 national survey (Delany 1992), originally suggested, and subsequently updated, by Ian Patterson of Aberdeen University.

Information required

- maximum number of territorial pairs
- total number of birds
- map, showing the survey area and boundary.

Number and timing of visits

Three visits, 25 April to 17 May.

Time of day

Within two hours either side of low tide, after midday.

Weather constraints

None.

Sites/areas to visit

Shallow coasts and estuaries, or inland seas and lakes, with access to fresh water and suitable nest-sites.

Equipment

- 1:25,000 field map of the area
- prepared recording forms (see example in Figure 1).

Safety reminders

Ensure that someone reliable knows of your whereabouts and when you will return. Ensure that they know what to do if you are late. Check that you are not liable to be cut off by the tide. Beware of slipping into saltmarsh creeks, which can be hidden by vegetation, and very soft sediments from which it may be difficult to extricate oneself.

Disturbance

Keep disturbance to a minimum.

Methods

Most shelducks have taken up territory by the beginning of May, so counts in the first week of May will minimise under-recording due to nest failure and desertion later in the year. The count period is extended over three weeks so that at tidal sites the scope for selecting the best tide on which to count can be maximised.

Mark the boundaries of the survey area clearly on a map. Shelducks are strongly territorial and maintain feeding territories if they have a nest nearby. The male defends the territory while the female incubates. This means that a carefully conducted count of birds on feeding territories can be strongly indicative of the number of pairs nesting in the area.

Ideally, the count should be conducted within two hours of low tide on an afternoon or evening in the first ten days of May. At big estuaries, afternoon heat haze may make counts of distant birds difficult, so a mid-tide count (when birds are less distant) or a morning count (when heat haze is less serious) may be better. Split large sites into manageable sections, each of which can be counted in about four hours or less. Ideally, these should be covered on the same tide by a number of observers.

For each visit, record:

(a) the number of lone well-spaced males (territorial males)
(b) the number of discrete pairs (territorial pairs, where the female is taking a break from incubation)
(c) the number of birds in groups (non-breeders).

Many non-breeding groups consist of paired birds, and care is needed to separate these non-breeding pairs from genuinely territorial pairs. If time permits, watch pairs for evidence of territorial behaviour, eg the male guarding the female as she feeds, both birds remaining on territory. Non-territorial pairs often occur in mobile groups which range widely as they feed.

Mark all the birds on a map, stating how many there are, whether they are male or female, and whether they are (a), (b) or (c) above.

At the end of each visit, transfer the results to a recording form (see example in Figure 1). Report the maximum number of 'total territories'

Site 1,Name:	Population Territorial males	Territorial pairs	Total territories	Non-breeders	Total birds
Visit 1 Date					
Visit 2 Date					
Visit 3 Date					
Maximum figures					

Site 1,Name:	Productivity No.of adults	No. of young	No.of viable young
Visit 1 Date			
Visit 2 Date			
Visit 3 Date			
Maximum figures			

Site 2,Name:	Population Territorial males	Territorial pairs	Total territories	Non-breeders	Total birds
Visit 1 Date					
Visit 2 Date					
Visit 3 Date					
Maximum figures					

Site 2,Name:	Productivity No.of adults	No. of young	No.of viable young
Visit 1 Date			
Visit 2 Date			
Visit 3 Date			
Maximum figures			

Figure 1
An example of a form which could be used for recording breeding shelduck. Modified from Delany (1992). See text for further details.

from any visit as the total number of breeding pairs at any one site. Calculate 'total territories' by adding (a) the number of territorial males to (b) the number of territorial pairs. Also report 'Total birds' by assuming that there were two birds for every 'total territory' and adding (c) the number of non-breeders.

Breeding season survey – productivity

Information required

- maximum number of viable young
- number of adults.

Number and timing of visits

Three visits, within a period of 7–10 days in late July and early August, the week before the oldest young fledge.

Time of day

Mid-tide (on a rising or falling tide) at tidal sites.

Weather constraints, Sites/areas to visit, Equipment, Safety reminders, Disturbance

As for the population survey (above).

Methods

Mark the boundary clearly on a map. Survey the same area as for the population counts in May. By the time the oldest juveniles are ready to fledge, the youngest (which could have hatched as late as early July) will have passed their most vulnerable first few weeks and are likely to survive to fledge after being counted. Surveying at mid-tide will help to avoid the difficulties of birds hiding in deep creeks at low tide and dense roosting groups of ducklings gathering at high tide.

Age the chicks in relation to the size of the adults, eg ½ , ¾, full-grown. For a guide to the plumage development of young waterfowl, see Figure 5 in *Dabbling and diving ducks* in the generic survey methods section.

Count the number of adults, young and viable young (three weeks old, about ¾-grown, size class IIA/IIB: Gollop and Marshall 1954) for each visit and transfer the results to the same recording form used for the population survey in May. Report the maximum number from any one visit. Report productivity as the maximum number of viable young divided by the maximum number of territorial pairs (from the population survey).

Winter survey

WeBS.

See *Inshore marine waterfowl* in the generic survey methods section.

References

Delany, S (1992) *Pilot Survey of Breeding Shelduck in Great Britain and Northern Ireland*. WWT, unpubl.

Gollop, J B and Marshall, W H (1954) *A Guide for Ageing Duck Broods in the Field*. Miss. Flyway Coun. Tech. Sect. mimeo.

Kirby, J S, Salmon, D G, Atkinson-Willes, G L and Cranswick, P A (1995) Index numbers for waterbird populations. III. Long-term trends in the abundance of wintering wildfowl in Great Britain, 1966/67–1991/92. *J. Applied Ecology* 32: 536–551.

Wigeon
Anas penelope

Status

Amber listed: WI, WL
Non-SPEC
Annex II/1 of EC Wild Birds Directive

National monitoring

WeBS.

Population and distribution

Only 300–500 pairs of wigeon breed in the UK, in Scotland and northern England (*Population Estimates*). However, the species is common in winter throughout the UK, both inland and on the coast (*Winter Atlas*). The estimated number of wigeon wintering in the UK is 291,000 individuals (*Population Estimates*), but the actual size of the wintering wigeon varies according to the severity of the weather (Kirby et al 1995).

Ecology

Breeds mainly along the shores of upland lakes and boglands, where it nests in marginal grass or shrub cover, or on islands (*88–91 Atlas*). A clutch of 8–9 eggs can be laid as early as mid-April but the main laying period is May. There is one brood and the young fledge in July (*Red Data Birds*). Outside the breeding season the feeding and roosting habitat is varied. On coastal mudflats the traditional diet is eel-grass *Zostera*, but birds also graze on land and will take grain in stubble fields. About 20% of British wintering wigeon are found on inland flooded grassland (*Red Data Birds*).

Breeding season survey

See *Dabbling and diving ducks* in the generic survey methods section.

Winter survey

See the generic methods section on WeBS counts.

Reference

Kirby, J S, Salmon, D G, Atkinson-Willes, G L and Cranswick, PA (1995) Index numbers for waterbird populations. III. Long-term trends in the abundance of wintering wildfowl in Great Britain, 1966/67–1991/92. *J. Applied Ecology* 32: 536–551.

Gadwall
Anas strepera

Status

Amber listed: WI, WL, SPEC 3 (V)
Annex II/1 of EC Wild Birds Directive

National monitoring

WeBS.

Population and distribution

There is a relatively small breeding population (790 pairs) of gadwall in the UK (*Population Estimates*) mainly concentrated in the south and east of England (*88–91 Atlas*). The increase in winter numbers of gadwall since the early 1970s is thought partly to be due to increasing numbers of continental immigrants and the increased use of gravel pits and reservoirs. However, it may also be due to an ability to feed in deeper waters by kleptoparasitising coots (Fox 1991, Kirby et al 1995). The estimated number of gadwall wintering in the UK is 8,400 (*Population Estimates*).

Ecology

Gadwall breed close to water on the shores or islets of lakes and slow rivers, or in marshes with reeds *Phragmites* or *Glyceria* on open pools; they can be solitary or social. Gadwall are primarily found on inland waters, mainly gravel pits and reservoirs. A clutch of 8–12 eggs is laid between late April and the end of June. There is a single brood, and young fledge by mid-August (*Red Data Birds*). In winter, they are found predominantly on inland fresh water, and feed on the leaves and stems of aquatic plants (*Winter Atlas*).

Breeding season survey

See *Dabbling and diving ducks* in the generic survey methods section.

Winter survey

See the generic survey methods section on WeBS counts.

References

Fox, A D (1991) The gadwall in Britain. *British Wildlife* 3: 65–69.

Kirby, J S, Salmon, D G, Atkinson-Willes, G L and Cranswick, P A (1995) Index numbers for waterbird populations. III. Long-term trends in the abundance of wintering wildfowl in Great Britain, 1966/67–1991/92. *J. Applied Ecology* 32: 536–551.

Teal
Anas crecca

Status

Amber listed: WI
Non-SPEC
Annex II/1 of EC Wild Birds Directive

National monitoring

WeBS.

Population and distribution

In the UK, teal are shy breeding birds, found predominantly in the north-west and preferring the undisturbed moorland pools, bogs and mires of upland areas (*88–91 Atlas*). In at least the last 20–30 years, teal numbers have dropped and there has been a contraction of their breeding range. There are a number of proposed causes including increased disturbance and afforestation (*88–91 Atlas*). Teal are very common in winter, with a south-east bias in the UK, in direct contrast to their distribution during the breeding season. Winter numbers were increasing until recent years, mirroring the trend in the north-east European population. There are an estimated 1,600–2,800 pairs of teal breeding, and 141,000 individuals wintering in the UK (*Population Estimates*).

Ecology

Teal breed in dense cover, usually (but not always) near water, both far inland and near coasts. A clutch of 8–11 eggs is laid during mid-April to early June. There is a single brood, and the young fledge in July. In winter, teal frequent areas of shallow water on estuaries, coastal lagoons, coastal and inland marshes, flooded pasture and ponds (*Red Data Birds*).

Breeding season survey

See *Dabbling and diving ducks* in the generic survey methods section.

Winter survey

See the generic methods section on WeBS counts.

Pintail
Anas acuta

Status

Amber listed: BR, WI, WL, SPEC 3 (V)
Schedule 1/II of WCA 1981
Annex II/1 of EC Wild Birds Directive

National monitoring

Rare Breeding Birds Panel.
WeBS.

Population and distribution

Pintails were first recorded breeding in Inverness-shire in 1869. They are an occasional breeding bird with a scattered distribution in the UK (*88–91 Atlas*). A high proportion of wintering pintails are found at a few important sites including the Dee estuary, the Mersey estuary, the Wash and Morcambe Bay (*WeBS 1992–93*). The wintering population both in the UK and in north-west Europe increased between the 1960s and the 1980s. Several factors contributed to the population increase, such as the development of man-made habitats, for example gravel pits and reservoirs, a series of relatively mild winters, increasing water eutrophication, increasing numbers of refuges and the shortening of the hunting season in some other countries (Kirby et al 1995). There are currently estimated to be 8–42 pairs of pintails breeding and 28,100 individuals wintering in the UK (*Population Estimates*).

Ecology

Pintails breed close to water, which can be shallow lowland lakes and marshes or upland lochs and moorland pools. A clutch of 7–9 eggs is laid between mid-April and late June. There is a single brood and the young fledge by early August. In winter, pintails occur mainly on estuaries, where their most important food is the mollusc *Hydrobia*, but they are also found on inland floodplains.

Breeding season survey

See *Dabbling and diving ducks* in the generic survey methods section.

Winter survey

See the generic methods section on WeBS counts.

References

Kirby, J S, Salmon, D G, Atkinson-Willes, G L and Cranswick, PA (1995) Index numbers for waterbird populations. III. Long-term trends in the abundance of wintering wildfowl in Great Britain, 1966/67–1991/92. *J. Applied Ecology* 32: 536–551.

Garganey
Anas querquedula

Status

Amber listed: BR, SPEC 3 (V)
Schedule 1 of WCA 1981
Annex II/1 of EC Wild Birds Directive

National monitoring

Rare Breeding Birds Panel.
WeBS.

Population and distribution

The only wildfowl species which is entirely a summer visitor to the UK, garganey are sparsely distributed mainly in England (*88–91 Atlas*). Numbers fluctuate from year to year, being highest in warm springs following wet winters, when shallow floods remain on the water meadows and marshes. Numbers fell between the 1940s and the late 1980s, since when there has been a slight increase in the numbers of pairs proven to breed (*Red Data Birds*). There are an estimated 15–125 pairs of garganey breeding in the UK (*Population Estimates*).

Ecology

Garganey arrive between mid-March and late May, and have left mainly by September. Their preferred breeding habitat is lowland wet grassland, marsh, fen or grassland with intersecting ditches often edged with reed *Phragmites*. A clutch of 8–11 eggs is laid between the end of April and late June. There is a single brood and the young fledge by mid-August (*Red Data Birds*).

Breeding season survey

See *Dabbling and diving ducks* in the generic survey methods section.

Shoveler *Anas clypeata*

Status

Amber listed: WI
Non-SPEC
Annex II/1 of EC Wild Birds Directive

National monitoring

WeBS.

Population and distribution

In the UK, breeding shovelers are usually found on shallow eutrophic waters, with highest numbers in the south-east and Midlands of England (*88–91 Atlas*). Breeding numbers have recently declined after an increase during the 1960s and 1970s (*88–91 Atlas*). In winter, shovelers are widespread on the inland waters of England up to the Midlands, but are more scarce in Wales and Scotland (*Winter Atlas*). There are an estimated 1,000–1,500 pairs of shovelers breeding and 10,300 individuals wintering in the UK (*Population Estimates*). Shovelers use man-made wetlands in the winter and this may be a key reason for the increase in numbers to the current relatively high level (Kirby et al 1995).

Ecology

Shovelers breed in or near wetland areas, feeding on small crustaceans, molluscs, insects and their larvae, and plant material. A clutch of 9–11 eggs is laid between mid-April and early June. There is a single brood and the young fledge in July. In winter, shovelers frequent shallow freshwater areas on marshes, flooded pasture, reservoirs and lakes with plentiful marginal reeds or emergent vegetation (*Red Data Birds*).

Breeding season survey

See *Dabbling and diving ducks* in the generic survey methods section.

Winter survey

See the generic method section on WeBS counts.

Reference

Kirby, J S, Salmon, D G, Atkinson-Willes, G L and Cranswick, PA (1995) Index numbers for waterbird populations. III. Long-term trends in the abundance of wintering wildfowl in Great Britain, 1966/67–1991/92. *J. Applied Ecology* 32: 536–551.

Pochard
Aythya ferina

Status

Amber listed: WI, WL
SPEC 4 (S)
Annex II/1 and III/2 of EC Wild Birds Directive

National monitoring

Rare Breeding Birds Panel.
WeBS.

Population and distribution

The pochard became established as a UK breeding species early in the 19th century, and it is now largely found in the east of England and Scotland, at lowland waters with significant emergent and submergent vegetation (*88–91 Atlas*). The wintering numbers expanded quickly up to the 1970s, with numbers then decreasing in line with those in north-west Europe (*Winter Atlas*). The reasons for the reduction in pochard numbers remain unclear (Kirby et al 1995). There are an estimated 250–400 pairs breeding and 81,200 individuals wintering in the UK (*Population Estimates*).

Ecology

Pochards breed on large pools, lakes or slow-moving streams in Britain. A clutch of 8–10 eggs is laid between early May and the end of July but mainly in June. There is a single brood and the young fledge by mid-August. In winter, pochards occur on lowland freshwater reservoirs, lakes, ponds, gravel pits, etc, usually with a good growth of submerged aquatic plants and small molluscs (*Red Data Birds*).

Breeding season survey

See *Dabbling and diving ducks* in the generic survey methods section.

Winter survey

See the generic methods section on WeBS counts.

Reference

Kirby, J S, Salmon, D G, Atkinson-Willes, G L and Cranswick, PA (1995) Index numbers for waterbird populations. III. Long-term trends in the abundance of wintering wildfowl in Great Britain, 1966/67–1991/92. *J. Applied Ecology* 32: 536–551.

Scaup
Aythya marila

Status

Amber listed: BR, SPEC 3^{W} $(L)^{W}$
Schedule 1 of WCA 1981
Annex II/2 of EC Wild Birds Directive

National monitoring

Rare Breeding Birds Panel.
WeBS.

Population and distribution

The scaup is the rarest of Britain's breeding ducks, there being no evidence of regular annual breeding attempts anywhere in the country. Britain does, however, have sites of national and international importance for wintering populations (*WeBS 1994–95*). This is one of the few wildfowl that has declined in numbers in the UK in recent years, and the number of scaup wintering around the UK is estimated to be 13,400 birds (*Population Estimates*).

Ecology

In winter, scaup occur mainly in coastal or estuarine areas, and a strong attraction to sewage outfalls has been noted. Food consists of mussels and also items in, or prey benefiting from, sewage outfalls (*Red Data Birds*).

Breeding season survey

No method is given here.

Winter survey

See the generic methods section on WeBS counts.

Detailed survey

Where a more detailed survey of wintering scaup than WeBS is required, two main generic methods are available. These involve counts either onshore or offshore (from an aeroplane or ship). Both methods are outlined in the generic methods section under *Inshore marine waterfowl* and *Waterfowl and seabirds at sea.*

Scaup feed mainly at night and tend to flock together to roost on the sea during the day. If daytime roosting areas in large estuaries are known or located, a variation on the onshore method specific to scaup involves approaching on foot at low tide where they can be counted, sometimes with greater accuracy than at high tide. In the Solway, for example, low-tide counts have picked up higher numbers of scaup than high-tide counts. See *Low Tide Counts* in the generic survey methods section.

Eider
Somateria mollissima

Status

Amber listed: WL
Non-SPEC
Annex II/2 of EC Wild Birds Directive

National monitoring

Intensive monitoring at specific breeding colonies eg Forvie Sands, Coquet Island.
WeBS (although poorly covered by this).
WeBS special surveys (eg NEWS in 1997–98).

Population and distribution

The exclusively coastal breeding distribution of eider in the UK is confined almost entirely to Scotland (including Orkney and Shetland) and Northern Ireland (*88–91 Atlas*). In most areas, birds are highly dispersed, but dense colonies occur in a few locations also, including Walney Island in south-west Cumbria. Numbers have risen gradually in recent decades to an estimated 31,000 breeding females in Britain and 1,000 in Ireland, most of the latter in Northern Ireland (*88–91 Atlas, Population Estimates*). Although, after mallard, eider is the second most common breeding duck in the UK, numbers are small in comparison with the 2.4–3.6 million individuals in the west Palearctic as a whole (Scott and Rose 1996), corresponding to an estimated 800,000–960,000 breeding pairs in Europe and Russia (Keller and Hario 1997). British birds are largely sedentary and, as a consequence, the 78,000 birds that are found in the UK over winter (*Population Estimates*) have a broadly similar distribution to that in the breeding season, although many birds congregate at key sites, notably the Tay estuary. Distribution away from these key sites is relatively poorly known since, although routinely included in WeBS, counting sea-ducks is problematic and much of the Scottish open coastline is not monitored.

Ecology

The morphology and sedentary nature of British eiders has led to the suggestion that birds in Shetland and Orkney constitute a separate population of the *faroeensis* subspecies, and those in the rest of Britain and Ireland constitute a separate population of the *mollissima* subspecies (Rose and Scott 1997). The distribution of breeding eiders is somewhat unusual in that they are highly dispersed along much of their range and yet in a few locations they occur in dense colonies. Nests are often located in the shelter of vegetation or rocks, and although islands and near-shore areas are favoured, nests can be up to 3 km inland. A clutch of 4–6 eggs is laid between mid-April and late May, with incubation lasting approximately 25–29 days. Young are cared for by the female and fledge after 65–75 days. First-year females do not breed (Baillie and Milne 1982).

Breeding season survey – population

At many sites, it is not possible to estimate the number of breeding and non-breeding female eiders without very detailed nesting studies of marked birds. From count data alone it is impossible to distinguish females which do not lay at all from those which laid but failed early. In the absence of detailed studies, the method presented here yields the most realistic estimate of the breeding population, and is based on the peak number of 'adult' females.

Information required

- peak number of second-year and older females
- peak numbers of all females, adult males and first-year males
- 1:25,000 OS map of the site boundary.

Number and timing of visits

Weekly visits from mid-April to the end of June at sites with large numbers of birds. Fortnightly visits should be sufficient at sites with smaller numbers.

Time of day

High tide.

Weather constraints

Counting is difficult during strong winds.

Sites/areas to visit

When monitoring a site with an existing boundary, cover all of the area within the boundary. Otherwise survey all suitable habitat.

Equipment

- 1:25,000 OS map of the survey area.

Safety reminders

Take care on any areas of rocky coastline as the going underfoot can be extremely slippery. Be aware of tide times; ensure that you do not become stranded by a rising tide.

Disturbance

Avoid disturbing nesting birds wherever possible.

Methods

Where sites are monitored by existing schemes (eg WeBS), use recognised site boundaries. Where this is not the case, mark the boundaries of the count area on a 1:25,000 OS map. Alternatively, counts can be carried out tetrad by tetrad (OS 2 × 2 km).

Survey the whole of the count area, ideally from elevated vantage points. Count birds under suitable tidal conditions (high tide is easier as birds are less likely to be actively feeding). Record numbers of adult males, first-year males and all females separately (adult and first-year females are normally very difficult to separate in the field).

At the end of the season, report the peak (maximum) count of each age-/sex-group separately, since variation in arrival dates, commencement of breeding and moult dispersal may mean that the peak numbers of each group occur at different times.

Assuming a 1:1 sex ratio of first-year males to females, the peak number of first-year females can be taken to be the same as the peak number of first-year males. Subtract the peak number of first-year females from the peak number of all females. The resulting figure is the peak number of second-year and older females. Report this figure as the most representative and comparable figure for the breeding population.

Breeding season survey – breeding success and productivity

Assessing the number of broods from counts of young birds is impossible since broods often amalgamate to form large crèches. More detailed nest counts at large colonies can be conducted in late June and early July, when the majority of young have hatched. However, precise methods will vary according to conditions at each site (size of colony, topography, nature of vegetation, etc). Experienced observers can obtain information on brood size, hatching success, predation, etc, from recently used nests. Counts of juvenile birds in late July or August prior to dispersal to wintering grounds (ie when birds are large enough to have passed the period of greatest juvenile mortality yet are still distinguishable from adult birds) can be undertaken to estimate recruitment. At some sites, different age/sex categories depart on different dates and this can be used to good advantage during such counts.

Winter survey

WeBS.

See *Waterfowl and seabirds at sea* and *Inshore marine waterfowl* in the generic survey methods section.

Contributed by Mark Pollitt and Ian Patterson

References

Baillie, S R and Milne, H (1982) The influence of female age on breeding in the eider *Somateria mollissima*. *Bird Study* 29: 55–66.

Keller, V and Hario, M (1997) Eider. Pp 110–111 in Hagemeijer, W J M and Blair, M J (eds) (1997) *The EBCC Atlas of European Breeding Birds: their distribution and abundance*. Poyser, London.

Rose, P M and Scott, D A (1997) *Waterfowl Population Estimates*. Wetlands International Publ. 44. Wageningen, Netherlands.

Scott, D A and Rose, P M (1996) *Atlas of Anatidae Populations in Africa and Western Eurasia*. Wetlands International Publ. 41. Wageningen.

Long-tailed duck *Clangula hyemalis*

Status

Non-SPEC
Schedule 1 of WCA 1981
Annex II/2 of EC Wild Birds Directive

National monitoring

WeBS.

Population and distribution

Long-tailed ducks last bred in Britain in 1926 (Spencer et al 1993). Winter flocks occur in open coastal waters on the east coast of Scotland and the Northern Isles. Counts from the shore only detect a small proportion of the total population, but more detailed counts have been carried out by the RSPB and BP in the Moray Firth. The UK wintering population is estimated to be 23,500 individuals (*Population Estimates*).

Ecology

The main winter flocks occur on open coastal waters. Shallow, sandy areas within the Moray Firth are particularly favoured, although small flocks may also occur off rocky coasts, within smaller estuaries or on brackish, coastal lochs (*Red Data Birds*).

Winter survey

WeBS.

See *Waterfowl and seabirds at sea* and *Inshore marine waterfowl* in the generic survey methods section.

Detailed survey

Long-tailed ducks are very difficult to count because they tend to remain well offshore. This is one of the few species where, in the UK at least, counting from the land will substantially underestimate the numbers present unless good flight-line movements at dusk can be detected. For an accurate count of winter long-tailed duck populations, aerial counts are probably essential. Aerial methods are outlined in the generic survey methods section under *Waterfowl and seabirds at sea*.

Long-tailed ducks come closer to shore to roost at sunset, so the next-best method for winter counts is an onshore method which depends on detecting good flight-line movements. The best time to do this is in the hours before and after sunset. Peak counts of flighting birds generally occur at sunset with arrivals trailing off after that. However, birds do still come in as darkness falls. All other recommendations for this type of survey are the same as for WeBS (see the generic survey methods section). When quoting figures obtained using this method, it should be clearly stated that the count was conducted on birds flying in to roost.

Reference

Spencer, R and RBBP (1993) Rare breeding birds in the United Kingdom in 1990. *British Birds* 86: 62–90.

Common scoter *Melanitta nigra*

Status

Red listed: BD, BR, WL
Non-SPEC
Schedule 1 of WCA 1981 and Wildlife Order NI 1985
Annex II/2 of EC Wild Birds Directive

National monitoring

National survey: 1995 (WWT/IWC/RSPB/JNCC/SNH), 2005.
Rare Breeding Birds Panel.
WeBS.

Population and distribution

This species started breeding in northern Scotland in the mid-19th century. The main populations are in the north and west of Britain and Ireland. The breeding population (particularly in Northern Ireland) has suffered both population decline and range contraction; breeding numbers in the UK are estimated to be 95 pairs (Underhill et al 1998). Wintering flocks are difficult to count, with wintering grounds in the North Sea, mainly offshore and in sheltered bays. The UK wintering population was estimated to be 37,550 individuals (*Population Estimates*).

Ecology

Common scoters breed in dense vegetation on the scrub-covered islands or uninhabited shorelines of lochs. In Ireland, these areas are usually large loughs (sometimes >100 km^2) with complicated shorelines and island systems. In Scotland, the breeding sites can be larger lochs, remote hill lochans, small moorland lochs and pool systems. Slow-flowing rivers are occasionally used in the main breeding areas (eg the Flow Country). A clutch of 6–8 eggs is laid between late May and mid-July. There is a single brood and most of the young fledge by the end of August. Winter flocks occur on open coastal waters where shallow bays with high bivalve populations are particularly favoured (*Red Data Birds*).

Breeding season survey – population

Most of the methods given here are based on work in the Scottish flow country (Hancock 1991) and on methods used for the British Isles breeding survey (Underhill et al 1998).

Information required

- maximum count of females
- details of presence and absence at all sites checked.

Number and timing of visits

Three visits: first visit 24 April to 7 May; second visit 8–21 May; third visit 22 May to 4 June.

Time of day

Any time.

Weather constraints

Avoid poor weather.

Sites/areas to visit

Sites where scoters are known to have bred and potential sites within the range of common scoter.

Equipment

- 1:25,000 map of the lake or water system to be surveyed
- telescope
- Schedule 1 licence.

Safety reminders:

When working alone in remote areas, ensure someone knows where you are and when you are due back. When working near water, work in a team of two wherever possible. A compass and maps should always be carried. Spare warm clothing, a plastic survival bag, first-aid kit and food supplies should be carried.

Disturbance

Common scoters are not readily put off their nests, but observers should not go close to, or on to, known nesting islands. Observers should avoid disturbing adult scoters with chicks, especially if the chicks are small. Any disturbance could distract adults and leave chicks more vulnerable to predation.

Methods

Scoters begin returning to their breeding lakes in mid- or late April, the main arrival usually being within a week of the first birds. Mark the area to be surveyed on a map. Scan the area systematically from one or more suitable vantage points. One vantage point will be sufficient for small moorland lochans, but for slightly larger lakes more than one observation point will be necessary. If parts of the shoreline are not visible, walk the shoreline to ensure complete coverage. Walking surveys are not useful along very complex loch shorelines where islands are also present; in this case, use a boat.

Divide the lough into sections and systematically traverse each section, scanning for birds. Ensure that there are two observers in the boat and carefully scan all of the shoreline and islands and all areas of water – scoters often feed very close inshore or loaf on banks.

Report negative results as well as positive ones. An example recording form (as used in the 1995 common scoter survey) is given in Figure 1; the following are definitions of the terms that should be recorded on the form.

Term	*Definition*
Site Name	Name of site.
Grid Ref	Two-letter and six-figure OS grid reference.
Pairs	The number of common scoters seen in pairs.
?Pairs	The number of common scoters where birds were probably paired but pairing was not obvious.
♂♂	The number of single adult male common scoters (over one year old).

Imm. ♂♂	The number of single first-year male common scoters.
♀♀	The number of single female common scoters (this will include first-year and older females).
Unsexed or unaged	It is possible that very poor views of flying common scoters will not enable the observer to determine the sex or age of the bird. Enter all such observations in this column, but give as much additional detail as possible regarding possible sex/age.

In some situations, for example on the larger loughs in Ireland, it is more difficult to classify birds as either pairs, probable pairs or single birds because they are more likely to form groups. If this problem is encountered and observations do not reveal any information on the likely number of pairs, note the number, sex and age of each bird and make clear that these birds formed one group.

Report the number of females, pairs and probable pairs (?pairs). The population estimate is expressed as the number of females present, regardless of pairing status.

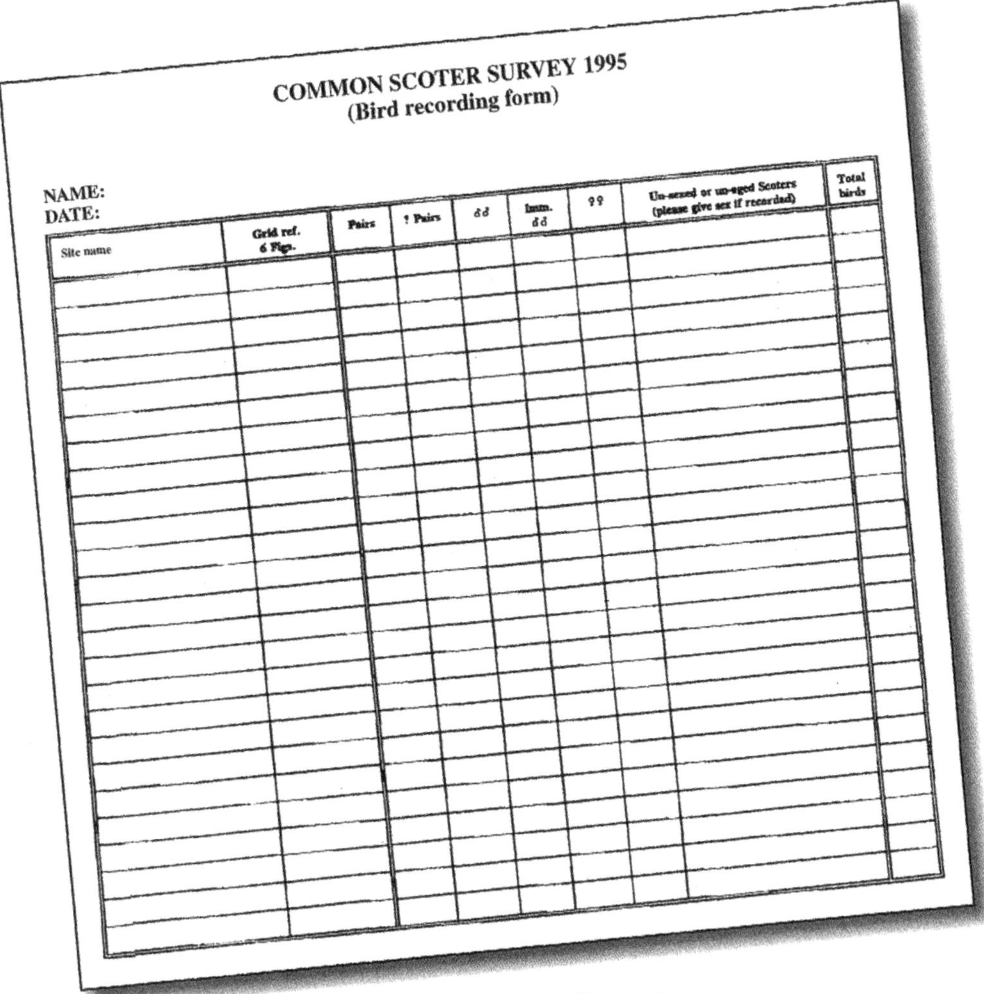

COMMON SCOTER SURVEY 1995
(Bird recording form)

NAME:
DATE:

Site name	Grid ref. 6 Figs.	Pairs	? Pairs	♂♂	Imm. ♂♂	♀♀	Un-sexed or un-aged Scoters (please give sex if recorded)	Total birds

Figure 1
The recording form used for the 1995 common scoter survey. The same type of information is relevant to smaller-scale common scoter surveys.

Breeding season survey – productivity

Information required

- number of fledged young per pair
- size and number of broods with ducklings classified into four approximate size-classes.

Number and timing of visits

Two visits, mid-July to mid-August (including one in the second week of August).

Weather constraints, Sites/areas to visit, Equipment, Safety reminders, Disturbance

As for the population survey (above).

Methods

Visit all sites where breeding pairs have been seen in order to look for any young. Some broods are attended by the female through to fledging but others are abandoned soon after hatching. Ducklings can be very difficult to spot. On smaller lochs, it is best to use two observers (standing 50–100 m apart) to locate chicks, especially on complex pool systems. As for the population survey, the larger and more complex sites should be surveyed by boat. Broods can be very difficult to find, but when approached they do tend to move out from the shoreline rather than seeking cover, and only dive when approached very closely.

The number of young, their size and the number of broods found on each site should be reported. Broods sometimes join together. While broods of different ages can be distinguished, this is impossible for combined broods of the same age.

It is important to report negative results, ie sites with breeding adults but where no young were seen on a visit. Record the size of the young seen on each visit as a fraction of adult size (eg about $\frac{1}{4}$, $\frac{1}{4}$–$\frac{1}{2}$, $\frac{1}{2}$–$\frac{3}{4}$, $>\frac{3}{4}$ or the same size), although this can be difficult if no adults are present. Estimating the size of the young (even roughly) can help to distinguish between different broods in subsequent visits and can be used to estimate duckling survival. The later the checks (ie into the first week in August) the more difficult it can become to distinguish adult females from large juveniles; good views of the birds are required to do this. Some extra checks on scoter broods after 6 August may help to assess survival of chicks and locate any later broods.

Report the number of broods, the size of the brood and the approximate size-class of each young scoter. Report fledging success as the estimated total number of fledged young divided by the peak count of paired birds.

Winter survey

See the generic survey methods section on WeBS counts. Winter counts are highly variable, with huge numbers in some years and many fewer in others (probably dependent on bivalve populations). Only coordinated aerial surveys would give a comprehensive picture of the UK population. Two main generic methods are outlined in the generic survey methods section: see *Waterfowl and seabirds at sea* and *Inshore marine waterfowl*.

References

Hancock, M (1991) *Common Scoter in the Flow Country, Results of the 1991 Survey: comparison with previous years and suggested monitoring method.* RSPB unpubl.

Underhill, M C, Gittings, T, Callaghan, D A, Hughes, B, Kirby, J S and Delany, S (1998) Status and distribution of breeding common scoters *Melanitta nigra nigra* in Britain and Ireland in 1995. *Bird Study* 45: 146–156.

Velvet scoter *Melanitta fusca*

Status

Amber listed: WL, SPEC $3^{W}(L)^{W}$
Schedule 1 of WCA 1981
Annex II/2 of EC Wild Birds Directive

National monitoring

WeBS.

Population and distribution

Winter, passage and moulting flocks all occur in open coastal waters, mainly along the east coast of Scotland. The number recorded in the UK is regularly 3,000 (*Population Estimates*) and occasionally up to 10,000 (*Red Data Birds*). Although there are differences in the counting methods, there does seem to have been a reduction in the number of wintering velvet scoter since the early 1980s.

Ecology

Shallow, sandy areas are particularly favoured, though smaller numbers may occur within smaller estuaries or on freshwater sites inland. The food of the velvet scoter is poorly known in Britain, but the wintering diet elsewhere is predominantly molluscs, crustaceans and small fish (*Red Data Birds*).

Winter survey

WeBS.

Winter counts are highly variable, with huge numbers in some years and many fewer in others (probably dependent on bivalve populations). Only coordinated aerial surveys would give a comprehensive picture of the UK population. Two main generic methods are outlined in the generic survey methods section: see *Waterfowl and seabirds at sea* and *Inshore marine waterfowl*.

Goldeneye
Bucephala clangula

Status

Amber listed: BR, WL
Non-SPEC
Schedule 1/II of WCA 1981
Annex II/2 of EC Wild Birds Directive

National monitoring

Rare Breeding Birds Panel.
WeBS.

Population and distribution

Goldeneyes were first proved to breed in Scotland in 1970. Since then, these tree-hole-nesting ducks have increased in Britain, helped by the provision of nest-boxes (*88–91 Atlas*). The numbers wintering in the UK are gradually increasing; favoured wintering grounds include Lough Neagh, the Forth estuary, Inner Moray Firth and Abberton reservoir (*WeBS 1992–93*). In the UK there are an estimated 83–109 pairs of goldeneye breeding and 32,000 birds wintering (*Population Estimates*).

Ecology

Goldeneyes breed in coniferous forest near water. The nest can be up to 1 km from water in adjacent forests, in natural tree-holes or woodpecker holes and under rocks or logs. The birds readily use nest-boxes, but rarely burrows. A clutch of 8–11 eggs is laid from early April. The young are taken to the water soon after hatching. In winter, the birds occur in fresh and salt water, sometimes in large concentrations, especially at waste outfalls (*Red Data Birds*).

Breeding season survey

See *Dabbling and diving ducks* in the generic survey methods section.

Winter survey

See the general guidelines for WeBS counts in the generic survey methods section.

Red-breasted merganser *Mergus serrator*

Status

Non-SPEC
Annex II/2 of EC Wild Birds Directive

National monitoring

National breeding survey (sawbills on rivers): 1987 (BTO, Gregory et al 1997).
WeBS.

Population and distribution

The red-breasted merganser has a wide breeding range across northern Eurasia and North America, breeding in forested tundra and temperate forest zones. In the UK, highest densities occur along Scotland's west coast, and the species has spread to north-west England, north-west Wales and Northern Ireland in recent times. There has been a slight range contraction since the late 1960s (*88–91 Atlas*), and there are currently estimated to be 2,200 pairs in Great Britain and a further 100 in Northern Ireland (*Population Estimates*). Birds move to the coast to moult before dispersing (eg Starling et al 1992, Marquiss and Duncan 1993), and around 10,000 winter in the UK (*Population Estimates*), almost all found on the coast. British birds are augmented by continental and Icelandic birds in winter (Scott and Rose 1996).

Ecology

The red-breasted merganser is predominantly a marine species, living in shallow coastal waters, and most of the British population breeds in sheltered sea lochs and estuaries, with a small proportion on the lower reaches of rivers (Marquiss and Duncan 1993). Breeding is late, the ducklings hatching in late summer to coincide with peak production of the fry of summer-breeding fish (Marquiss and Duncan 1993). Peak hatching occurs in July, and most deaths occur within a week of hatching. Ducklings fledge in September. Most drakes move to moulting sites on the sea in late July, begin moult by mid-August and have completed it by early September (Marquiss and Duncan 1993).

Breeding season survey – population

Survey methods are based largely on those for goosander, modified to take account of the largely coastal distribution of red-breasted merganser and its later breeding season.

Information required

- peak number of paired and unpaired birds
- peak number of adult males present at the time first eggs hatch.

Number and timing of visits

Two visits, one in the last week of April to record the peak number of paired and unpaired birds and another at the end of May or early June to record adult males. Ideally, repeat each of these counts within a week (Marquiss and Duncan 1993).

Time of day

From one hour after sunrise. When surveying rivers, do not count too early or too late in the day before the end of April, otherwise you may miss some birds because they have not yet flown upstream from communal roosts at river mouths or because they have commuted downstream in the last 2–3 hours of daylight (Marquiss and Duncan 1993).

Sites/areas to visit

Red-breasted mergansers prefer shallow inshore waters, particularly estuaries and sea lochs, but avoid rocky coasts. In Orkney, brackish lochs are also used. They are found along rivers, though densities are lower at greater distances from the river mouth, and on narrower, steeper and higher-altitude rivers (Marquiss and Duncan 1993). Birds nest in rank vegetation, especially on islands, but also on shingle spits, river banks and under cliff ledges, with records from gullies also (M Marquiss pers comm).

Equipment

- photocopied 1:25,000 OS field maps of the survey area.

Safety reminders

When working alone in remote areas, ensure someone knows where you are and when you are due back. When working near water you should do so in a team of two wherever possible. Always carry a compass and maps, spare warm clothing, a plastic survival bag, first-aid kit and food supplies.

Disturbance

Red-breasted mergansers flush easily, especially on rivers, and this can reduce the accuracy of the count. Avoid disturbing birds wherever possible (see *Methods*, below).

Methods

Divide the river/coast into roughly equal sections, each of which can be covered by a single observer, bearing in mind that one observer can cover about 10–13 km in half a day. Further divide these sections into 1-km subsections using easily discernible features, such as bridges and islands. This will allow a more precise assessment of bird distributions. Mark the section and subsection boundaries on a map to aid comparison with future surveys. Ideally, if there are sufficient observers, survey all sections of contiguous habitat on the same day, with all observers beginning the survey of their section at the same time (one hour after sunrise). All observers should walk in the same direction (upstream on rivers), thus maintaining the same distance between one another and reducing the chance that any flushed birds will be recorded by another observer. When a river narrows to less than 10 m there is no need to continue further upstream.

Since red-breasted mergansers on rivers are likely to flush readily, do not walk too close to the river if possible. At suitable vantage points, stop to scan as far ahead as possible (ideally using a telescope) prior to walking along the banks. If no birds are seen, walk alongside the river for a short distance and then scan ahead again. Once a bird is seen, attempt to move away from the river and pass around the bird to avoid disturbing it, scanning downstream again on regaining the river to check that the bird has not flown off and that other birds have not been missed. Adopt a similar approach on coastal sites where there is a risk of disturbance.

Where possible, sex and age birds and record them as paired or unpaired (see *Sexing and ageing*, below). Record the precise time, exact location and movements of all birds on prepared maps of the study area. This helps eliminate double registrations, especially where birds are recorded close to section boundaries by two separate observers, or where the same bird repeatedly flies in front of the observer. Ideally, repeat both surveys (those of paired birds and adult males) within a week to verify the earlier counts.

Eliminate obvious double registrations when estimating the number of birds in each 1-km section. This may sometimes prove difficult, so alternatively report the minimum and maximum with the maximum based on the assumption that all registrations were independent, and the minimum that all birds flushed ahead were seen again later.

After the first visit, sum the counts of paired (and unpaired) birds across all 1-km subsections and sections, and report these totals. Where the count was repeated a week later, report the maximum of the two counts. Provide further details, as outlined in *Sexing and ageing* (below), where possible. After the second visit, sum the number of adult males and report this total. Again, if a repeat count was made, report the maximum of the two counts.

Breeding season survey - breeding success

Information required

- number of young from brood counts in early August.

Number and timing of visits

One, preferably two, in early August; the second visit is to verify the results of the first. If possible, make an additional visit later in August to locate late-hatching broods.

Time of day

As for the population survey (above).

Sites/areas to visit

As for the population survey (above). Ideally, survey all suitable breeding habitat. Alternatively, only survey areas which held adult males on territory in spring.

Equipment, Safety reminders, Disturbance

As for the population survey (above).

Methods

Use the same boundaries as for the population surveys earlier in the year and follow similar methods.

Age and sex all birds recorded (see *Sexing and ageing*, below), and count the number of ducklings in each brood and estimate their size-class. Carefully map broods, noting especially the number of ducklings and their age-/size-class, so that, if more than one visit is made, movements between counts can be traced. Relating these observations to the position of males in late spring may also help.

Sexing and ageing

Wherever possible red-breasted mergansers should be sexed, and aged as adult or immature (in their first year). Record adult birds as paired or unpaired. Thus, birds can be classified into one of six groups: paired birds, unpaired adult males, immature males, unpaired redheads (see below), or females with young or broods. In some cases, where only poor views are obtained and birds are not sexed or aged, it may be necessary to record birds as unaged males or simply as unsexed birds. All groups (paired birds, unpaired males, immature males, etc) should be recorded if present, even if the survey was conducted primarily to assess the abundance of a single cohort.

Adult drakes are distinguished from ducks by their white forewings, and adults from first-years by their vermilion-coloured legs. It is rarely possible to sex juveniles in the field until spring, when most young drakes have pale tertials and often other adult feathers (Marquiss and Duncan 1993).

In general, broods will be discrete parties of ducklings at a similar stage of development, attended by a single female. Occasionally, however, broods may join crèches, more commonly so than in goosander, and particularly on larger waterbodies such as sea lochs. Record the number of young in each brood and classify the young into one of four size-classes, relative to the size of the adult bird:

Class 1 up to ¼ adult size
Class 2 ½ adult size
Class 3 ¾ adult size
Class 4 fully grown.

As red-breasted mergansers take 50–60 days to fledge, each size-class corresponds roughly to a two-week age period (though duckling growth is more rapid during the first few weeks).

Winter surveys

WeBS.

See *Inshore marine waterfowl* in the generic survey methods section.

River surveys

Mergansers are generally found near river mouths during winter. Up to the end of April, they may roost communally at the mouth, commuting daily upstream to feed from an hour after daybreak, and returning in the last 2–3 hours of daylight (Marquiss and Duncan 1993). During winter surveys, aim to count all rivers in the watershed simultaneously. Adopt methods similar to those used for the population survey in the breeding season (above).

Contributed by Peter Cranswick

References

Marquiss, M and Duncan, K (1993) Variation in the abundance of red-breasted mergansers *Mergus serrator* on a Scottish river in relation to season, year, river hydrography, salmon density and spring culling. *Ibis* 135: 33–41.

Scott, D A and Rose, P M (1996) *Atlas of Anatidae Populations in Africa and Western Eurasia*. Wetlands International Publ. 41. Wageningen.

Starling, A E, Kirby, J S, Pickering, S P C, Gilburn, A S and Howell, R (1992) *British Status and Distribution of Wintering Cormorants, Goosanders and Red-breasted Mergansers in Great Britain with Specific Reference to the Welsh Region*. WWT report for the Wetlands Advisory Service under contract to NRA Welsh Region. NRA, Cardiff.

Goosander *Mergus merganser*

Status

Non-SPEC
Annex II/2 of EC Wild Birds Directive

National monitoring

National survey of breeding sawbills 1987 (Gregory et al 1997).
Surveys of breeding birds on selected river catchments in Wales in 1981, 1985, 1990 and 1993 (RSPB 1981, Tyler et al 1988, Griffin 1990, Underhill 1993).
WeBS.
Winter surveys of the River Wye, Wales, in 1993 and 1995–96 (Underhill 1993, Jennings et al 1997).

Population and distribution

Goosanders occur throughout the northern hemisphere, breeding mainly between 50°N and the Arctic Circle (*BWP*). Most drakes are absent from May to October. Evidence from ringing recoveries suggests that most, if not all, drake goosanders from western Europe leave their breeding grounds in late May or June to moult on the Tana estuary in Finnmark, Norway (Little and Furness 1985). This is corroborated by the observation of a gathering of up to 30,000 drakes on the Tana estuary in 1983, which agrees well with the estimate of 70,000 individual goosanders of both sexes in Europe during the winter (Ogilvie 1975).

Goosanders were first recorded breeding in Britain in Perthshire in 1871, since when they have spread steadily southward. The first confirmed breeding record for Wales was in 1972, with records from Derbyshire, Devon and Shropshire more recently illustrating the continuing expansion. The British population was estimated at 1,000–2,000 pairs at the time of the *68–72 Atlas*, increasing to 2,600 at the time of the 1987 survey (*Population Estimates*). Distributed widely throughout Scotland, northern England and Wales, breeding populations are most concentrated in the southern uplands and the northern English counties (*88–91 Atlas*).

The number of goosanders wintering in Britain increased between 1960–61 and 1981–82, and was particularly rapid during 1972–82 (Owen et al 1986). Around 8,900 birds are estimated to winter in Britain (*Population Estimates*). Scott and Rose (1996) and Rose and Scott (1997) suggest that birds in Britain constitute a separate population, although winter numbers are swollen by birds from the continent, especially during cold weather (Cranswick et al 1997). Goosanders occur only very occasionally in Ireland.

Ecology

In Britain, wintering birds are found mainly on the lower reaches of rivers, on reservoirs and on gravel pits, with large numbers also on some estuaries (Starling et al 1992). Many birds feed on rivers by day, but gather on still waters (eg reservoirs), often in large numbers, to roost (eg Marquiss and Duncan 1994). Goosanders move to the upper reaches of rivers in spring where they breed, typically on upland rivers, preferring wider stretches of shallow gradient, ie with fewer areas of riffle or white water (Gregory et al 1997). Breeding birds have, however, been found nesting on lowland reaches and upland lakes. During the breeding season, goosander foraging activity peaks in the first two hours of the day (Marquiss and Duncan 1994). Early in the breeding season, however, some birds may still be using communal winter roosts and may arrive at the feeding areas later.

Breeding season survey – population

Survey methods are based on Underhill (1993) and subsequently developed by staff at WWT (WWT unpubl).

Information required
- peak number of paired birds
- peak number of adult males at time that first eggs hatch.

Number and timing of visits

At least two visits, ideally three, at weekly intervals from the middle week in March. Where possible, two further visits, one week apart in mid- or late April.

Time of day

From one hour after first light until midday.

Weather constraints

Avoid days with poor weather.

Sites/areas to visit

Rivers (usually upland stretches with wider stretches of shallow gradient) and upland lakes.

Equipment

- photocopied 1:25,000 OS maps of the survey area on which sections and subsections are marked (see below).

Safety reminders

When working alone in remote areas, ensure someone knows where you are and when you are due back. When working near water, work in a team of two wherever possible. A compass and maps should always be carried. Spare warm clothing, a plastic survival bag, first-aid kit and food supplies should be carried.

Disturbance

As river habitats are feeding areas adjacent to breeding sites (hollows in trees and rocks) it is unlikely that this method will cause significant disturbance to adult birds. Disturbance may, however, cause adults on the river to fly. This may lead to the same individuals being double-counted and leave dependent young vulnerable to predation. Once a bird is seen, move away from the river and pass around the bird to avoid disturbing it.

Methods

Wherever possible, survey the whole site in half a day on each visit, if necessary using several observers to survey adjacent sections simultaneously. This avoids problems caused by birds moving between different parts of the river over a period of days. This is particularly important when assessing potential numbers of breeding birds, since birds may be relatively mobile at this time. It is less important when counting adult males since they are likely to remain loyal to their territories. Divide the site into roughly equal sections for individual observers to walk (a maximum of about 10–13 km per observer per visit). Subdivide sections further into 1-km lengths using easily discernible features such as bridges and islands. This allows a more accurate assessment of bird distributions on the site. Mark the full details of section and subsection boundaries carefully on a map to aid comparison with future surveys. Use sufficient fieldworkers to cover the entire site in half a day, all beginning the survey of their section at the same time (one hour after sunrise).

All observers should walk upstream, thus maintaining the same distance between one another and reducing the chance of any flushed birds being recorded by another observer.

Goosanders are easily flushed, so do not walk too close to the river if possible since this may drive birds upstream ahead of you. At suitable vantage points, scan as far ahead as possible (ideally using a telescope) prior to walking along the banks. If no birds are seen, walk alongside the river for a short distance and then scan ahead again. Once a bird is seen, try to move away from the river and pass around the bird to avoid disturbance, scanning downstream again on regaining the river to check that the bird has not flown off or that other birds have not been missed.

Where possible, birds should be sexed, aged and recorded as paired or unpaired (see *Sexing and ageing*, below). Record the precise time, exact location and movements of all birds on prepared maps of the study area. This helps eliminate double registrations of the same bird, especially where birds are recorded close to section boundaries by two separate observers, or where there are multiple sightings of the same birds flying upstream in front of you, and allows comparison with surveys of broods.

Make two, preferably three, visits, each a week apart, starting in mid-March, to record the number of paired birds (the potential breeding population). Make two visits in mid- to late April (peak hatching of first clutches), about 30 days after the second March visit, to record the number of adult males (number attempting to breed).

Calculate the total number of paired and unpaired birds per 1-km length of river, eliminating all obvious double registrations of the same birds. In practice, this may be difficult and it is probably worth also reporting the maximum number (ie assuming that all registrations are independent) and minimum number (assuming that any birds that flew upstream were seen later), as this will give a range within which the true estimate will lie.

Note that a peak count of pairs/unpaired birds may still underestimate the true number of pairs, as some birds may already have started incubating in mid-March.

Breeding season survey - breeding success

Information required
- number of young.

Number and timing of visits

Two visits, separated by a week in early to mid-July (ideally, second week). If two additional visits can be made, one should be earlier (in late June) and one later (third week of July).

Time of day

As for the population survey (above).

Sites/areas to visit

As for the population survey. Ideally, survey all areas of suitable breeding habitat. Alternatively, survey areas which held adult males on territory in spring.

Equipment, Safety reminders, Disturbance

As for the population survey.

Methods

Use the same boundaries and methods of avoiding double-counting and disturbance as for the population survey.

Record, age and sex all birds. Count the number of ducklings in each brood and estimate their size-class (see *Sexing and ageing*, below).

Most young goosanders will be on the river from late June until the end of July and if only two visits are possible, make them in the middle of this period. However, since some broods will hatch earlier than this and others later, some may be missed, so make two additional visits if possible, one early (late June) and one late (third week of July). Note, however, that broods and females gradually move downstream as the season progresses. Careful mapping of broods, especially the number of ducklings and their age-/size-class, should allow such movements to be traced. Relating these observations to the position of males in late spring may also help to determine the total number of broods.

Sexing and ageing

Where possible, goosanders should be sexed, and aged as adult or immature (in their first year). Adult birds should be recorded as paired or unpaired. Thus, birds can be classified as one of six groups: paired birds, unpaired adult males, immature males, unpaired redheads (see below), or females with young or broods. In some cases, where only poor views are obtained and birds are not sexed or aged, it may be necessary to record birds as unaged males or simply as unsexed birds. All six groups should be recorded if present, even if the survey is being conducted primarily to assess the abundance of a single group.

Immature males (in their first winter) are usually easy to distinguish from females by their large rounded head profile, strikingly white breast and neck, black feathers on the mantle, more white in the wing and, as winter progresses, an increasing amount of white on their flanks. Juvenile females can be distinguished from adults by their brown or dull orange legs and feet; adults have bright orange or scarlet legs and feet. However, since these features are not always visible, some birds may have to be recorded simply as 'redheads'. During winter and spring, this category may contain adult and first-year females, and during summer may also include recently fledged male and female birds.

Adult males leave the breeding areas around May, most migrating to Norway, although some moult in the Moray Firth. Thus, moulting males should not be encountered during any survey. Young males are identifiable in their first spring, but are unlikely to be present in breeding areas at this time.

Paired birds are often obvious since the male closely attends the female. However, communal courtship in early spring may make identification of pairs difficult.

Broods are defined as discrete parties of ducklings at a similar stage of development, attended by a single female. However, crèches of ducklings of different ages may occasionally be found. The number of young in each brood should be recorded and young classified into one of four size-classes, relative to the adult bird: (1) up to ¼ adult size; (2) ½-grown; (3) ¾-grown; (4) fully grown. As goosanders take 60–70 days to fledge, each size-class corresponds to roughly a two-week period (although duckling growth is more rapid during the first few weeks).

Winter surveys

WeBS.

See *Inshore marine waterfowl* in the generic survey methods section.

River surveys

Although goosanders are recorded by WeBS, few rivers are covered by the scheme and most counts are made during the day, often in late morning. Thus, a large proportion of the goosander population, particularly that which feeds on rivers by day and only visits still waters to roost, will be missed by WeBS.

Peak winter numbers of goosanders generally occur in mid- to late winter (Cranswick et al 1997), following the return of males from their moulting areas and the influx of winter immigrants from north-west Europe.

Studies of the diurnal activity patterns of goosanders in Scotland (Marquiss and Duncan 1994) and anecdotal information from surveys carried out in Wales (Tyler et al 1988) suggest that, during the winter, most goosanders feed on the lower reaches of rivers during the day, returning to roost on large waterbodies at night. This diurnal activity pattern means winter surveys should aim to count all rivers in the watershed simultaneously. The upper reaches may hold fewer birds, but the results of Underhill (1993) indicate that some birds may be found well upstream on the main river.

Due to the difficulties involved in covering large stretches of river, coordinated roost counts may give the most accurate figures and offer a less labour-intensive method. However, it is not known if some birds remain on rivers during the night, and some may arrive at the roost very close to dusk or even after dark, making roost counting difficult.

For a winter survey, make a minimum of two visits in February using similar methods to those used in the breeding season.

Contributed by Mark Underhill and Peter Cranswick

References

Cranswick, P A, Waters, R J, Musgrove, A J and Pollitt, M S (1997) *The Wetland Bird Survey 1995–96: wildfowl and wader counts.* BTO/WWT/RSPB/JNCC, Slimbridge.

Gregory, R D, Carter, S P and Baillie, S R (1997) Abundance, distribution and habitat use of breeding goosanders *Mergus merganser* and red-breasted mergansers *Mergus serrator* on British rivers. *British Birds* 44: 1–12.

Griffin, B (1990) *Breeding Sawbills on Welsh Rivers 1990.* Report prepared by RSPB Wales.

Jennings, P, Langworthy, J, Purveur, R A J, Sellers, R M, Smith, I D and Wells, C (1997) *Wye Wintering Bird Survey 1995/96.* Report WWBS-001.

Little, B and Furness, R W (1985) Long-distance moult migration by British goosanders *Mergus merganser. Ringing and Migration* 6: 77–82.

Marquiss, M and Duncan, K (1994) Diurnal activity patterns of goosanders *Mergus merganser* on a Scottish river system. *Wildfowl* 45: 209–221.

Ogilvie, M A (1975) *Ducks of Britain and Europe.* Poyser, Berkhamsted.

Owen, M, Atkinson-Willes, G L and Salmon, D G (1986) *Wildfowl in Great Britain.* Cambridge University Press.

Rose, P M and Scott, D A (1997) *Waterfowl Population Estimates.* Wetlands International Publ. 44. Wageningen.

RSPB (1981) *Sawbills in Wales: a short survey of selected rivers.* MSC Special Project (8/31/0578/81); River Site Evaluation Project. Report prepared by RSPB Wales, Newtown, Powys.

Scott, D A and Rose, P M (1996) *Atlas of Anatidae Populations in Africa and Western Eurasia.* Wetlands International Publication 41. Wageningen.

Starling, A E, Kirby, J S, Pickering, S P C, Gilburn, A S and Howell, R (1992) *British Status and Distribution of Wintering Cormorants, Goosanders and Red-breasted Mergansers in Great Britain with Specific Reference to the Welsh Region.* Report prepared by WWT for Wetlands Advisory Service under contract to NRA Welsh Region. NRA, Cardiff.

Tyler, S J, Stratford, J O and Lucas, R M (1988) Goosanders in Wales. *Welsh Bird Report,* 146–93.

Underhill, M C (1993) *Numbers and Distribution of Cormorants and Goosanders on the River Wye and its Main Tributaries in 1993.* WWT.

Red kite
Milvus milvus

Status

Red listed: HD, BR
SPEC 4 (S)
Schedule 1 of WCA 1981
Annex I of EC Wild Birds Directive

National monitoring

Known sites currently monitored annually by the RSPB/SNH in Scotland, EN/RSPB and the Red Kite Study Group in England and the Red Kite Study Group in Wales; this level of monitoring will be reviewed as the red kite population expands.
Rare Breeding Birds Panel.

Population and distribution

Breeding red kites were almost extinct in Britain at the start of the twentieth century, with only a small population existing in Wales. In recent years this population has gradually increased, and birds have been successfully reintroduced to Scotland and England. The current breeding population is estimated at 160 pairs (*Population Estimates*).

Ecology

Red kites in Wales breed in mature woodland, chiefly of oak, often on the steep sides of valleys. In Scotland, a wide variety of trees is used, including mature conifers. The nest is usually high in the main fork of a broadleaved tree, made from sticks and turf, and lined with wool, although sometimes rags and other rubbish are used (*Red Data Birds*). Kites are known to adapt the disused nests of corvids and buzzards. Feeding areas usually encompass pastoral agricultural land and can be up to 10 km from the nest-site. Eggs are laid from late March to May. There are usually 2–3 eggs in a clutch; there is a single brood, and young fledge by mid-July (*Red Data Birds*).

Breeding season survey – population

The survey procedures for red kite monitoring are intense, and are currently followed for all known sites (Bainbridge et al 1995).

Information required

- number of occupied territories
- number of breeding pairs
- a map showing all registrations
- completed nest recording and sighting forms.

Number and timing of visits

Three visits, February to late March, for evidence of occupation. Every 3-4 days to occupied territories from late March to the end of April, for signs of nest-building and incubation.

Time of day

Any time of the day.

Weather constraints

Avoid visiting in persistent rain, strong winds or poor visibility.

Sites/areas to visit

Woodland, usually mature broadleaved but other woodland possible.

Equipment

- 1:25,000 map of the area
- A4 photocopied field maps
- nest recording forms (Figure 1)
- sighting forms (Figure 2)
- Schedule 1 licence.

Safety reminders

Always tell someone where you are going and when you expect to return. Carry a compass and know how to use it. In more remote upland areas, always carry extra clothing, food, a survival bag and first-aid kit.

Disturbance

Keep disturbance to a minimum. Egg-collectors are a threat to this species.

Methods

During the winter, check winter roosts to see which birds are associating as pairs, noting the details of any wing-tags seen.

Map the boundary of the survey area on to a field map. Make three visits between the beginning of February and late March and map all potential nest-sites, ie areas with suitable nesting habitat – particularly areas where kites have been reported, areas where single birds or pairs have been seen previously, and territories which have been occupied in the past. Defended territories can be small, and neighbouring nests can be as little as 300 m apart, although 3–5 km is more usual.

In late March and April, visit potential territories every 3-4 days and watch for 1–2 hours for evidence of breeding behaviour, ie nest-building or egg-laying. Nest-building behaviour is usually seen before 10 April and includes birds carrying nest material, both male and female entering the wood at the same point, or birds being aggressive towards crows and buzzards. Egg-laying usually takes place from 10 to 25 April when the male may be seen on his own in a potential territory, or the female might be seen coming off a nest. Estimating when the eggs were laid is important for estimating breeding success because it can be used to work out the predicted hatching date. Always watch from outside the wood. However, if the wood is open, it may be possible to observe the nest without disturbing the birds.

Map all the potential red kite nest-sites and annotate the map as follows:

× Potential nest-sites located in the survey area
⊗ Territories occupied by red kite (red kites seen, no evidence of breeding behaviour)
● Red kite breeding sites (breeding behaviour seen)
■ Sites lost (destroyed/developed) during the survey

Outline and hatch shade any areas which were not searched because they were unsuitable. As well as mapping the required information, fill out a nest recording form (Figure 1) for each occupied territory, ie each site at which red kites are seen. Each time a red kite is seen away from the breeding site fill out a red kite sighting form (Figure 2) and report details of any wing tags seen.

The number of occupied red kite territories is defined as the total number of suitable breeding sites at which red kites were seen. The number of breeding pairs of red kites is defined as the number of occupied territories at which there was evidence that nest building took place and/or eggs were laid.

RED KITE NEST RECORDING FORM

SITE/PAIR

Observer	Region/district	Year
Site name	Grid reference	Altitude (m)
Tree species	Tree height (m)	Aspect
Other site details		
Female (tag, etc)	Male (tag, etc)	

SITE VISITS

Date	Time	Adults present	Eggs	Young	Age	Comments

continue overleaf if necessary

RINGING DATA

Nest						Date		
Ring no.	Tag colour	Tag mark	Wing	Tarsus	Leg	Weight	Blood	Sex

Figure 1
A red kite nest recording form.

RED KITE SIGHTING FORM

DATE OF MESSAGE:
RECEIVED BY:

OBSERVER DETAILS

NAME:
ADDRESS:

TELEPHONE:

DETAILS OF SIGHTING

PLACE:

GRID REFERENCE:
DATE/TIME:
NUMBER OF BIRDS:

	BIRD 1		BIRD 2		BIRD 3	
ACTIVITY						
HABITAT						
TAGS:	L	R	L	R	L	R
Tags looked for?						
Possible to see tags?						
Tags seen?						
Main colour						
Bar at end of tag (colour?)						
Diagonal bar (colour?)						
Spot (colour?)						
Horizontal bar (colour?)						
Letter/number only (what?)						
IDENTITY (Office use)						

Figure 2
A red kite sighting form.

Breeding season survey – breeding success and productivity

Information required

On each visit:

- breeding status of the birds, eg confirm sitting birds, young or old chicks or failure
- clutch size
- number of eggs hatched
- brood size, early June
- brood size, late June/early July
- brood size at fledging.

Number and timing of visits

Numerous. One between 22 and 28 April to record the clutch size, one to record number of chicks hatched, one in early June to record the brood size early in the chick-rearing period, and one in late June/early July to record the brood size and to ring the chicks. Then, every 3–4 days throughout July and possibly into August, to check progress.

Time of day

Any time of day.

Weather constraints

As for population survey (above).

Sites/areas to visit

All occupied territories and breeding sites recorded earlier in the season (see above).

Equipment

- 1:25,000 field maps of the survey area
- nest recording forms (Figure 1)
- sighting forms (Figure 2)
- a mirror attached at 45° on a long (up to 20 m) extendable pole (you should be sure that this works before approaching the nest)
- ringing, tagging and measuring equipment
- Schedule 1 licence.

Safety reminders

Always tell someone where you are going and when you expect to return. Only climb trees with the assistance of a ladder and ensure that you are accompanied. You will need help to ring, tag and measure the chicks quickly and safely.

Disturbance

Keep disturbance to an absolute minimum. Egg-collectors are a serious threat. Always make initial observations from a distance.

Methods

To record the clutch size, the number of eggs that have hatched and the brood size early in the chick-rearing period, use a mirror on an extendable pole to look at the nest contents and count the eggs and/or young. Fill in the details on the nest recording form (Figure 1) and each sighting of adults away from the breeding site on a sighting form (Figure 2). Each visit should take no more than 5–10 minutes.

To record the brood size later in the chick-rearing period and ring the young, visit the nest-site in late June/early July (dependent on the progress of the young and the predicted best time for ringing). Climb to the nest, secure the chicks and lower them down to the ground for ringing, tagging, blood sampling (if undertaken) and measurements. While at the nest, record any food items present. Record appearance/progress of chicks and details of rings, tags and measurements on the nest recording form (Figure 1). This visit should take no more than 30–40 minutes.

To record the brood size at fledging, visit the vicinity of the nest every 3–4 days in July and watch for 2–3 hours. The approximate fledging date can be worked out if the hatching dates are known. Visits should start about 40 days from the hatching date; red kites can leave the nest between 45 and 60–70 days after hatching. Again, fill in the nest recording form for each site after each visit and continue to fill out sighting forms for any other records.

In order to calculate nesting success, visit at least once every 10 days to ensure that the nest is still occupied. If, during a visit, the nest appears to be unoccupied, search for causes of failure, dead adult, blown-out nest, predated eggshells, partially eaten chicks, etc. Regular visits are important for the accurate calculation of daily survival rates of clutches and broods.

Reference

Bainbridge, I P, Summers, R W and O'Toole, L (1995) *Red Kite Monitoring in Scotland: Annual Red Kite Monitoring Procedures*. RSPB unpubl., Edinburgh.

White-tailed eagle *Haliaeetus albicilla*

Status

Red listed: HD, BR, SPEC 3 (R)
Schedule 1 of WCA 1981
Annex I of EC Wild Birds Directive

National monitoring

All white-tailed eagle sites are monitored annually by the RSPB and/or SNH.
Rare Breeding Birds Panel.

Population and distribution

Once relatively common in the far north and west of Britain, white-tailed eagles became extinct as a British breeding species in 1916. They bred in the more remote coastal areas on cliffs and islands, and since 1975 a long-term reintroduction programme has successfully brought the birds back as a breeding species to western Scotland (Love 1983). In 1996, there were estimated to be 13–14 occupied home-ranges in Scotland.

Ecology

Breeding is largely confined to coastal areas. On cliffs, nests may start as simple scrapes which are added to in subsequent years. On ledges, cliff tops, slopes and in trees, these can become bulky structures of branches, twigs and driftwood lined with grass, lichens or seaweed. Eggs are laid from March to May and there are usually two eggs in a clutch; there is one brood and the young fledge from mid-July (*Red Data Birds*).

Breeding season survey – population

These procedures (taken from Bainbridge 1995) are considered the minimum standard of information required from each breeding site. They are carried out annually.

Information required

- number of occupied territories
- number of breeding pairs
- completed nest recording forms (Figure 1) and sighting forms (see Figure 2 in red kite).

Number and timing of visits

One visit in December or January to check for the presence of a pair (not essential). For high-priority sites, at least one visit (preferably more) between 15 and 28 February to determine territorial occupancy. Weekly visits from 5 March to 5 April to confirm incubation, unless confirmation has been obtained earlier. For low-priority sites, at least one visit during the season.

Time of day

Any time.

Weather constraints

Avoid visiting in persistent rain, strong winds or poor visibility.

Sites/areas to visit

Priority 1 All nest-sites where breeding has been proved/attempted in the last five years.

Priority 2 All nest-sites at which eagles have been recorded in the last five years, but where breeding has not been proved.

Priority 3 All nest-sites at which eagles have been recorded, but not in the last five years should be visited at least once between January and August.

Priority 4 Any other potential white-tailed eagle nest-sites should, if time allows, receive at least one visit between January and August.

If birds are recorded at priority 3 or priority 4 sites, those sites should be treated as of higher priority and visited more regularly.

Equipment

- 1:25,000 field maps of the survey areas
- recording forms
- Schedule 1 licence.

Safety reminders

Always tell someone where you are going and when you expect to return. Carry a compass and know how to use it. In more remote upland areas, always carry extra clothing, food, a survival bag and first-aid kit.

Disturbance

Keep to a minimum. Egg-collectors are a serious threat.

Methods

Visit all potential nest-sites between 15 and 28 February, in order of priority (see above), and watch from a distance for 3–4 hours. Look for evidence of territoriality and nest-building, ie a pair seen together, birds carrying nest material, copulation, courtship behaviour, display, or aggression towards other birds.

Make weekly visits from 5 March to 5 April to watch for evidence of incubation. Watch the nest without disturbing the birds; a sitting bird

and/or the eggs may be visible; otherwise, watch for an incubation changeover.

Record all breeding information on the nest recording form (Figure 1). Any sightings away from the nest should be recorded on the sighting form (see Figure 2 in red kite). There are other forms which must be filled in as part of annual sea-eagle monitoring, to record nest-site characteristics, home-ranges and individual bird history. Please contact RSPB Edinburgh for details.

The number of occupied white-tailed eagle territories is defined as the total number of suitable breeding sites at which white-tailed eagles were seen. The number of breeding white-tailed eagles is defined as the number of occupied territories at which there was evidence that nest-building took place and/or eggs were laid.

WHITE-TAILED EAGLE NEST RECORDING FORM

Record all checks to all eyries even if nothing is found. If continuous watches are being made record summary information.

Observer: | Region/District: | Year: | Home range code number:

Site name: | Grid reference: | Other site setails:

SITE VISITS

Date	Weather	Eyrie Code	Start Time	Finish Time	Adult(s) Present	Eggs	Young	Age	Comments	Prey Remains	Pellets Collected

(continue overleaf if necessary)

RINGING DATA NEST: DATE:

Ring No.	Tag Colo	Tag Mark	Wing	Tarsus	Leg	Weight	Blood	Sex

Figure 1
Example of a white-tailed eagle nest recording form.

Breeding season survey – breeding success and productivity

Information required

- breeding status on each visit
- clutch size (if possible)
- brood size (if possible)
- when and how many chicks fledge
- presence of fledged juveniles (not essential).

Number and timing of visits

Weekly from 5 March until chicks have fledged or nest has failed.

Time of day

Any time.

Weather constraints

As above.

Sites/areas to visit

All active nest-sites recorded earlier in the season (see above).

Equipment

- 1:25,000 field maps of the survey area
- Schedule 1 licence and separate licence to collect addled eggs
- nest recording and sighting forms
- a mirror attached at 45° on an extendable pole up to 20 m long (dependent on site)
- equipment for ringing, tagging and measuring chicks, and for collecting prey remains, moulted feathers and addled eggs.

Safety reminders

Always tell someone where you are going and when you expect to return. Only climb trees if trained to do so, and ensure that you are accompanied. Don't attempt to reach cliff or ledge sites if this poses any danger to yourself. Do not climb alone or without appropriate safety equipment. You will need help to process the chicks quickly and safely.

Disturbance

Keep disturbance to an absolute minimum. Egg-collectors are a threat to this species.

Methods

On the first visit, establish whether incubation is taking place by watching the site for 3–4 hours. An incubation changeover should take place during this time. If possible, record the clutch size. Try to do this by observing the contents from a distance. If this is not possible, but the nest is accessible, visit the nest-site. Use a mirror on a pole to look at the nest contents and count the eggs. On subsequent visits, record the clutch size, the number of eggs that have hatched and the brood size early in the chick-rearing period. Visits to the nest should only last 5–10 minutes and, whenever possible, should be made when the adults are absent.

During each observation record and identify (stating whether alive or dead) any food items that are brought in by adults. If making visits to the nest, record any prey items or pellets found in the nest. If supplementary feeding is taking place, record the prey species and the number of items involved. Fill in the details on the nest recording form (Figure 1).

Young should only be ringed and tagged if the site is accessible and the process can be completed without danger to birds or people. Visit the nest-site in June to record the brood size later in the chick-rearing period and to ring the young. On the nest recording form (Figure 1) give details of the progress of the young (eg size, feather stage), rings, tags, measurements and blood samples taken.

Throughout this period, if failure is suspected it must be confirmed. If all the chicks are lost, there is no need to monitor the site further.

Visit the vicinity of the nest every week from 1 July to 10 August to record the brood size at fledging. The approximate fledging date can be worked out if the hatching dates are known. Visits should start about 60 days from the hatching date. White-tailed eagles fledge at 70–75 days and may be fed at the nest for a further 30–40 days before becoming independent (*BWP*). Again fill out the nest recording form for each site after each visit and continue to fill out sighting forms for any birds seen away from the nest-site.

During August–October, if time is available, continue to make weekly visits to record the activities of any fledging or post-fledging chicks. In particular, record the size and appearance of the young and the sex of young flying with adults.

Apart from nest recording and sighting forms there are other forms which must be filled in as part of annual sea-eagle monitoring; these are to record nest-site characteristics, home-ranges and individual bird history. Please contact RSPB Edinburgh for details.

References

Bainbridge, I P (1995) *Sea Eagle Monitoring Procedures and Recording Forms*. RSPB unpubl., Edinburgh.

Love, J A (1983) *The Return of the Sea Eagle*. Cambridge University Press.

Marsh harrier
Circus aeruginosus

Status

Red listed: HD, BR
Non-SPEC
Schedule 1 of WCA 1981
Annex I of EC Wild Birds Directive

National monitoring

National surveys: 1995, 2005.
Rare Breeding Birds Panel.

Population and distribution

The UK stronghold for breeding marsh harriers is the coast of East Anglia where there is the highest concentration of large reedbed sites, the favoured breeding habitat. In 1971, breeding marsh harriers had declined to only one pair at Minsmere in Suffolk, predominantly due to the effects of organochlorine pesticides (*88–91 Atlas*). Since these chemicals were banned, numbers have increased to an estimated 157–160 breeding females in the UK (*Population Estimates*).

Ecology

Although most nest in reedbeds, even quite small ones, marsh harriers have also nested in arable fields planted with wheat, barley, and oil-seed rape. The first young hatch about 30–35 days after the first egg is laid. The mean hatching date is 4 June and most eggs hatch between mid-May and mid-June, although hatching has been observed as late as the end of June. Young from the same brood fledge over a period of a few days, usually in the second and third week of July. The mean period between hatching and fledging is 40 days (Underhill-Day 1990).

Breeding season survey – population

This method is based on that used in the 1995 survey devised by John Underhill-Day.

Information required

- number of probable nesting sites
- number of confirmed nesting sites.

Number and timing of visits

At least three visits mid-April to mid-May to locate 'probable' nest-sites. A minimum of two mid-May to mid-June to confirm nesting.

Time of day

Any time of day.

Weather constraints

No weather constraints, although it is best to avoid extreme wet and windy conditions.

Sites/areas to visit

Reedbeds. Adjacent and/or coastal arable fields planted with tall-growing crops.

Equipment

- 1:25,000 field map of the survey area
- Schedule 1 licence.

Safety reminders

Nothing specific. See general health and safety advice in the *Introduction*.

Disturbance

Do not attempt to get close to the nest-site as marsh harriers are very easily disturbed and may desert if the nest is approached. Make all observations from a distance, preferably from a hide.

Methods

Make three visits during mid-April to mid-May to count the number of probable nesting sites. On each visit, observe potential nest-sites from a suitable vantage point for at least four hours.

A 'probable' nest-site is one at which one or both of the following observations are made:

- a female carrying nest material stays at a potential nest-site for an hour or more.
- a female stays at the nest-site for more than four hours and receives a prey delivery from the male during this time.

Observations of males carrying nesting material cannot be taken as evidence of probable nesting, as non-breeders will also carry nest material during this period.

Revisit all probable nest-sites during mid-May to mid-June to confirm breeding. Two observations of 2–3 hours at least two weeks apart at probable nest-sites should be sufficient, but make more if necessary. During these visits only, a 'confirmed' nest-site is one to which adults are observed bringing prey.

Breeding season survey – productivity

Information required

- number of fledged juveniles.

Number and timing of visits

At least two visits after 10 July.

Time of day

Any time of day.

Weather constraints

Avoid adverse weather conditions.

Sites/areas to visit, Equipment, Safety reminders

As for the population survey (above).

Disturbance

Do not attempt to get close to the nest-site as marsh harriers are very easily disturbed and may desert if the nest is approached. Make all observations from a distance, preferably from a hide.

Methods

Make two visits of at least 2–3 hours to each confirmed nest-site. Count the number of fledged young flying with the adults.

Visit probable nesting sites from late August onwards to look for eggs that failed to hatch, prey remains or any other clues to assess the outcome of the nesting attempt and possible reasons for failure.

Reference

Underhill-Day, J C (1990) The status and breeding biology of marsh harrier *Circus aeruginosus* and Montagu's harrier *Circus pygargus* in Britain since 1900. PhD thesis. RSPB and ITE.

Hen harrier *Circus cyaneus*

Status

Red listed: HD, SPEC 3 (V)
Schedule 1 of WCA 1981
Annex I of EC Wild Birds Directive

National monitoring

National breeding surveys: 1988–89, 1998 (RSPB/SNH/EN/CCW/DETR/Raptor Study Groups).
National Winter Roost Survey: annual (recording forms and details available from Roger Clarke and Donald Watson, c/o New Hythe House, Reach, Cambridge CB5 0JQ).

Population and distribution

Hen harriers mainly breed in the north and west of Britain, the Isle of Man and Ireland, with strongholds in Orkney, the east Highlands and Strathclyde north and west of the Clyde (*88–91 Atlas*). Between the 1968–72 and the 1988–91 breeding atlases the Scottish population remained relatively stable, although those in England and Wales declined. The lack of expansion of the breeding population to former levels is thought to be the result of human persecution (*88–91 Atlas*). There are an estimated 630 breeding pairs of hen harriers in Britain and 180 in Ireland (Bibby and Etheridge 1993, *88–91 Atlas*). In winter, birds disperse southwards, and are much more widely distributed throughout the UK. Breeding birds are resident and most young winter in Britain. The winter population is estimated at 1,500–2,000 birds (B Etheridge pers comm).

Ecology

Hen harriers breed on moorland, especially where there is old, deep heather and also in heather within young conifer plantations. Eggs are laid between May and June, there is a single brood, and the young fledge from the end of June (*Red Data Birds*). In winter, hen harriers roost on platforms of trampled vegetation. These can be small gaps in the vegetation just wide enough to take a harrier, or wider gaps up to 1 m or more across, surrounded by standing vegetation. They can be scattered over 1–2 ha, but are often grouped 1–2 m apart within a small favoured area. The birds will move around locally during the winter and between winters, but tend to return to the same favoured spots.

Breeding season survey – population

These methods are based on those used in the survey of breeding hen harriers in 1988–89 (Bibby and Etheridge 1993).

Information required

- minimum number of breeding pairs
- maximum number of breeding pairs
- a map showing the boundary of the survey area and summarised registrations.

Number and timing of visits

At least two visits, between early April and the end of May. If breeding is not confirmed, a third visit should be made between late June and the end of July.

Time of day

0700–1900 BST.

Weather constraints

Avoid poor weather.

Sites/areas to visit

Heather-dominated moors from sea level (in west Scotland) to 550 m (in central Highlands), usually on well-drained slopes where heather grows longer. Young plantations in upland areas (particularly north and west Britain).

Equipment

- 1:25,000 map (1:50,000 will do)
- A4 photocopied map of the survey area for use in the field
- Schedule 1 licence.

Safety reminders

Ensure someone knows where you are going and when you expect to return. Take a compass and always carry a survival bag, waterproofs, whistle, extra clothing, food and a first-aid kit in remote areas.

Disturbance

Keep disturbance to a minimum. Try to confirm breeding without a visit to the nest. Do not disturb any bird during the laying period in late April/first half of May. If survey visits are made during this time, observe the nest-site from a distance. Do not cold search for hen harrier nests. Human persecution is a serious threat to this species. Keep site information confidential.

Methods

Take a new field map on each visit to the site and clearly mark the boundary of the survey area on the map.

During the first visit to the site, walk to within 250 m of each spot. Mark on your visit map all the areas that are unsuitable for hunting and breeding (see above). These can be excluded from further searching. Search for hen harrier activity and signs of hen harrier presence.

During the second visit (until the end of May) search all suitable areas, recording all hen harrier observations, and attempt to confirm breeding by watching all potential nesting sites from a distance. Continue searching even if a pair was confirmed as breeding on the first visit, as there may be other birds breeding in fairly close proximity (800–1,000 m) or more than one female with the same male.

If breeding is not confirmed, make a third visit between late June and the end of July to watch birds and to try to locate a nest-site and confirm breeding behaviour. Again, try to achieve this from a discreet distance, without a visit to the nest itself.

During each visit, record on the map the presence or absence of hen harriers, their activity, sex and details of any nests encountered, and cross-refer to any more-detailed comments. By the end of the survey the observer should know the number of confirmed, probable and possible breeding pairs in the survey area.

Breeding is confirmed if:

- a nest containing eggs or young is found
- an adult is seen carrying food for the young
- a used nest or eggshells are found (occupied or laid within the period of the survey)
- recently fledged young are seen
- agitated behaviour or anxiety calls are given by the adults.

Breeding is probable if:

- a pair of hen harriers is seen in suitable nesting habitat in the breeding season
- a permanent territory is presumed through the registration of territorial behaviour on at least two different days, separated by more than one week
- courtship and display behaviour is witnessed
- birds are seen visiting a probable nest-site.

Breeding is possible if:

- a hen harrier is observed in the breeding season in possible nesting habitat.

The minimum number of breeding pairs of hen harrier is the number of confirmed + probable pairs; the maximum number is the confirmed + probable + possible pairs.

Winter survey

This method is based on that used for the annual survey of wintering hen harriers in Britain (Clarke and Watson 1990).

Information required

- a map of the survey area, including its boundary and the location of hen harrier roost(s)
- numbers of hen harriers using the roost(s) on the survey dates
- the proportions of grey adult males to 'ringtails' (first-winter males and females).

Number and timing of visits

Make an initial visit to the site during the day to familiarise yourself with it. Recommended visit dates are Sundays in the periods 20–26 October, 24–30 November, 15–21 December, 12–18 January, 15–21 February and 15–21 March. The national organisers will inform you of any changes to these dates.

Time of day

From 1½ hours before sunset to half an hour after sunset (or until it becomes too dark to see). The morning after the watch date is acceptable, but you must be on station at first light and know the exact roosting area, if you are not to miss birds as they leave.

Weather constraints

Windy and dry weather conditions are best. Some wind will keep the birds up and active, making them easier to count.

Sites/areas to visit

All sites known or suspected to be winter roost sites or potential roost sites. Usually areas of rank ground vegetation, in a variety of open habitats, particularly saltmarsh, reedbeds and other marshes, rough grassland, heather moor, lowland heath and (more unusually) conifer plantations.

Equipment

- 1:25,000 map
- torch and compass
- Hen Harrier Winter Roost Survey recording forms.

Safety reminders

You will be finishing in the dark and may be working near water, so make sure someone knows where you are and when you expect to return. A torch, compass and map are essential.

Disturbance

Do not get too close to the (potential) roost, or birds will desert the site. Do not draw attention to the roost site.

Methods

Mark the boundary of the survey area on the map. During a daytime visit, mark areas of suitable habitat within the survey area on the map. If possible, find a concealed vantage point overlooking the suitable habitat and watch for incoming birds from there.

It may be impossible to get an accurate count. Make the best estimate possible and beware of double-counting birds flying in and out of view, settling and then rising again. Figure 1 is a sample of the recording form used in the National Survey. The recording form columns are open-ended, please rule off after each watch.

On each visit record the following:

- Date.
- Time of start and finish.
- Weather conditions.
- Times of the first and last hen harriers arriving/settling and, if possible, of any birds in between (in the 'others' column). 'Arrival' is defined as when a hen harrier first arrives in the vicinity of the site. 'Settling' is defined as a hen harrier landing at a presumed roosting place and not immediately rising again.
- Number of hen harriers.
- Number of ringtails and grey males. First-winter brown males can sometimes be distinguished by their small size, but only if other harriers are present for comparison. They must, however, be recorded as 'brown' in the relevant columns. Record all older males as 'grey', even if their plumage has substantial amounts of brown.

Further notes in the comments column should include the direction of arrival and departure and the presence of wing tags. Wing tags, if present, will be on both wings. It is important to note on the recording form the colour of the wing tag and which colour on which wing. If possible note the letters/number on the tags as well.

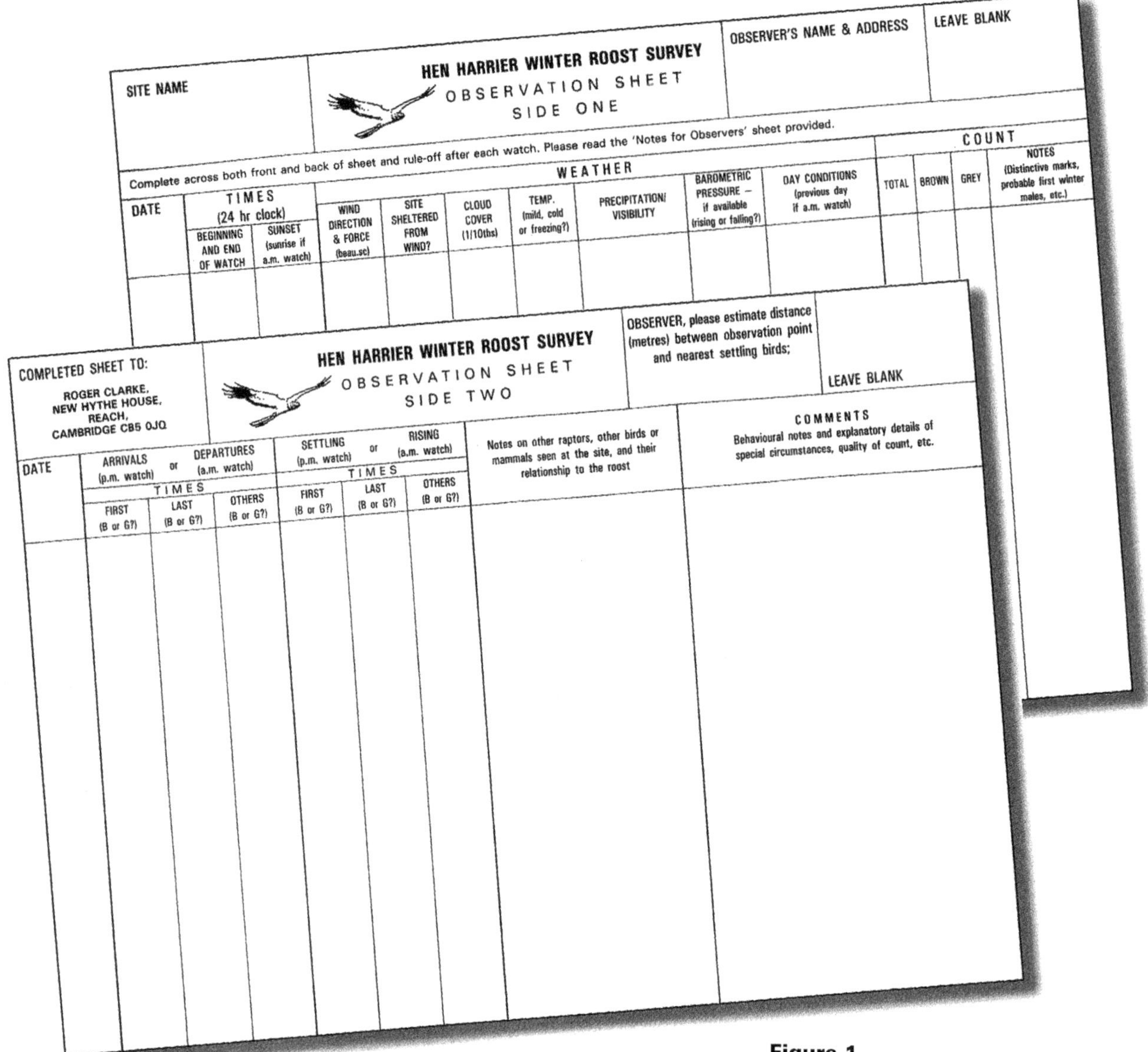

SITE NAME

HEN HARRIER WINTER ROOST SURVEY
OBSERVATION SHEET
SIDE ONE

OBSERVER'S NAME & ADDRESS

LEAVE BLANK

Complete across both front and back of sheet and rule-off after each watch. Please read the 'Notes for Observers' sheet provided.

DATE	TIMES (24 hr clock)		WEATHER							COUNT			
	BEGINNING AND END OF WATCH	SUNSET (sunrise if a.m. watch)	WIND DIRECTION & FORCE (beau.sc)	SITE SHELTERED FROM WIND?	CLOUD COVER (1/10ths)	TEMP. (mild, cold or freezing?)	PRECIPITATION/ VISIBILITY	BAROMETRIC PRESSURE – if available (rising or falling?)	DAY CONDITIONS (previous day if a.m. watch)	TOTAL	BROWN	GREY	NOTES (Distinctive marks, probable first winter males, etc.)

COMPLETED SHEET TO:
ROGER CLARKE,
NEW HYTHE HOUSE,
REACH,
CAMBRIDGE CB5 0JQ

HEN HARRIER WINTER ROOST SURVEY
OBSERVATION SHEET
SIDE TWO

OBSERVER, please estimate distance (metres) between observation point and nearest settling birds;

LEAVE BLANK

DATE	ARRIVALS (p.m. watch) or DEPARTURES (a.m. watch) TIMES			SETTLING (p.m. watch) or RISING (a.m. watch) TIMES			Notes on other raptors, other birds or mammals seen at the site, and their relationship to the roost	COMMENTS Behavioural notes and explanatory details of special circumstances, quality of count, etc.
	FIRST (B or G?)	LAST (B or G?)	OTHERS (B or G?)	FIRST (B or G?)	LAST (B or G?)	OTHERS (B or G?)		

Figure 1
Example recording form (front and back) as used in the Hen Harrier Winter Roost Survey.

Report the figures for each month, highlighting the maximum number, and provide a map of the site with summarised registrations, the boundary of the survey area and a brief description of the habitat. Return completed national survey forms to the organisers at the end of the season.

References

Bibby, C J and Etheridge, B (1993) Status of the hen harrier in Scotland 1988–89. *Bird Study* 40: 1–11.

Clarke, R and Watson, D (1990) The hen harrier *Circus cyaneus* winter roost survey in Britain and Ireland. *Bird Study* 37: 84–100.

Buzzard
Buteo buteo

Status

Non-SPEC

National monitoring

BBS.
Survey of Britain and Ireland, 1983 (BTO).
Population changes between 1983 and 1996 in West Midlands investigated (RSPB).

Population and distribution

Buzzards are fairly common in western and northern Britain, but are absent from much of central and eastern England and Scotland and most of Ireland (*88–91 Atlas*). Once present throughout most of Britain, the present westerly distribution has been linked with several factors, notably persecution, pollution and a decline in rabbit populations due to myxomatosis (*88–91 Atlas*). There are an estimated 12,000–17,000 territories in the UK (*Population Estimates*).

Ecology

Buzzards are linked with a variety of landscapes including forest clearings, woodlands (amidst pasture, meadow, arable or wetland), cultivated landscapes with groves or tall isolated trees, and hilly slopes, ridges or uplands with some or much tree cover. They prefer hunting over open tracts of land with low vegetation, but nest in trees. A diverse diet enables them to use a wide variety of habitats in which some kind of prey is abundant and accessible. A clutch of 2–4 eggs is laid from early April and most pairs will have full clutches by mid-April. Hence incubating females are unlikely to be seen soaring during April.

Breeding season survey – population

This method was used to estimate regional densities of the British and Northern Irish populations in 1983 (Taylor et al 1988) and to estimate the change in numbers between 1983 and 1996 in an area of the west Midlands (Sim et al in prep). The methods are designed to obtain fairly crude estimates of population size over wide geographical areas with relatively little time spent in the field in any particular area.

Information required

- maximum number of soaring buzzards seen on any one visit per tetrad (2 × 2 km)
- map showing the boundary of the survey area and the locations of soaring birds.

Number and timing of visits

One visit in March. A minimum of 1 hour (preferably two) must be spent in each tetrad.

Time of day

1000–1400 BST.

Weather constraints

Counts should ideally be carried out in good soaring conditions (ie dry, cloud cover less than 50%, wind speed less than 30 mph). However, as such conditions are uncommon in March, simply try to avoid counting in wet, misty or windy conditions.

Sites/areas to visit

Any suitable habitat (see *Ecology*, above).

Equipment

- 1:25,000 map.

Safety reminders

No specific advice. See general safety reminders in the *Introduction*.

Disturbance

This survey will not disturb buzzards.

Methods

Mark the boundary of the survey area on to a map and split it into tetrads (2 × 2 km). Cover the whole of the area searching for soaring buzzards. This does not require great expertise and can be carried out from roadsides or public footpaths. The best way of surveying a tetrad will depend on local topography, but often a single vantage point offering good all-round visibility will suffice. In hilly or well-wooded areas, it will probably be necessary to find two or more survey points to obtain a reliable estimate. Buzzards can soar to great heights and move large distances in a short time, and this needs to be taken into account when carrying out soaring surveys. Movements of soaring birds between tetrads should be marked on to the map and taken into account when analysing results. Separate groups of soaring buzzards should be indicated by dotted lines between the groups. Movements of individual birds between tetrads should be represented by solid lines.

Report the maximum number of soaring buzzards seen in each tetrad (relate this to the map) on a single visit, but make allowances for birds seen to move between tetrads (ie ensure they are not counted twice). Report the density as the mean number (± standard error) of soaring buzzards per tetrad visited.

References

Sim et al (in prep) Correlates of the population increase of common buzzards *Buteo buteo*.

Taylor, K, Hudson, R and Horne, G (1988) Buzzard breeding distribution and abundance in Britain and Northern Ireland in 1983. *Bird Study* 35: 109–115.

Golden eagle
Aquila chrysaetos

Status

Amber listed: SPEC 3 (R)
Schedule 1 of WCA 1981
Annex I of EC Wild Birds Directive

National monitoring

National surveys: 1982, 1992, 2002 (RSPB/SNH).
Raptor Study Groups may be coordinating monitoring in your area; contact them to find your nearest regional co-ordinator.

Population and distribution

In the UK, golden eagles breed throughout the uplands of Scotland and at a site in northern England. No birds breed in Ireland. In the last few decades there have been some modest changes in the species' breeding distribution (*88–91 Atlas*), although its population is remarkably stable. The greatest threats to golden eagles in Britain today are the persecution of adult birds and human interference at nests, including egg-collecting (*Birds in Europe*). There are an estimated 420 breeding pairs of golden eagle in Britain (Green 1996).

Ecology

Breeding habitat is mainly mountainous upland. Eyries are usually on ledges or in hollows of rock faces. In Scotland, nest heights range from 16 m to 900 m above sea-level. Generally, golden eagles will avoid inland waters, wetlands and dense forests, preferring open areas with rough grass moors and smooth grassland, especially slopes or plateaux allowing a wide view and use of air currents. Perches are on crags, trees or other suitable look-out points. The home range size varies from 30 to 120 km^2, averaging 50 km^2 and is determined by prey density, topography and the availability of nest-sites. A clutch of 1–3 eggs is laid in March–April and the young fledge from the beginning of July (*Red Data Birds*). By about 15 days old, the young begin to develop their primaries and have their second coat of down which is long, coarse and creamy-white. The fledging period is about 65–70 days, but the chicks may fledge from the age of 50 days onwards, by which time they have many brown adult feathers. The female stays with the young for up to about 40 days.

Breeding season survey – population

This survey method is based on the instructions for the 1992 national golden eagle survey (Green 1996).

Information required

- number of occupied home ranges
- a map showing the survey area boundary and all eagle registrations.

Number and timing of visits

As many visits as are required to confirm occupancy of the home range, between January and early April.

Time of day

Any time of day.

Weather constraints

Wherever possible make visits in good weather, but do not cancel visits because of poor weather, except when this is so extreme as to be hazardous to personal safety.

Sites/areas to visit

Any site known to have a previous history of golden eagle nesting. Ledges and hollows on rock faces.

Equipment

- 1:25,000 OS map of the survey area
- telescope (optional)
- Schedule 1 licence.

Safety reminders

Always carry a compass and ensure that someone knows where you are going and when you are due back. Carry spare warm clothing, a plastic survival bag, whistle, first-aid kit and food supplies. If possible, surveyors should not work alone in more remote upland areas.

Disturbance

During laying or incubation, keep disturbance to an absolute minimum. If you make visits during this time, observe the nest-site from a discreet distance. No visits to nests are necessary to confirm home range occupancy. Egg-collectors are a serious threat to this species.

Methods

This is a difficult bird to survey. Despite their size, golden eagles can be quite elusive, breeding in fairly inaccessible mountainous habitat where the weather is frequently poor. Also, their use of widely distributed alternative eyries can make locating the actual breeding site difficult.

Use past information on breeding sites as much as possible. Raptor Study Groups regularly monitor many sites; if you have not been in contact with them already, do so before starting fieldwork.

Mark the boundary of the survey area on a map. On initial visits (in January) find all those areas that might be potential nest-sites and mark them on the map. You can exclude from the survey any other areas not suitable as hunting grounds (eg dense forest, plantations, populated areas) and mark them on the survey map as such. More experienced observers and observers with good information on the likely location of eagles may be able to concentrate on particular sites and cover much less ground.

Early visits are a good time to find displaying golden eagles, to record the presence of immature birds and to record nest-building activities. There are usually several alternative eyries within a home range, but as nest building progresses it should become more obvious which site will be used for breeding. Usually a deep, thickly lined bowl signifies that a hen will lay, and usually in that nest. However, it is possible to have several well constructed nests within one home range, and females (especially young females) may lay eggs in a poorly constructed nest (Watson et al 1989).

On subsequent visits (January to early March) work out a route that ensures that you visit all potential nest-sites (scanned with binoculars or telescope, no need for rock climbing) and that you have approached all suitable hunting areas to within 500 m. It should be possible to cover about 2 km^2 in one day, depending on the accessibility of the area.

If you do not see two golden eagles together during visits to the survey area until late March, visit potential eyries (late March or April) to locate an incubating bird. Always approach these areas with caution at this time and only watch from a discreet distance so that eagles are not flushed early in incubation.

Proof of occupancy of a home range by a pair requires that you either see two eagles together or witness incubation by one of the adults (usually the female).

Record registrations of any other eagle activity on each visit. Mark eagle registrations on a map, using standard BTO notation (Appendix 1). Use a new map for each visit. Even if you do not see any eagles, note the location of any evidence of their presence, such as fresh pellets, moulted feathers or fresh food remains in an eyrie.

Report whether home range occupancy was confirmed and how. Provide a summary map of the survey area, showing eagle registrations, eyrie and alternatives. If you are not able to confirm occupancy, report all eagle registrations and evidence of eagle presence on a summary map with additional explanatory notes if necessary.

Breeding season survey – breeding success and productivity

Information required

- confirmation of fledged young (number fledged if possible).

Number and timing of visits

Two, one in mid-June and one in late June/early July.

Time of day

Any time of day.

Weather constraints

Avoid poor weather conditions.

Sites/areas to visit

Sites where the home range and the nest-site have been confirmed.

Equipment

- telescope
- Schedule 1 licence.

Safety reminders

As for the population survey (above).

Disturbance

Be discreet during your observations. Nest visits are not necessary to determine if there are any chicks in the eyrie. Do not loiter if the adult is still brooding the young. Keep the location of the nest confidential.

Methods

Make the first visit in mid-June to check on the progress of any young. Do not go near the nest-site if an adult is still brooding: leave, and return a week later.

If young are present, note their progress. If they are near to fledging stage (about 50 days old) no further visits are necessary. If they are still downy, make another visit at the end of June/beginning of July. When observing the nest-site and trying to ascertain the number of young, it may be necessary to wait until one of the adults brings food to the nest and feeds the young before a clearer view is possible. This may mean observing for several hours. If this is not possible, just state what you saw.

References

Green, R E (1996) The status of the golden eagle in Britain in 1992. *Bird Study* 43: 20–27.

Watson A, Payne, S and Rae, R (1989) Golden eagles *Aquila chrysaetos*: land use and food in northeast Scotland. *Ibis* 131: 336–348.

Osprey
Pandion haliaetus

Status

Red listed: HD, BR, SPEC 3 (R)
Schedule 1 of WCA 1981
Annex I of EC Wild Birds Directive

National monitoring

Monitored annually by the Scottish Osprey Study Group (amateur and RSPB observers; results collated by Roy Dennis).

Population and distribution

Historically, ospreys bred throughout the British Isles, but due to long-term persecution over many centuries the species became extinct in England in 1847 and in Scotland in 1918. Successful recolonisation occurred in Scotland in 1954 (Dennis 1995) at Loch Garten, and the population has slowly increased to 111 pairs in 1997 (Dennis 1997).

Ecology

In Scotland, ospreys breed in a wide variety of wooded landscapes, including native Scots pine forests, and by freshwater lochs and rivers, which provide nesting and feeding sites. Ospreys also feed along estuaries. The nest is a bulky platform of sticks, typically in the crown of a live or dead Scots pine or other coniferous tree, rarely a deciduous tree; artificial platforms are also used. A clutch of usually three eggs is laid mid-April to mid-May, and the young fledge between mid-July and August.

Breeding season survey – population

At present all known sites are regularly monitored by the Scottish Osprey Study Group and their contacts.

Information required

- number of occupied eyries
- map showing the boundary of the survey area and all osprey registrations.

Number and timing of visits

As many visits during April and May as are required to confirm eyrie occupancy.

Time of day

Any time of day.

Sites/areas to visit

All known sites, artificial nests and suitable areas should be visited.

Equipment

- 1:25,000 map of the survey area
- Schedule 1 licence
- prepared field maps
- telescope.

Safety reminders

Ensure someone knows where you are and when you are due back.

Disturbance

Do not make unnecessary visits to nest-sites. During the laying or brooding period, disturbance should be kept to an absolute minimum. If visits are made during this time then observe the nest-site from a discreet distance. Disturbance distances vary from 100 to 500 m. No visits to nests are necessary to confirm occupancy.

Egg-collectors are a threat to this species. Keep site information confidential, carry minimal site information with you, and be alert to any suspicious activity.

Methods

Ospreys are generally faithful to successful nest-sites, so most of the annual monitoring involves visits to known sites. Finding new sites is more difficult and involves cold searching, following birds with fish and using local information.

Mark the boundary of the survey area on a map. Make one or more initial visits in early April, to find all those areas which could hold potential nest-sites and mark them on the map. Also mark any vantage points which overlook suitable feeding habitat. Any other areas not suitable as nesting or feeding sites can be excluded from the survey and marked as such on the survey map. Search for osprey activity.

For subsequent visits (until the end of May) work out a route which ensures that all potential nest-sites are visited (scanned with binoculars or telescope) and that all suitable feeding habitat is watched from vantage points. Ensure that you get to within 500 m of all points in suitable habitats. Record all osprey observations and attempt to confirm occupancy (without visiting nests) at all potential nesting sites.

Proof of occupancy by a pair requires that either (a) two ospreys are seen together at an eyrie on more than one occasion separated by a week or (b) incubation by one of the adults is witnessed or parents are seen feeding chicks. If occupancy is not thus proven by the presence of two adults by the end of April, revisit potential eyries in May in order to check for incubating birds or parents feeding chicks. Approach these areas with caution at this time and only watch from a discreet distance so that ospreys are not flushed early in incubation. There is no need to revisit eyries where occupancy has been confirmed.

Record all osprey activity on to a new map for each visit using standard BTO notation (Appendix 1).

Report the following:

1. The number and location of eyries with a pair present for more than a week.
2. Any incubating birds present (or parents feeding chicks).
3. Any pairs present for less than one week.
4. Any single birds present.

5. The identity of adults (rings seen or plumage characteristics noted, particularly on the head).

Provide a summary map of the survey area showing all osprey registrations. The number of occupied eyries is the sum of 1 and 2 above.

Breeding season survey - breeding success and productivity

The Scottish Osprey Study Group also collects valuable information on clutch size, the number of chicks hatched, the numbers of chicks ringed, the numbers of chicks fledged and the causes of any failures. Please contact the group or the RSPB for further instructions on how to gather this information safely.

As part of a long-term study of osprey breeding biology, most young ospreys are ringed with coloured plastic leg rings. Please use a telescope to check for these on migrating ospreys. Note which leg is colour-ringed, the colour, and the repeated inscription on the ring. Please report this to the Study Group.

Osprey Study Group contacts: Roy Dennis, Inchdryne, Nethybridge, Inverness-shire; RSPB Offices for South and West Scotland, East Scotland and North Scotland (see *Useful addresses* in the *Introduction*)

Contributed by Roy Dennis

References

Dennis, R (1995) Ospreys *Pandion haliaetus* in Scotland – a study of recolonization. *Vogelwelt* 116: 193–195.

Dennis, R (1997) *Osprey News 1997*. Unpublished newsletter.

Kestrel
Falco tinnunculus

Status

Amber listed: BDM, SPEC 3 (D)

National monitoring

CBC, BBS.

Population and distribution

The kestrel is the most common raptor breeding in the UK, able to adapt to most habitats, including city centres. It is scarce only in north-west Scotland, the Western Isles and Shetland (*88–91 Atlas*). There has been a decline in kestrel numbers in the west of the British Isles, the reasons for which are unclear (*88–91 Atlas*). There are an estimated 52,000 pairs of kestrel breeding in the UK (*Population Estimates*).

Ecology

Kestrels nest in a wide variety of sites. In the uplands they use mainly old stick nests and crags. In open moorland, likely sites are old crows' nests in conifer shelterbelts or rock ledges over 3 m high. In some years (of high vole numbers), some shelterbelts may have more than one nesting pair, and even small crags might be used. In thick plantations nests are hard to find from the ground and are most easily found by observing displaying kestrels. In lowland areas, holes in trees (especially ash and elm) are the most likely sites, although isolated barns and other buildings (eg church towers, disused windmills and industrial sheds) may also be used. Kestrels will habituate to human disturbance, and sites near human habitation cannot be ruled out without checking.

Breeding kestrels are difficult to survey accurately because of the irregular spacing of their nests and the difficulty in finding them. Where nests are clumped, the density within the clump may be high with very few nests in the surrounding area (Village 1990). The total number of territorial pairs will include a number of pairs that will mate but will fail to produce eggs, and some which lay eggs but subsequently fail. If an area is only surveyed late in the nesting season, pairs will be missed because only successful pairs will be found (Village 1990). The densities of breeding pairs of kestrel vary according to habitat, eg grassland 32 pairs/100 km^2; mixed farmland 19 pairs/100 km^2; arable farmland 12 pairs/100 km^2. In Britain, kestrel breeding activity begins in late February and March. Laying starts between mid-April and late May, and young fledge during June and July. Juveniles remain in their natal territory for about a month after fledging, which means that most pairs finish their breeding cycle by mid- or late August.

Breeding season survey – population

Two methods are given: measuring density (*Method 1*) and monitoring occupancy (*Method 2*). Use Method 1 to make an accurate density estimate. This can only be achieved with a study area of 50–200 km^2 which is thoroughly searched to find all (or at least 90%) of the territorial/laying pairs. It is the only accurate way of comparing kestrel numbers in different areas in the same year. Use Method 2 to monitor changes in numbers from year to year in small survey areas or in cases where not all the breeding pairs are likely to be found.

Information required

For both methods:
- minimum and maximum number of pairs
- a map showing the area searched.

Number and timing of visits

Method 1 – At least one visit to all potential sites before mid-May. Make a later visit to any unoccupied sites checked very early in the season, or any site where occupancy may have been missed. Further visits as necessary up to mid-July to record breeding performance at occupied sites. If the 50–200 km^2 study area is in open country with few possible nest-sites, this could be covered by a single fieldworker over a season. A team would be needed to cover a similar area in lowland Britain.

Method 2 – Five visits between April and mid-July with at least a week between visits.

Visit 1: 1–15 April
Visit 2: 16–30 April
Visit 3: 1–15 May
Visit 4: 16–31 May
Visit 5: 15 June to 15 July

Time of day

Both methods – 0700 to 1900 BST.

Weather constraints

Both methods – Avoid poor weather.

Sites/areas to visit

Any suitable habitat. Kestrels nest in a wide variety of sites (see *Ecology*, above).

Equipment

- 1:25:000 OS map
- prepared field maps.

Safety reminders

When working in more remote or upland areas, ensure someone knows where you are and when you are due back. Always carry a compass. If possible, surveyors should not work alone in remote upland areas, and spare warm clothing, a plastic survival bag, whistle, first-aid kit and food supplies should be carried.

Disturbance

The disturbance caused by this method is minimal. Try to confirm breeding without visiting the nest. Do not disturb birds during the laying period (late April and first half of May). Observe the nest-site from a discreet distance at this time.

Methods

Method 1: Measuring density

Define the survey area prior to searching; avoid concentrating effort in areas where kestrels are known to breed. A 10 × 10 km square would be a suitable study area.

Spend the first visits searching for possible nest-sites (old stick nests, rock ledges, suitable tree holes, etc) and mark them on a map.

All potential nest-sites should be visited from early April to mid-May to check for occupancy, with a second check of sites found unoccupied early in the season. In lowland England, most nesting territories are occupied year-round, but in upland areas most breeders leave in winter and some territories may not be occupied until May. Occupied sites should be further visited as necessary to record breeding performance.

Checking occupancy

Kestrels are often seen at occupied sites, or can be heard calling, in the pre-laying period (April and early May for most pairs). Fluttering-wing flight (level flight with shallow wing-beats) is common near the nest. Signs of kestrel presence include pure white, creamy droppings (not yellowish as with owls), pellets (20–30 mm × 12–17 mm, tapered) and piles of feathers or vole fur. The substrate of nest-ledges or tree-holes is usually scraped before eggs are laid into a shallow cup. Later in the season look for moulted feathers, droppings under stick nests or down ledges, and listen for young calling.

Some stick nests and tree-holes may eventually become unsuitable (due to collapse, etc), so check the surrounding area carefully for alternatives if sites become unoccupied.

During each visit, record the presence (or absence) of kestrels, their activity, sex and details of any nests located on the map and cross-reference these to any more-detailed comments. By the end of the survey you should know the number of confirmed, probable and possible breeding pairs in the survey area.

Method 2: Monitoring occupancy

Ideally the survey area should contain at least ten territorial pairs of kestrels. Mark the boundary on a map and keep it the same from year to year. Spend the first two visits thoroughly searching the whole area for potential nest-sites and noting kestrel behaviour. Find as many of the potential sites in the area as possible. Check occupancy as for Method 1, above. Although a few sites may be missed each year, as long as most of the potential breeding sites are known, changes in occupancy should give an index of the variation in numbers over time. The third, fourth and fifth visits should concentrate on proving occupancy at all the potential nest-sites mapped. All potential nest-sites should be visited at least once each year.

Breeding is *confirmed* if:

- a nest containing eggs or young is found
- an adult is seen carrying food for the young
- a used nest or eggshells are found (occupied or laid within the period of the survey)
- recently fledged young are found.

Breeding is *probable* if:

- agitated behaviour or anxiety calls are given by the adults
- a pair of kestrels is seen in suitable nesting habitat in the breeding season
- a permanent territory is presumed through the registration of territorial behaviour on at least two different days separated by more than one week
- courtship and display behaviour are witnessed

Breeding is *possible* if:
- a kestrel is observed in April and/or May in possible nesting habitat.

Report the minimum number of breeding pairs of kestrels as the number of confirmed pairs; and the maximum number as the confirmed + probable + possible breeding pairs. Report the size of the survey area and provide a map of summarised registrations.

Contributed by Andrew Village

Reference

Village, A (1990) *The Kestrel.* Poyser, London.

Merlin
Falco columbarius

Status

Red listed: HD
Non-SPEC
Schedule 1 of WCA 1981

National monitoring

National surveys: 1983–84, 1993–94, 2003–04.

Population and distribution

Merlins in Britain breed on moorland, notably in the Scottish Highlands and islands, the Welsh mountains, the Pennines and the southern uplands. In the last 20 years there has been some evidence of a decline in south-west England, south-west Scotland and the Borders (*88–91 Atlas*). There are estimated to be 1,330 pairs of merlin breeding in the UK (*Population Estimates*).

Ecology

Merlins require open ground for hunting, preferably heather moorland. They may nest on the ground in heather, on a rocky outcrop or in crows' old stick nests. Some pairs nest well into woodland, including conifer plantations, but always with open moorland nearby. A clutch of 3–5 eggs is laid in late April or May. There is usually a single brood, but early egg losses are sometimes replaced (*Red Data Birds*).

Breeding season survey – population

These methods are based on those developed for the 1993–94 national survey (Rebecca and Bainbridge 1994, 1998).

Information required
- number of breeding pairs
- a map showing the boundary of the survey area and location of breeding pairs.

Number and timing of visits

At least four visits, one each month (April–July), with follow-up visits where necessary to confirm breeding.

Time of day

Any time of day.

Weather constraints

Avoid surveying in wet or windy conditions (winds above Beaufort force 4).

Sites/areas to visit

Areas of heather moorland or bracken, young plantations with small trees and heather cover, plantation edges, including the edges of open areas within afforested blocks and the edges of rides within 100 m of the plantation/moorland boundary and open woodland (eg birchwood and Caledonian pine forest), usually below 600 m altitude.

The following can be excluded from the survey: towns and villages, enclosed in-bye pastures, arable farmland, areas of grassy moorland which lack trees, bushes, crags and stream banks (all areas of heather moorland must be checked), land above the 600 m contour (where merlins are known or suspected to nest above 600 m, they should be included with a note made of the altitude).

Equipment

- 1:25,000 OS map of survey area
- A4 photocopies of the map for use in the field
- prepared recording forms (see Figure 1 for example)
- Schedule 1 licence.

Safety reminders

Ensure someone knows where you are going and when you will return. Do not climb trees or crags. Always carry a compass. If possible, do not work alone or in more remote upland areas. Carry spare warm clothing, plastic survival bag, whistle, first-aid kit and food supplies.

Disturbance

Disturb the birds as little as possible and do not visit nests. If your presence does disturb a bird, withdraw immediately and observe from a distance. Always try to observe birds from a discrete vantage point.

Methods

Make sure the survey boundary is clear and is mapped.

The first visit in April will require more detailed gathering of information than subsequent visits:

- get to within 500 m of every spot in the survey site
- map all those areas that are unsuitable as potential nest-sites (see above) so that these areas can be excluded from further searching
- locate all crows' nests, as these may be used by nesting merlins
- search for merlin activity and signs of merlins.

In May, June and July:

- search all suitable areas, recording all merlin observations and attempt to confirm breeding (without nest visits) at all potential nesting sites (it is not necessary to revisit any sites at which breeding has already been confirmed in a previous month).

Work out the most effective route beforehand and mark it on the map. Alternate the direction the route is travelled between visits. On each visit to the site, whether the observer is visiting the whole area as on the first visit, or only suitable nesting habitat, the following search methods should be used:

In moorland habitat, check along fence-lines and around crags, steep stream banks, rocks, grouse butts, stone walls/dykes, hummocks, recently burned areas, isolated trees and other perches for whitewash (faeces), prey remains, pellets and moulted merlin feathers. Check old crows' nests.

In open birch, pine or alder woods, tree-lined gorges and rivers, search the woodland for crows' nests and check for birds leaving the site as you approach.

At sites with forest edge and open areas within afforested blocks, inspect fence posts, etc, for feathers and whitewash. Check plantation edges along rides for about 100 m from the plantation/moorland boundary. Check perches such as boulders and walls in moorland near the forest edge.

On the May–July visits, alternate searching with watching from a vantage point. Look for displaying birds and males bringing food. Listen for calls. Watch crows, herons, harriers and other raptors to see if they provoke alarm calls or mobbing from merlins. Be prepared to watch for at least two hours if there is a good view over a large area.

If you find merlins on the first visit, note the location on a map and retreat to a safe distance to watch. No further searching is required within a 1-km radius of that location. On subsequent visits, return to watch from a vantage point to see whether the birds are nesting there. If breeding is not yet confirmed, return again on another occasion.

Signs of merlin presence such as moulted merlin feathers, prey remains and pellets can only be taken as indicating presence by those experienced enough to be able to distinguish merlin signs from those of

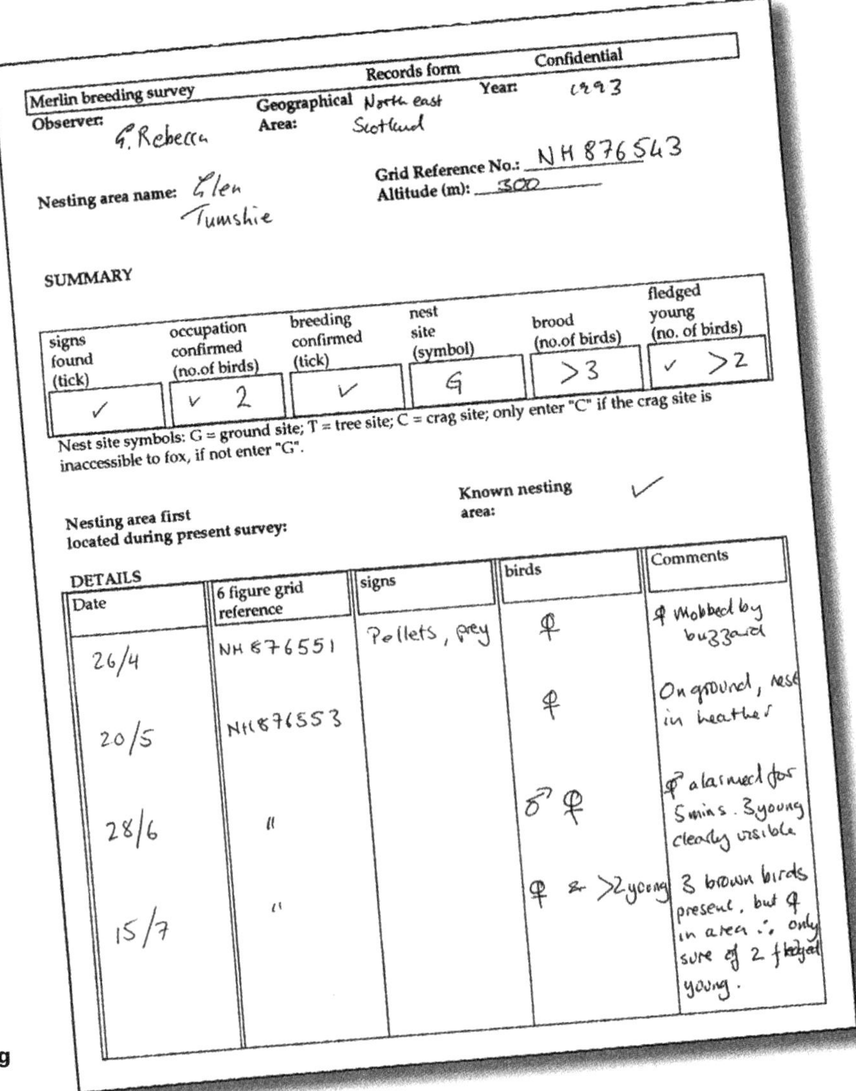

Confidential

Records form

Merlin breeding survey

Observer: G. Rebecca

Geographical Area: North east Scotland

Year: 1993

Grid Reference No.: NH 876543

Altitude (m): 300

Nesting area name: Glen Tumshie

SUMMARY

signs found (tick)	occupation confirmed (no.of birds)	breeding confirmed (tick)	nest site (symbol)	brood (no.of birds)	fledged young (no. of birds)
✓	✓ 2	✓	G	>3	✓ >2

Nest site symbols: G = ground site; T = tree site; C = crag site; only enter "C" if the crag site is inaccessible to fox, if not enter "G".

Known nesting area: ✓

Nesting area first located during present survey:

DETAILS

Date	6 figure grid reference	signs	birds	Comments
26/4	NH 876551	Pellets, prey	♀	♀ mobbed by buzzard
20/5	NH876553		♀	On ground, nest in heather
28/6	"		♂ ♀	♀ alarmed for 5mins. 3 young clearly visible
15/7	"		♀ & >2 young	3 brown birds present, but ♀ in area ∴ only sure of 2 fledged young.

Figure 1
An example of a completed recording form used in a breeding merlin survey.

other species. Experienced observers can use these signs to confirm the continued presence of merlins in an area by removing the signs and then checking to see if they are present again a week or more later.

You can confirm occupation of a new site and count the observations as evidence of a probable breeding pair if you see or hear at least one merlin, or find fresh signs (experienced observers only) on two visits separated by one week or more.

You can confirm breeding if you see an adult returning to a nest, if you find eggs or young, if the adults are repeatedly alarm calling at the appropriate time or if the signs of occurrence indicate that a pair has probably bred and failed.

Use a new map for each visit. Each map should have the areas of excluded habitat marked. Mark suspected nest-sites, observations of birds and signs of occupation on the maps. Transfer the information to the recording forms for that site (see Figure 1 for an example), using a new form for each pair. Put all merlin sightings or signs on a recording form.

If you see any chicks or fledged young, record this on the form. Do not disturb birds to get this information; it is likely that during more lengthy observations you may see young.

On the recording form there is a box in the summary section labelled 'nest-site'. Enter a letter 'G' for a ground site, 'T' for a tree site or 'C' for a crag site. Only enter 'C' if the crag site is inaccessible to foxes, if not enter 'G'.

At the end of the season, the number of confirmed breeding pairs of merlin in the survey area and the number of other registrations should be obvious from the recording forms. Draw a summary map that includes the survey boundary, the habitat searched, the habitat excluded, the locations of all breeding pairs (with the suspected nest-site also marked) and the locations of any other registrations.

References

Bibby, C J and Nattrass, M (1986) Breeding status of merlin in Britain. *British Birds* 79: 170–185.

Rebecca, G and Bainbridge, I (1994) *Survey of Merlins in Great Britain 1993–1994: survey instructions*. RSPB unpubl.

Rebecca, G and Bainbridge, I (1998) The breeding status of the merlin *Falco columbarius* in Britain in 1993–1994. *Bird Study* 45: 172–187.

Peregrine *Falco peregrinus*

Status

Amber listed: SPEC 3 (R)
Schedule 1 of WCA 1981
Annex I of EC Wild Birds Directive

National monitoring

National surveys: 1961, 1971, 1981, 1991 (BTO/JNCC and regional raptor groups), 2001.
Annual surveys of known sites (Raptor Study Groups).

Population and distribution

The peregrine breeds across much of Scotland, Northern Ireland and Wales, although in England its breeding distribution is concentrated in the north and west and on the south-west coast (*88–91 Atlas*). The peregrine breeding population was severely depleted in the 1950s and 1960s due to the effects of pesticide contamination (Crick and Ratcliffe 1995). Recently the population has recovered in many areas and the present UK breeding population is estimated at 1,285 pairs (*Population Estimates*).

Ecology

Peregrines breed mainly on coastal, moorland and mountain terrain with undisturbed cliffs and crags providing nesting sites (*Red Data Birds*). Other alternative sites used are: quarries, nests on the ground, industrial waste-tips with scarped faces, open-cast coal workings and mine excavations, tall chimneys, bridges, warehouses and low rocks on top of moorland hills. The recent recovery in peregrine numbers has led to an increase in the use of more unusual nest-sites, eg worked quarries and buildings (Crick and Ratcliffe 1995). A clutch of usually 3–4 eggs is laid from mid-March to May; there is one brood and the young fledge from the end of May (*Red Data Birds*).

Breeding season survey – population

This method is based on that used for the 1991 Peregrine survey (BTO 1991, Crick and Ratcliffe 1995).

Information required

- number and location of all potential peregrine nest-sites
- number and location of all occupied territories.

Number and timing of visits

One visit in late March. If no birds are recorded, a further visit may be necessary one month later. (This does not include visits to assess breeding success, mentioned below.)

Time of day

Any time of day.

Sites/areas to visit

All potentially suitable peregrine nesting sites within the survey area, including traditional nest-sites and all possible alternative sites.

Equipment

- 1:25,000 map of the survey area
- recording form
- Schedule 1 licence.

Safety reminders

If you are going to undertake extended coastal or mountain walks, tell someone where you are going and when you intend to return. Do not attempt to climb cliffs.

Disturbance

There is no need to get close to a nest-site when recording the breeding population in March and April; observations from a distance will suffice. Carry minimal site information, keep it confidential and be alert to any suspicious activities.

Methods

Mark the boundary of the survey area on to the map. Map all the potential peregrine nest-sites as follows:

- × Potential nest-sites located during the survey.
- *f* Sites occupied by peregrines during the survey.
- ● Sites at which peregrines bred during the survey.
- ■ Sites lost (destroyed/ developed) during the survey.

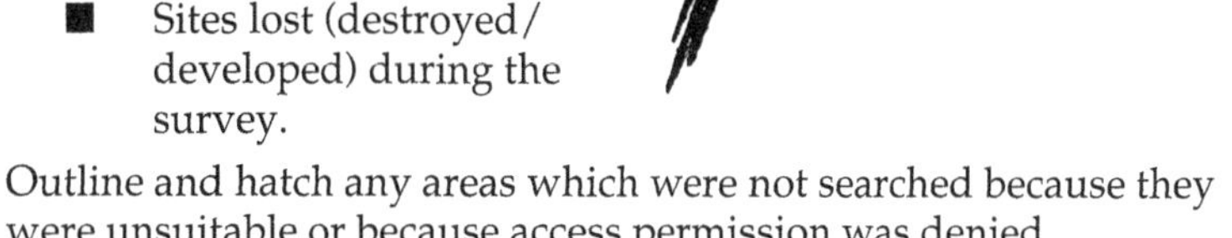

Outline and hatch any areas which were not searched because they were unsuitable or because access permission was denied.

Thoroughly search the study area for suitable nesting sites in late March. Record all potential sites, both known sites and those not previously known to have held breeding peregrines. Record the number of peregrines present at these sites. A site recording form (see example in Figure 1) should be completed for each site (including empty or unoccupied sites). Ensure that the grid references given allow cross-referencing with mapped information.

If any peregrines are seen at a potential nest-site on the first visit, consider it occupied and do not make any more visits to that site (except to check breeding success later in the season; see below). If no peregrines are seen on the first visit, make another a month later (end of April) to try to confirm occupancy.

The presence of one or more peregrines at a suitable site is regarded as evidence of territorial occupation. Report the number of potential peregrine nest-sites and the number of occupied peregrine nest-sites.

CONFIDENTIAL PEREGRINE SURVEY SITE RECORDING FORM

COUNTY:

SITE NAME:

GRID REF(6 fig):

OBSERVER:

SITE DESCRIPTION (Please tick boxes): Traditional (pre-1939) | Recent (1939-81 | New (post 1981) | Uncertain of site age

Regularly used (in most years) | Occasionally used | Not sure how often used

Brief site description: coastal: | Inland:

Main land-use in area:

Please fill in the date of the visit in the space provided

Visit summary (give numbers or tick box if unsure	Visit 1 Date:	Comments	Visit 2 Date:	Comments	Visit 3 Date:	Comments	Visit 4 Date:	Comments	Visit 5 Date:	Comments	Visit 6 Date:	Comments
Adults present												
Eggs laid												
Small young												
Large young												
Fledged young												

Alternative sites checked (dates and grid references if applicable)

Additional Comments

Please also pass on this, and more detailed habitat information, to the British Trust for Ornithology's Nest Record Scheme, write to them for record cards and more information.

Figure 1
An example of the peregrine survey site recording form used in 1991.

Breeding season survey – breeding success

Please try to contribute this and relevant habitat information to the BTO's Nest Record Scheme.

Information required

- number of fledged young or evidence of fledged young.

Number and timing of visits

At least two visits, in June.

Time of day

Any time during the day.

Sites/areas to visit

Occupied sites found in March or April

Equipment

- recording form (same as used for population survey)
- Schedule 1 licence.

Safety reminders

If you are in any doubt about the safety of getting near to a nest, watch it from a distance and do not attempt to get close.

Disturbance

Keep disturbance to an absolute minimum; some pointers on how to achieve this are given below. Remember that a Schedule 1 licence is required to approach a peregrine nest.

Methods

In June, revisit those sites found to be occupied in March and April in order to look for large young or evidence that young have fledged from the site. Wherever possible, observe the nest from a good, distant vantage point to try and obtain this evidence. If chicks are present, a further visit will be required unless the chicks are well-grown and close to fledging. If you have to visit the nest, a licence will be required and great care must be taken. Do not cause any accidental damage to the nest, as this may lead to desertion or reveal the nest to predators. It is especially important to make your initial observations from a distance because startled brooding adults can inadvertently carry small young to the edge of the nest (or even out of it) by their rapid departure.

Record the presence of any adults, eggs, or young on the site recording form.

References

BTO (1991) *Guidelines and Site Record Form for the Fourth BTO Peregrine Survey 1991*. BTO, Norfolk.

Crick, H Q P and Ratcliffe, D A (1995) The peregrine *Falco peregrinus* breeding population of the United Kingdom in 1991. *Bird Study* 42: 1–19.

Black grouse *Tetrao tetrix*

Status

Red listed: BD, HD, SPEC 3 (V)
Annex II/2 of EC Wild Birds Directive

National monitoring

National survey: 1995–96 (RSPB/GCT/JNCC/EN/SNH), 2005–06. Perthshire black grouse study group counts leks annually.

Population and distribution

The British strongholds of black grouse are Grampian, Tayside and parts of the northern Pennines, with its range extending into northern Scotland, south-west Scotland and Wales (*88–91 Atlas*). Long-term evidence from records of birds shot, and more recent evidence from counts of birds at leks undertaken by the Game Conservancy Trust (Baines and Hudson 1995), have shown a dramatic decline in the population and a contraction in the breeding range in Britain.

There are estimated to be 6,510 male black grouse in Britain (Hancock et al 1999). The main causes of the population decline are habitat loss, fragmentation and degradation (Baines and Hudson 1995). The population is fairly sedentary and the distribution in the winter is similar to that in the breeding season.

Ecology

The number of males attending leks (communal display areas) varies between sites from only one or two birds, to more than 30. Although male black grouse are present throughout the year at lek sites, peak numbers occur during the breeding season in early mornings in spring.

Lek arenas are typified by having short vegetation and good all-round visibility: in-bye pastures on moorland and woodland edges, foddering sites on short heather, patches of young heather resulting from muirburn, and tracks and clearings in young plantations are all frequently used.

Several females may nest in the vicinity of a lek and each female may lay a clutch of 6–11 eggs from late April to early June. There is usually one brood. The young can fly at two weeks old but are not fully independent for 2–3 months (*Red Data Birds*).

Breeding season survey – population

These instructions are based on those written for the 1995–96 national survey by Etheridge and Baines (1995), which were based on the studies of Robinson et al (1993) and Baines and Hudson (1995).

Information required

- maximum number of males attending each lek
- a map showing the location of each lek.

Number and timing of visits

One visit (after several preparatory visits), between the last week in March and mid-May.

Time of day

Preparatory visit(s) to locate suitable habitat can be at any time of day. Preparatory visit(s) to locate leks should be up to two hours after dawn (preferably) or in the evening, before dusk. The visit to count males at the lek should be between one hour before and one hour after sunrise.

Weather constraints

Visit(s) to locate suitable habitat can be made in almost any weather. Visits to locate leks and count males must be in good visibility, in dry and calm conditions (wind not exceeding Beaufort force 3).

Sites/areas to visit

The preferred habitats for black grouse include mosaics of moorland or heathland, woodland, plantations, rough grazing, in-bye land and meadows. They are transitional or marginal between the enclosed fields on valley slopes and the lower edges of heather moorland. These habitats correspond to a distinct altitudinal range of 200–550 m. Previous work in Perthshire (Robinson et al 1993) suggests that more than 95% of leks are located within this range.

Moorland. Within northern Britain, heather moorland, often managed for red grouse, is the main habitat for black grouse. They tend to be found on the edges of moorland from which they have access to other habitats such as scrub or woods, rough grazing and herb-rich in-bye pastures.

Native woodland. Black grouse favour two types of native woodland in the uplands: birch and birch/scots pine mixes. They prefer either small woods, woodland edges or even rows of shelterbelt trees. Open canopied woods are preferred as these allow sufficient light to reach the forest floor and create a field rich in herbs and dwarf scrubs. They avoid closed-canopy woods.

Forestry. The recent spread of afforestation in the uplands has resulted in short-term benefits for black grouse. Under relaxation from grazing and heather burning in the early stages of afforestation, heather, bilberry and scrub will form a luxuriant layer that can provide increased food and nesting cover. Black grouse numbers can thus be high in young forestry. However, the benefits are short-lived, and conditions rapidly deteriorate on canopy closure 10–15 years after planting.

Unsuitable areas. The following areas are generally unsuitable for black grouse leks and may not be occupied: ground above 550 m (600 m in England); ground below 200 m in northern England and southern Scotland (however, ground below 200 m can be occupied regularly if near to the coast, particularly in western and north-west Scotland); built-up areas; enclosed arable farmland; the interiors of unbroken post-thicket stage forest blocks and dense native woodland.

Equipment

- 1:25,000 OS map of the area
- photocopied A4 field maps
- prepared recording forms (see example Figure 1)
- telescope (if possible).

Safety reminders

Ensure someone knows where you are going and when you expect to return. Carry a compass and know how to use it. In more remote upland areas, carry extra clothing, food, a survival bag and first-aid kit. Always obtain prior permission from owners, tenants or their agents for entry to private land.

Disturbance

Do not disturb black grouse, particularly at a lek. Observe from a discreet distance. A lek can often be counted from a distance of several hundred metres (even from inside a vehicle) using a telescope.

Methods

Make sure the boundary of the survey area is marked clearly on a map. Make one or more initial visits to locate suitable habitat. Plan a route that takes you to within at least 500 m of every point. During these visits, mark on the map the areas of suitable and unsuitable habitat. Question locals (gamekeepers, agents, etc) for knowledge of potential lek sites.

Make one or more further visits to locate leks, following a planned route. Confirm the locations of leks you have been told about and note the positions of any others you find. The number of visits required to do this will depend on the size of the area and the number of leks. Consider the best way to approach each lek to avoid disturbing birds. Under relatively calm conditions on bright mornings, the distinctive

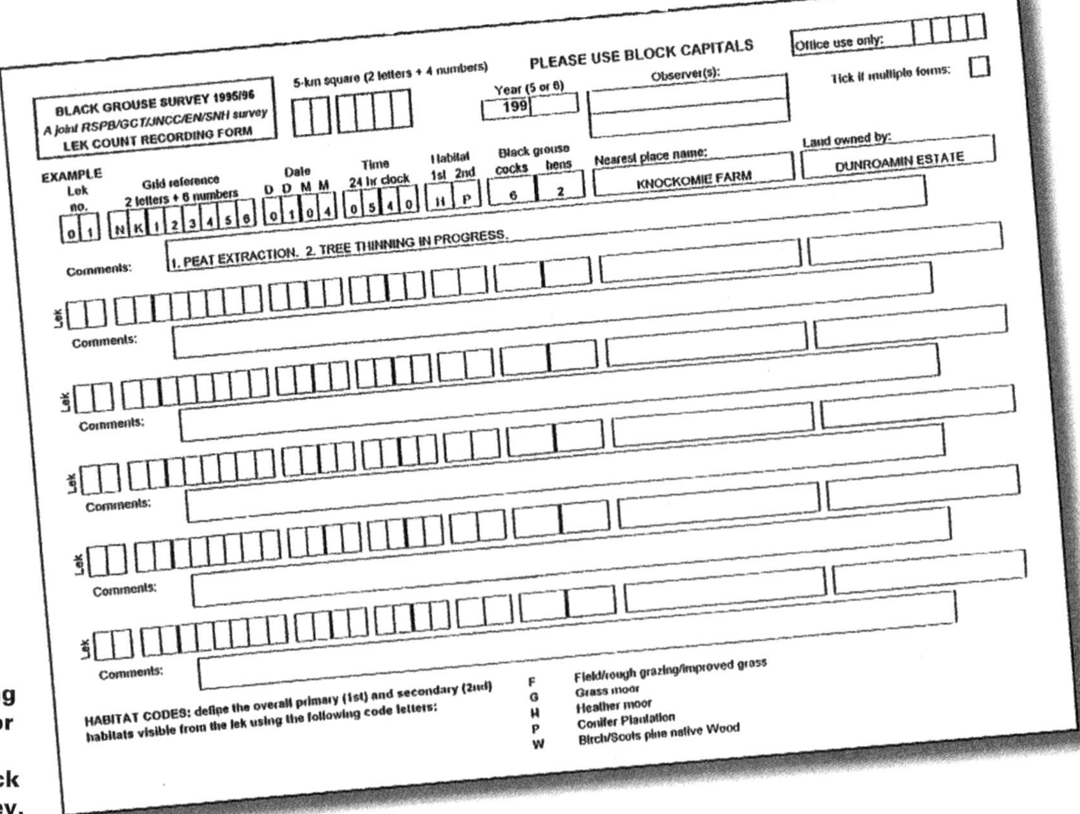

BLACK GROUSE SURVEY 1995/96
A joint RSPB/GCT/JNCC/EN/SNH survey
LEK COUNT RECORDING FORM

5-km square (2 letters + 4 numbers)

PLEASE USE BLOCK CAPITALS

Office use only:

Year (5 or 6) 199

Observer(s):

Tick if multiple forms:

EXAMPLE

Lek no.	Grid reference 2 letters + 6 numbers	Date D D M M	Time 24 hr clock	Habitat 1st 2nd	Black grouse cocks	Black grouse hens	Nearest place name:	Land owned by:
0 1	N K 1 2 3 4 5 6	0 1 0 4	0 5 4 0	H P	6	2	KNOCKOMIE FARM	DUNROAMIN ESTATE

Comments: 1. PEAT EXTRACTION. 2. TREE THINNING IN PROGRESS.

Lek
Comments:

Lek
Comments:

Lek
Comments:

Lek
Comments:

Lek
Comments:

HABITAT CODES: define the overall primary (1st) and secondary (2nd) habitats visible from the lek using the following code letters:

F Field/rough grazing/improved grass
G Grass moor
H Heather moor
P Conifer Plantation
W Birch/Scots pine native Wood

Figure 1
The recording form used for the 1995–96 national black grouse survey.

sound made by lekking males can be heard over a distance of almost 1 km. Early morning listening from a good vantage point overlooking suitable habitat can help locate leks. However, the audibility of leks varies immensely, so do not use this as a substitute for visually searching all potential ground. If you hear a lek in the distance, always confirm its exact location. Depending on the terrain, one person should be able to cover a 5-km grid square in two or three mornings.

Carry out a dawn lek count within three days of a lek being located. To avoid disturbing birds as they arrive, be in position at least an hour before sunrise, when it is still quite dark. If during initial visits to the site you are able to find a good vantage point from where you will not disturb any birds, you can be more flexible about your arrival time, as long as the count takes place within the count period.

Black grouse arrive at the lek in the half-light of dawn. Count the maximum number of males present in the period between one hour before and one hour after dawn. Count all males, not just those displaying. Not all males will be visible on the display ground at the same time, other males may be perched quietly in nearby trees or hidden by vegetation. Also count the number of females. Enter the counts on a standard recording form (see Figure 1 for an example).

Treat leks that are 200 m or more apart as separate leks. Count all displaying birds you encounter within the survey area; note that some males 'lek' on their own, often at transient, non-traditional sites. If a displaying male is separated from any other leks by 200 m or more, record it as a lek of one male.

References

Baines, D and Hudson, P J (1995) The decline of black grouse in Scotland and northern England. *Bird Study* 42: 122–131.

Etheridge, B and Baines, D (1995) *Instructions for the Black Grouse Survey 1995/6: a Joint RSPB/GCT/JNCC/SNH Project.* Unpublished.

Hancock, M, Baines, D, Gibbons, D, Etheridge , B and Shepherd, M (1999) The status of male black grouse *Tetrao tetrix* in Britain in 1995–6. *Bird Study* in press.

Robinson, M C, Baines, D and Mattingley, W (1993) A survey of black grouse leks in Perthshire. *Scottish Birds* 17: 20–26.

Capercaillie *Tetrao urogallus*

Status

Red listed: BD
Non-SPEC

National monitoring

National surveys: 1992–93 (RSPB, ITE and Game Conservancy Trust – Pinewoods Birds Project), 1998–99.

Population and distribution

The capercaillie became extinct in Britain in the mid-18th century but was successfully re-introduced to Scotland in the mid-19th century. After becoming more common in the 20th century, the numbers of breeding capercaillie declined dramatically and its range has contracted in recent years. It is now found from central Scotland to the Dornoch Firth. The decline in numbers and range has been attributed in part to habitat loss and overhunting, but also to increased mortality from birds flying into deer fences, decreased productivity as a result of a run of poor summers and, perhaps, increased predator numbers. The current breeding population is estimated to be 2,200 birds (Catt et al 1994).

Ecology

The capercaillie is a bird of open mature pinewoods on hills and valleys with an undergrowth of heather (*Calluna vulgaris* and *Erica* spp) and bilberry *Vaccinium myrtilus*. They are solitary birds, but males will gather at traditional sites to display (lek) to females. Eggs are laid mid-April to early July, there is a single brood and the young fledge by mid-June to late July (*Red Data Birds*).

Breeding season survey – population and breeding success

The breeding population is counted at the same time as breeding success is estimated.

Information required

- total number of hens
- average number of chicks per hen
- average brood size.

Number and timing of visits

One visit in late July or early August.

Time of day

Any time of day.

Weather constraints

Avoid cold or wet conditions.

Sites/areas to visit

All areas known or suspected to hold capercaillie. Capercaillie are usually found in mature, open, semi-natural pinewoods with extensive heather and bilberry/blaeberry. Populations are sometimes found in other types of woodland such as mature Scots pine and other plantations, especially non-native pinewoods which have wind-blown areas and equivalent undergrowth.

Equipment

- trained dogs (pointers or setters) and a trained dog-handler – it may be necessary to hire these
- compass
- 1:25,000 map of the area to be visited
- prepared recording form.

Safety reminders

Nothing specific, see *General health and safety advice* in the *Introduction*.

Disturbance

Inevitable and unavoidable. However, there is no reason to suspect that the disturbance caused by the survey method has any effect on numbers or breeding success. Breeding capercaillie are not monitored at the nest because of the risk of desertion.

Methods

Use a team of people walking with dogs in a line. The dogs locate any chicks by scent. In July the chicks, which have hatched in early June, will be big enough for dogs to find easily. In early August, the broods should still be together. Cover the survey area systematically.

Record the total number of chicks and hens (including hens with no chicks). Use these figures to calculate the average number of chicks per hen. Calculate the average brood size as the number of chicks found divided by the number of hens with at least one chick.

Winter survey

There are two methods:
1. Transect counts – less disruptive and require fewer participants.
2. Winter drives – have been used at a number of sites for many years.

1. Transect counts

For a detailed method used by the RSPB over a number of winters – the pinewood bird survey – see the generic survey methods section.

2. Winter drives

Information required
- total number and density of birds of both sexes within set areas counted annually.

Number and timing of visits
One per annum per count area, between November and March.

Time of day
0900–1600 BST.

Weather constraints
None.

Sites/areas to visit
As for population survey (above).

Equipment
- 1:25,000 map of the area
- compass
- whistles, notebooks and walking sticks.

Safety reminders, Disturbance
As for population survey.

Methods
This method is mainly used for annual comparisons in areas which are counted this way every year. Up to 70 people can be involved.

The size of the sections to be counted varies according to the site's topography and the number of personnel available. A line of beaters should walk through the section of woodland to be counted, walking a few metres apart and maintaining a slightly curved line with left and right edges a little further forward. They should make much noise by shouting, clapping, using whistles and tapping trees with sticks as they walk through the section being counted. End beaters should be alert to flushed birds flying out and back round into areas already counted.

A line of counters should stand at the opposite end of the section being counted, spaced in such a way as to count all birds being 'driven' from the wood. The counters should be in position before the beaters start. Each counter should record any birds flying over their right shoulder noting the number of males, the number of females and the time. At the end of the 'drive' one person should collate all the records.

Reference

Catt, D C, Baines, D, Moss, R, Leckie, F and Picozzi, N (1994) *Abundance and Distribution of Capercaillie in Scotland 1992–1993.* Final Report to SNH and RSPB from ITE and Game Conservancy Trust.

Grey partridge
Perdix perdix

Status

Red listed: BD, SPEC 3 (V)
Annex II/1 of EC Wild Birds Directive

National monitoring

CBC, BBS.
Annual March pair counts and August brood counts (Game Conservancy Trust).

Population and distribution

Grey partridges are distributed over most of Great Britain except west and north Scotland and west Wales; the Irish population is very small. They breed on farmland, particularly cereals and pasture, but are also found less usually in habitats such as fringes of rush-covered moorland. Grey partridge numbers have generally decreased over the last 40 years (*88–91 Atlas*). The main causes of the decline are threefold: reduced chick survival because of reduced insect availability through increased use of herbicides, loss of nesting habitat, and increased predation during the breeding season (Potts 1986). The number of grey partridges in Great Britain is estimated at 140,000–150,000 pairs (*88–91 Atlas*).

Ecology

By mid-March birds should have paired. They are difficult to see except at dawn and dusk when they come out into the open to feed. Nests are on the ground, often at the bottom of a hedge at the edge of a field. Clutches of about 15 eggs are laid from the end of April through to August. Young hatch from mid-June onwards and fly after 10–11 days. Grey partridges are single-brooded though repeat clutches are common (*Red Data Birds*).

Breeding season survey – population

This method is equivalent to the 'March pair count' organised by the Game Conservancy Trust for the last 65 years (Potts 1986, Karen Blake pers comm, Anon 1996). For recording forms and instructions, contact the Partridge Count Co-ordinator, Game Conservancy Trust, Loddington House, Fordingbridge, Hampshire SP6 1EF.

Information required
- total number of pairs
- total number of single birds
- map of the locations of the pairs and single birds.

Number and timing of visits

One visit in mid-March.

Time of day

During the two hours after dawn and/or the two hours before sunset.

Weather constraints

Visit in calm and, preferably, dry conditions.

Sites/areas to visit

All habitats except for woodland older than five years, particularly open farmland with very few trees, field margins, arable crops and grass leys.

Equipment

- 1:10,000 OS field maps of the survey area
- 4WD vehicle.

Safety reminders

Nothing specific. See general guidelines in the *Introduction.*

Disturbance

You may need to flush partridges in dense grassland and set-aside. Always get permission from all landowners before gaining access to private land. Never enter crop fields without first obtaining permission.

Methods

Map the boundary of the area. Survey all suitable grey partridge feeding areas equally, even if this means continuing in the evening or the following morning. It is acceptable to report the combined counts from morning and evening surveys.

Survey fields from suitable vantage points, preferably one that is elevated. It may be possible to survey from a vehicle. Scan all fields and field margins systematically with binoculars. The birds should show up well on recently sown fields. In well grown grass leys and in set-aside in which birds cannot be seen, criss-cross the field to get a closer look. If birds are flushed, make a note of where they land and try to avoid double-counting. Plot on a map all pairs and single birds seen. Use a number to denote number of pairs seen and an X to denote single individuals. Calculate separately, from the completed maps, the total number of pairs and the total number of single individuals seen.

Breeding season survey - breeding success

This method is equivalent to the 'August brood counts' organised by the Game Conservancy Trust for the last 65 years (Potts 1986, Anon 1996). Contact address as for the population survey.

Information required

For each covey (family group):
- number of pairs
- number of single birds
- the sex of adults
- presence of a brood
- number of young per brood
- map showing the location of the birds.

Number and timing of visits

One visit per field, between the end of July and mid-September, as soon as possible after harvest.

Time of day, Weather constraints, Sites/areas to visit, Equipment, Safety reminders

As for the population survey (see above).

Disturbance

Should not be a problem using this survey method.

Methods

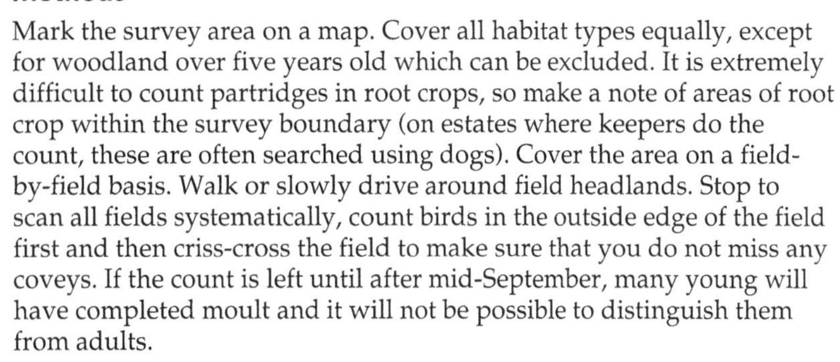

Mark the survey area on a map. Cover all habitat types equally, except for woodland over five years old which can be excluded. It is extremely difficult to count partridges in root crops, so make a note of areas of root crop within the survey boundary (on estates where keepers do the count, these are often searched using dogs). Cover the area on a field-by-field basis. Walk or slowly drive around field headlands. Stop to scan all fields systematically, count birds in the outside edge of the field first and then criss-cross the field to make sure that you do not miss any coveys. If the count is left until after mid-September, many young will have completed moult and it will not be possible to distinguish them from adults.

Pairs of grey partridge without young, as well as successful pairs, will be seen feeding on stubbles or other patches of open ground. Distinguish males from females and young birds from old; if necessary, use illustrations (see Potts 1986). Ageing should be possible even when young are 9–10 weeks old.

Record any pairs or single birds seen without broods as well as pairs seen with broods. For each covey, record the number of adult males, number of adult females and the number of young. There are often more than two older birds in each covey when young birds are present. Small coveys frequently comprise older birds only.

Plot the position of each covey on the map. For example: gp♂♂♀ + 10, corresponds to grey partridge – 2 males, 1 female, 10 young. If you cannot be sure of the number of birds in a covey (eg some run into cover), note this on the map and when quoting the final figures.

Calculate the final figures from the map and report the number of adult males and females and the number of young in each covey, eg as follows:

Covey number	Adult male	Adult female	No. of young
1	1	1	7
2	1	1	4 *incomplete*
3	1	1	
4	2	1	9

In upland areas, grey partridges are best counted using trained dogs to locate coveys on moorland fringes.

References

Anon (1996) *The Annual Partridge Count Scheme: August brood count recording form and instructions*. Game Conservancy Trust, Fordingbridge.

Potts, G R (1986) *The Partridge: pesticides, predation and conservation*. Collins, London.

Quail
Coturnix coturnix

Status

Red listed: HD, SPEC 3 (V)
Schedule 1 of WCA 1981
Annex II/2 of EC Wild Birds Directive

National monitoring

Rare Breeding Birds Panel.

Population and distribution

Quail were common in the UK up to the beginning of the twentieth century but now are relatively scarce. They arrive in waves during the spring and summer, their numbers varying considerably between years (*88–91 Atlas*). Some quails do not reach the UK until the second half of July (these are probably birds born in Africa or southern Europe in April/May of the same year). Their UK distribution varies between years, although their strongholds are in Wiltshire and Dorset and they are absent from upland areas (*88–91 Atlas*). There were an estimated 2,600 calling males in Britain in 1989, but this was an influx year and was particularly well covered by county recorders. Usually, there are records of about 300 calling males in Britain (*88–91 Atlas*).

Ecology

In the UK, quails breed in cereal and hay fields, particularly winter cereals and meadow-like wild grasslands. As with some other game birds, crop structures which allow good movement, provide protection from avian predators and are a source of insect food are most likely to be associated with successful breeding (Potts 1986, Stoate 1989). Many males probably remain unmated in the UK. However, recording singing males is the only practical way of censusing this elusive species, therefore the survey method given below provides an index of numbers present, not an accurate estimate of the number of potential breeding pairs (Moreau 1951). The advertising song of the male is a staccato, far-carrying (but difficult to locate) 'whIC whic-whIC', given usually 3–10 times in close succession (*BWP*).

Breeding season survey – population

Information required

- maximum number of singing males heard on any one of six visits
- a map of the survey site with the boundary clearly marked.

Number and timing of visits

Six visits, roughly a fortnight apart, from mid-May to the end of July.

Time of day

At dusk, between about half an hour before and an hour after sunset.

Weather constraints

Avoid rainy and/or windy conditions (greater than Beaufort force 3).

Sites/areas to visit

Cereal fields and meadow-like grasslands, eg hay fields.

Equipment

- 1:10,000 OS field maps of the survey area, detailed enough to show field boundaries
- compass
- torch
- Schedule 1 licence.

Safety reminders

Areas to be visited at night should first be visited during the day. Ensure someone knows where you are and when you are due back. Always seek permission for entry to private land, especially crop fields (avoid walking through the crop), and always warn the police that you propose to drive around and survey at night. Always carry a compass, map and torch.

Disturbance

Avoid disturbing singing birds. Approach only as close as is necessary to pinpoint singing positions to within 100 m. If possible, remain at the field edge or on the path/road.

Methods

Mark the boundaries of the survey area on a map. If you are unfamiliar with the area, visit it by day to plan a route that will take you to within 250 m of all potential quail habitat. Also mark on the map 'listening points', distributed so that you can listen from within 250 m of all potential quail habitat. Even if you know the area well, plan the route carefully in advance. Listen for calling birds for at least five minutes at each 'listening point'. It may be possible to survey most of the area by car from the public road. When surveying by car, make sure you get out of the car and switch off the engine to listen. Map the location of singing males to within 100 m. If you find it difficult to pinpoint a bird with this degree of accuracy, try listening to it from several different points.

References

Moreau, R E (1951) The British status of the quail and some problems of its biology. *British Birds* 44: 257–276.

Potts, G R (1986) *The Partridge: pesticides, predation and conservation.* Collins, London.

Stoate, C (1989) Some observations on habitat and censusing of territorial quail *Coturnix coturnix. Hampshire Bird Report*: 77–78.

Water rail
Rallus aquaticus

Status

Amber listed: BDM
Non-SPEC
Annex II/2 of EC Wild Birds Directive

National monitoring

Breeding surveys: Wales 1996.
WeBS.

Population and distribution

Water rails breed throughout the UK but accurate censusing of this elusive species is difficult. In 1968–72, the UK breeding population was estimated at 2,000–4,000 pairs (*68–72 Atlas*), and in 1988–91 it was estimated at 1,300–2,400 pairs (*88–91 Atlas*). However, the difference in mean population size is more likely to be related to difficulties in censusing this species than to a real decline in the population. There is a need for more accurate monitoring of this species to ascertain whether numbers are in decline.

Ecology

Water rails breed in areas that include tall emergent vegetation, normally reeds, sedges or rushes. In the UK, incubation starts in the last week of March (*BWP*). This will be later in Scotland, but it is not known how laying dates vary north to south. In winter, Britain and Ireland receive birds from north and east Europe (Flegg and Glue 1973). Migrants arrive from September, set up winter territories and some may still be present in the following March (*BWP*). The 'sharming' call given by the male and female is likened to the sound of piglets squealing. The male and female of a pair regularly answer one another with this call, alternating in a duet, both at close range or over considerable distances (*BWP*).

Breeding season survey – population

The method given here is based on that devised by Moyes and Robertson (1991).

Information required

- minimum number of breeding pairs
- a map showing the boundary of the survey area and location of calling birds.

Number and timing of visits

One visit, late March (southern England) to late April (northern Scotland).

Time of day

Early morning, after sunrise.

Weather constraints

Avoid days with winds stronger than Beaufort force 3.

Sites/areas to visit

Sites with standing or slow running fresh water, some flat muddy ground and dense tall aquatic vegetation, which is not too prone to flooding.

Equipment

Essential

- 1:10,000 OS map
- A4 photocopied map of survey area for use in the field
- A4 photocopied map of survey area for compiling final visit map
- compass
- high-quality continuous loop-tape recording of a pair of water rails giving territorial calls
- moderately powerful portable cassette player.

Useful

- 1:25,000 OS map, for locating rights of way
- waders, for walking through wetter areas
- long stick
- boat, if water too deep to walk through.

Safety reminders

Always inform someone where you are working and what time you expect to return. Take extra care when working close to water; if any boat trips are necessary, wear life-jackets.

Disturbance

Use existing paths, banks or causeways (if on foot) or dykes (if travelling by boat) as much as possible. Avoid areas containing rare breeding birds such as bitterns and marsh harriers (access will probably be restricted anyway). The use of playback for this survey method is purely to elicit a vocal response from birds, and not to lure birds from one area to another. Follow the instructions for playback carefully.

Methods

At least two people are required for this survey method. Complete the survey in the shortest possible time to avoid double-counting birds. To ensure quicker coverage for large areas that require more than a few days surveying, you may need more than one team of people. In this situation, make sure that all teams use similar equipment and work in similar weather conditions.

Map the site and split large areas into manageable sections of up to about 40 ha each. Familiarise yourself with the site beforehand. Work out the best route through each section or site which will take you to within 200 m of all available habitat within the site.

Water rails respond to the 'sharming' call made by both male and female. Stand about 20 m apart to assess more accurately the location of any responding birds. One person should hold the playback equipment and both should have maps. Play the tape for 60 seconds, then switch it off. Remain stationary and quiet, listening for 60 seconds. If there is no response, switch the tape on for a further 30 seconds. Stand still and listen for 30 seconds. If there is still no response, assume that there are no water rails present in the area, and move on 100 m as quickly as you can and repeat the playback procedure. Moving quickly is important to prevent birds following the tape or being counted more than once.

Both observers should mark on their map the location of any birds that respond, distinguishing between pairs and single birds. Established pairs will generally respond more readily and move quickly towards each other while calling. When the pair meet they will call together and continue to do so for some time after the observer has moved away. Single birds are more reluctant to respond, and where there is a response it will be brief. If undecided, make extra notes cross-referenced with mapped symbols to resolve later. Combine the maps from both observers into a final visit map as soon as each survey section or site is finished. Draw putative boundaries around registrations thought to be from the same pair.

Estimate separately the number of breeding water rail pairs and the number of single birds. If there is any doubt about the number of pairs responding to the tape, always report the lowest figure – double-counting of the same pair is more likely than pairs being overlooked.

References

Flegg, J J, Flegg, M and Glue, D E (1973) A water rail study. *Bird Study* 20: 69–79.

Moyes, S B and Robertson, D (1991) *A Census of Breeding Water Rails in the Tay Reedbed and the Effects of Reedbed Management on Breeding Densities, 1991*. Rep. from Tay Ringing Group to Nature Conservancy Council for Scotland, South East Region.

Spotted crake
Porzana porzana

Status

Amber listed: BDM, BR
SPEC 4 (S)
Schedule 1 of WCA 1981
Annex I of EC Wild Birds Directive

National monitoring

Rare Breeding Birds Panel.
WeBS.

Population and distribution

Spotted crakes are rare breeding birds in the UK. They inhabit fens and larger wetlands and before the extensive drainage of wetlands in the 18th and 19th centuries, it is thought that they were much more common (*88–91 Atlas*). Between one and 20 pairs of spotted crake currently breed in Britain (*Population Estimates*).

Ecology

The spotted crake breeds in swamps and fens with a high water table. The advertising call of the male is a loud, nasal, 'quek', likened to the sound of a whiplash or water dripping into an empty barrel. The call is usually heard at night, is given repetitively and can continue until after dawn. In calm weather, this call is sometimes audible up to 5 km away (*BWP*). Spotted crakes are monogamous and once a male has mated he will no longer call (N Schaffer pers comm). Eggs are laid mid-May to early August and the last young will fledge by mid-September (*Red Data Birds*).

Breeding season survey – population

The following survey method will only give an indication of population size, rather than the actual number of breeding pairs.

Information required

- maximum number of birds heard calling on any one visit
- maximum number of different birds heard across all visits
- a map showing the boundary of the survey area and location of calling birds.

Number and timing of visits

Three visits, one in the first half of May, the second in the second half of May, and the third in the first half of June.

Time of day

From half an hour after sunset up to 0200.

Weather constraints

Avoid cold, wet and windy conditions.

Sites/areas to visit

Carex beds, *Juncus*/*Scirpus* marsh, *Glyceria* beds, wet grassland and occasionally reedbeds.

Equipment

- 1:25,000 OS map of the area to be visited
- Schedule 1 licence.

Safety reminders

Always inform someone where you are working and what time you expect to return. Take extra care when working close to water and, if any boat trips are necessary, make sure you wear life-jackets.

Disturbance

Use existing paths, banks or causeways (if on foot) or dykes (if travelling by boat) as much as possible.

Methods

Establish enough listening points for the listener to be within 500 m of any suitable habitat. Listen for five minutes at each point and record as accurately as possible the position of any calling bird on a map and the time at which it was heard. The far-carrying call may be heard from more than one listening post, so double-counting is a potential problem. If you are unsure of a crake's location, revisit some of your listening points.

Report the maximum number of birds heard calling on any one visit, and the maximum number of different birds heard across all visits. Calculate the latter by assuming that birds recorded 500 m or more apart on different nights are separate individuals.

Corncrake
Crex crex

Status

Red listed: BD, HD, BL, SPEC 1 (V)
Schedule 1 of WCA 1981
Annex I of EC Wild Birds Directive

National monitoring

Annual censuses of the main UK areas since 1991 (RSPB).
National surveys: 1988, 1993, 1998, 2003.
Rare Breeding Birds Panel.

Population and distribution

The corncrake is one of three globally threatened birds in Britain. Its distribution within Britain and Ireland has contracted towards the north-west (Cadbury 1980, Hudson et al 1990). The decline in population and the contraction in range of this species are thought mainly to be linked to the intensification of farming practices (Stowe et al 1993). Corncrakes require vegetation cover and can be found in a variety of vegetation types although they typically breed in hay meadows and silage fields. In 1993, there were 480 calling male corncrakes in Britain (Green 1995). About 90% of potential breeding areas are monitored by the RSPB and Birdwatch Ireland annually.

Ecology

Male corncrakes sing to attract females. They can sing from late April until the end of August. Once males have mated they sing less until the female starts incubating, after which they sing more again.

Breeding season survey – population

In areas of Britain and Ireland where the RSPB and Birdwatch Ireland survey corncrakes annually, surveys are closely linked to corncrake initiative (grant) schemes. These survey instructions therefore incorporate some measures that are essential for corncrake protection.

Information required
- estimated number of males
- maps showing locations of all craking males

Number and timing of visits

A minimum of two, between 20 May and 30 June.

Time of day

0000–0300 BST.

Weather constraints

Avoid conditions that are windy (over 3 on the Beaufort scale) and wet.

Sites/areas to visit

Meadows and grassland but also gardens, nettlebeds, crops and any vegetation cover taller than 20 cm.

Equipment

- 1:10,000 OS map of the survey area
- Schedule 1 licence
- a compass with an internal light
- a clipboard and headtorch
- warm clothing.

Safety reminders

Visit the areas to be visited at night first during the day. Ensure someone knows where you are and when you are due back. Always carry a compass and know how to use it. Always obtain permission to enter private land, especially crop fields (avoid walking through the crop). Warn police that you propose to drive around in the middle of the night.

Disturbance

Do not disturb corncrakes while counting. It is not necessary to get closer than 100 m to pinpoint a male's singing position. If he stops singing you may be too close – stand still and make no noise until he starts singing again, then walk slowly away. Avoid flushing females by using field edges and paths to pinpoint bird positions. Never use playback of tapes, etc, to try to get a bird to crake.

Methods

Visit the survey area by day to plan a route that will take you within 500 m of all potential corncrake habitat. It may be necessary to reduce this to 200 m if windy conditions are unavoidable (eg Malin Head and the Uists). Mark suitable stopping places on the map. Note any features that will be identifiable at night (eg signs) close to stopping places and make sure you know where all the stopping places are on the map. Even if you know the area well, plan the route carefully in advance.

It is possible to survey many areas by car from public roads. At each stopping place, turn off the engine and get out of the car. Spend at least 1–2 minutes listening. If a corncrake happens to be near the road, it may stop singing when the car engine is turned off. If your car windows are down you will hear this and should wait to confirm the presence of the bird.

The position of more distant corncrakes can be assessed by cupping your hands behind your ears and scanning around. If you hear a corncrake, work out where it is by triangulation. Ensure this is accurate to within 100 m. If the landscape is such that the positions of singing birds need to be estimated over longer distances, use a compass with an internal light to improve accuracy. In the dark it is often difficult to relate actual positions to mapped positions. By marking your own position at night with a marker (eg tied to a fence or tree) and noting the relative position of the bird, you can return when it is light to map its position more accurately (all markers must later be removed).

Note that the strength of the sound varies depending on the direction the bird is facing. Beware of corncrakes calling near objects which reflect sound such as walls and buildings. This can cause you to misjudge distance and direction, or to record two or more birds singing where there is only one.

To avoid confusion, give each bird a unique code starting with the 10-km square reference, eg NB01/1, NB01/2, etc. It is important to list the 1-km squares searched during each visit in the specified night-time period, regardless of whether you heard anything. Note the time and the number of listening stops.

Update your records immediately after your last visit to a site, while the details are fresh in your mind. As well as mapped information, record the census date, time, grid reference (letters plus six numbers) and location name for every bird on each visit. Surveys organised by the RSPB record such information on standard forms (Figure 1). Note whether you saw or heard the bird and anything distinctive about the way it called (eg speed or pitch). Additional records of males singing during the day within the census season, young seen or nests found should also be noted. It does not matter if you survey different parts of the area on different dates.

When interpreting the results, it is sometimes not clear if a bird heard calling from different places on two visits is the same or a different bird. If the second location is within 200 m of the first, then it is almost certainly the same bird. If the second location is separated by 200 m or more, then they should be recorded as two separate birds unless (a) there is something distinctive about the call that tells you they are

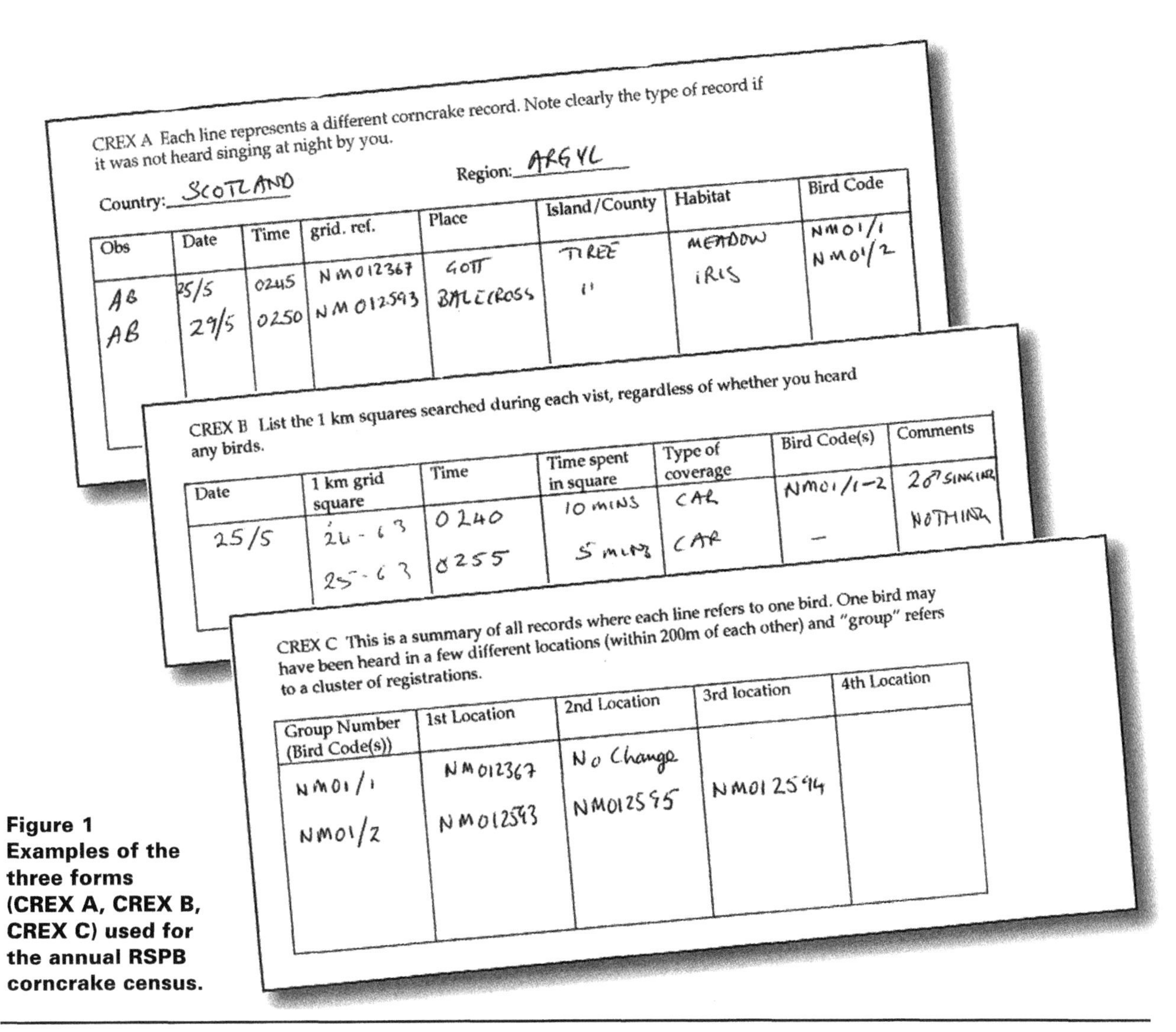

CREX A Each line represents a different corncrake record. Note clearly the type of record if it was not heard singing at night by you.

Country: SCOTLAND Region: ARGYL

Obs	Date	Time	grid. ref.	Place	Island/County	Habitat	Bird Code
AB	25/5	0245	NM012367	GOTT	TIREE	MEADOW	NM01/1
AB	29/5	0250	NM012593	BALLCROSS	"	IRIS	NM01/2

CREX B List the 1 km squares searched during each vist, regardless of whether you heard any birds.

Date	1 km grid square	Time	Time spent in square	Type of coverage	Bird Code(s)	Comments
25/5	24-63	0240	10 MINS	CAR	NM01/1-2	2♂ SINGING
	25-63	0255	5 MINS	CAR	–	NOTHING

CREX C This is a summary of all records where each line refers to one bird. One bird may have been heard in a few different locations (within 200m of each other) and "group" refers to a cluster of registrations.

Group Number (Bird Code(s))	1st Location	2nd Location	3rd location	4th Location
NM01/1	NM012367	No Change		
NM01/2	NM012593	NM012595	NM012594	

Figure 1 Examples of the three forms (CREX A, CREX B, CREX C) used for the annual RSPB corncrake census.

probably the same bird, (b) something has happened to the habitat at the first recorded location (eg it has been mowed) and the new location is the nearest place to which the bird could go or (c) local people who seem reliable tell you that they often listen to corncrakes and have not heard calling from both places at the same time.

Include records of nests or flightless young in the totals if they are more than 200 m from the location of a singing male detected in the census period. Include records of singing males from outside the census period if this is the only information available from a remote area where corncrakes are thought to breed, eg an offshore island.

Breeding season survey – productivity

These methods are based on those used as part of the RSPB/Birdwatch Ireland corncrake initiative.

Information required

- number of corncrake nests found in searches after mowing
- details of all areas searched
- number of corncrake chicks seen during mowing
- the number of chicks killed during mowing
- details of all mowing watched.

Number and timing of visits, Time of day

Depends on date and timing of mowing by individual farmers.

Weather constraints

As for the population survey (above).

Sites/area to visit

All fields that are within a 250-m radius of a singing male corncrake that has been present at the same site for a week or more.

Equipment

As for the population survey.

Safety reminders

Do not stand close to tractors as they mow: there is a real risk of being hit by stones or detached mower blades. Do not watch for birds from the tractor.

Disturbance

Corncrakes will be reluctant to come out of a field being mown and you may inadvertently scare them back in. Stand in a position that will avoid this (see Figure 3). However, if a bird appears not to be escaping to safety, try chasing it out of the field.

Methods

One of the conditions of the corncrake initiative scheme is that the landowner must inform an RSPB or Birdwatch Ireland fieldworker when they will be mowing their eligible fields so that someone from one of the conservation organisations can be present.

Record the grid reference of the centre of the field, the field dimensions in metres (pace it out), the type of mower used, and the mowing method (eg from the outside of the field inwards or from the centre of the field outwards). An example of a recording form is shown in Figure 2.

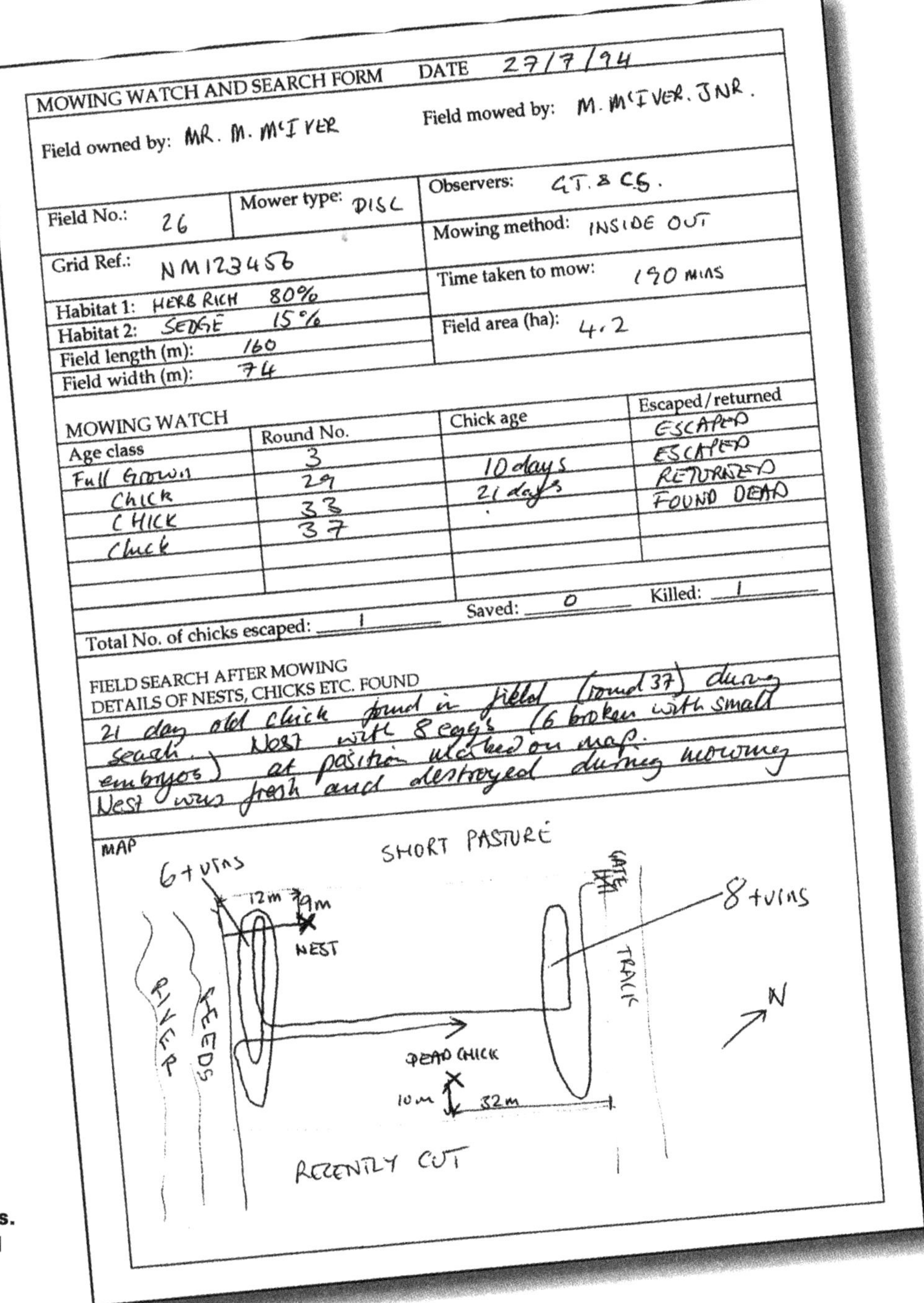

MOWING WATCH AND SEARCH FORM DATE 27/7/94

Field owned by: MR. M. McIVER

Field mowed by: M. McIVER. JNR.

Field No.: 26

Mower type: DISC

Observers: GT. & CG.

Grid Ref.: NM123456

Mowing method: INSIDE OUT

Habitat 1: HERB RICH 80%

Habitat 2: SEDGE 15%

Time taken to mow: 190 MINS

Field area (ha): 4.2

Field length (m): 160

Field width (m): 74

MOWING WATCH

Age class	Round No.	Chick age	Escaped/returned
Full Grown	3		ESCAPED
Chick	29	10 days	ESCAPED
CHICK	33	21 days	RETURNED
Chick	37		FOUND DEAD

Total No. of chicks escaped: 1 Saved: 0 Killed: 1

FIELD SEARCH AFTER MOWING
DETAILS OF NESTS, CHICKS ETC. FOUND

21 day old chick found in field (round 37) during search. Nest with 8 eggs (6 broken with small embryos) at position marked on map. Nest was fresh and destroyed during mowing

MAP

Figure 2
An example of the recording form used during mowing watches and searches. These forms are used during RSPB annual surveys.

Ideally, two people should be present at each mowing attempt, although a lone observer could be helped by the tractor driver. Figure 3 shows the positions in which observers should stand to watch fields being mown. In those being mown from the centre outwards, it is important to watch the edges of fields where they adjoin areas of short

OUTSIDE IN

Tractor

Observers watch cut edge opposite tractor, and move in as field is mown

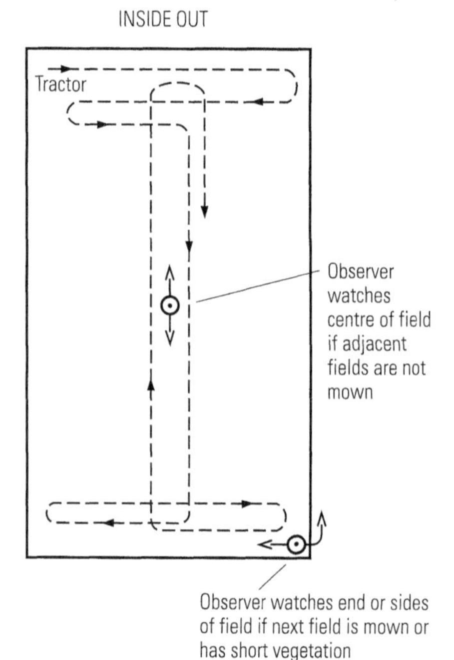

Figure 3
Recommended positions for observers (preferably two) and the directions in which they should look, during mowing procedures – with the tractor moving from the outside in and from the inside out.

vegetation as adults and chicks may appear there. If there is no adjoining short vegetation (or cut edge), one person should watch from the centre of the field and one should watch and listen for chicks or adults at the edge. If a field is being mown from the outside inwards, watch the cut edge of the grass opposite where the mower is cutting, as most birds will flee away from the mower (see Figure 3).

Every time birds are seen, even if it may mean that the same birds are recorded more than once, record the number of adults and chicks and the number of times the mower had travelled round the field when they were seen. If possible, indicate which records probably refer to the same individual(s). Estimate the age of birds running out of the crop by using the 'chick chart' (available from RSPB). Note whether the bird returns to the field or whether it escapes.

Immediately after a field has been mown, search for the remains of nests or adults and chicks that might have been killed during mowing. If the field is left for any length of time (hours) scavengers and predators will remove the evidence. Check the field by walking between the rows of cut grass looking between the rows and in the cut grass. It is not feasible to turn all the cut grass while searching for remains, but if you see a clue, such as possible nest material or eggshells, turn the grass to check for further signs. Dead chicks can be aged using the 'chick chart' to provide an estimate of hatching dates, and destroyed nests can provide information on clutch size.

Whenever destroyed nests are found, record whether they contain no egg remains, shell fragments from hatched eggs (usually have papery membranes attached to the shell), shells (indicating predation) or freshly broken eggs (destroyed by mowing), etc. Make detailed notes of the appearance of the remains. If embryos are present, record their stage of development by noting their size and the amount of down feathers. Record the approximate age of any dead chicks on a sketch map of the field. Record the position in metres (pace them) from the two nearest sides of the field, this can later be related to mowing procedure.

References

Cadbury, C J (1980) The status and habitats of the corncrake in Britain 1978–79. *Bird Study*, 27: 203–218.

Green, R E (1995) The decline of the corncrake *Crex crex* in Britain continues. *Bird Study* 42: 66–75.

Hudson, A V, Stowe, T J and Aspinall, S J (1990) Status and distribution of corncrakes in Britain in 1988. *British Birds* 83: 173–187.

Stowe, T J, Newton, A V, Green, R E and Mayes, E (1993) The decline of the corncrake *Crex crex* in Britain and Ireland in relation to habitat. *J. Appl. Ecol.* 30: 53–62.

Oystercatcher *Haematopus ostralegus*

Status

Amber listed: WI, WL
Non-SPEC
Annex II/2 of EC Wild Birds Directive

National monitoring

BBS.
WeBS.

Population and distribution

Oystercatchers breed throughout most of mid and northern England, Scotland and around the coasts of Wales, Ireland and southern England. These are adaptable birds which can occupy diverse inland as well as coastal habitats (*88–91 Atlas*). The species has increased in numbers over the last 30 years: an estimated 34,000–44,000 pairs breed, and 378,000 birds winter in the UK (*Population Estimates*).

Ecology

Coastal oystercatchers nest on shingle beaches, dunes, saltmarshes and rocky shores including cliffs, skerries and vegetated tops of islands. Inland nests are mainly on riverine shingle beds, lake shores and in fields, especially arable (*88–91 Atlas*). A clutch of usually three eggs is laid between mid-April and mid-July, and there is a single brood. Outside the breeding season, the oystercatcher is associated with predominantly sandy estuaries, beds of cockles and mussels, rocky shores and coastal pastures.

Breeding season survey – population

Details of breeding wader survey methods are in the generic survey methods section.

Winter survey

WeBS.

See *Generic wintering bird monitoring methods* in the generic survey methods section.

Avocet
Recurvirostra avosetta

Status

Amber listed: BL, WL, SPEC 3^{W}(L^{W})
Schedule 1 of WCA 1981
Annex I of EC Wild Birds Directive

National monitoring

Rare Breeding Birds Panel.
WeBS.

Population and distribution

Avocets became extinct as a breeding species in Britain in the mid-19th century, eventually recolonising Norfolk, Essex and Suffolk in the 1940s. They now also breed in Kent, and generally the population has slowly increased (*88–91 Atlas*). Avocets require a specialised habitat of shallow, brackish coastal lagoons with bare or sparsely vegetated low islands (*BWP*). The breeding population in England is 450–492 pairs with an estimated 1,270 wintering in Britain (*Population Estimates*).

Ecology

Avocets prefer to nest in the open on bare, low muddy islands. They will, however, also nest on shingle and short grass, and consequently by the end of the incubation period it is not unusual to find birds nesting in what appears to be totally unsuitable habitat (eg among dense grass or reed where they are barely visible). Avocet breeding seasons can be very variable. Nesting can start within a very short period after the birds have returned to their breeding areas, leading to closely synchronised hatching and fledging. On the other hand if the colony is large and regularly disturbed, birds can take a long time to settle and start incubating, so the last birds may still be going down on eggs as the first nests start to hatch. Although avocets only rear one brood a year, they will re-nest if they lose eggs or young. The latest successful nesting attempt recorded at Minsmere started on 29 June. Under these circumstances young from early nests will be fledging when late nests are hatching and re-nests are just starting (H Welch pers comm). A clutch of 3–5 eggs is laid from mid-April onwards (*Red Data Book*). Further details of avocet breeding behaviour are given at the end of this entry.

Breeding season survey – population

Information required
- number of breeding pairs
- map of the site with observation points marked.

Number and timing of visits

Three visits in May, with a week between visits.

Time of day

Preferably early morning (0600–0900 BST) and/or late afternoon or early evening (1600–1900 BST), although other times will do. Take as long as necessary. At sites where the numbers of adults are greatly affected by the tide, count individuals on the date when the tide is high; this makes birds easier to see.

Weather constraints

Do not survey if visibility is restricted by rain, mist or high winds.

Sites/areas to visit

Areas of shallow brackish standing water surrounded by bare or sparsely vegetated ground, eg man-made saline lagoons/scrapes, floodlands and flashes.

Equipment

- up-to-date large-scale OS map of the area (at least 1:10,000) showing all islands and waterbodies as accurately as possible
- clipboard
- telescope (20 × magnification)
- Schedule 1 licence.

Safety reminders

Take care when working close to water.

Disturbance

Do not disturb the colony while counting as this could affect breeding success. Carry out all observations from a discreet distance and from an unseen position or hide.

Methods

Before counting, assess the extent of the whole colony and mark it on the map. Find vantage points giving good views of all parts of the colony, checking that there are no blind spots. The number of vantage points required will depend on the size of the colony and the geography of the area. Split the colony into convenient sections, depending upon which areas can be seen from which vantage points, and mark and number them on the map.

Be absolutely clear (both in your mind and on the map) which sections are to be counted from which vantage points. You may find it useful to carry out a 'practice run' to familiarise yourself with the site and to establish whether birds visible at, for example, point 'A' are the same as those seen from hide 'B'.

Visit each vantage point in turn, in a logical sequence dependent on the layout of the site and on light conditions. Count the number of adult avocets in each section and note the count on the map.

Record also the following on the map: observer, site, date, start and finish times of observations and weather (wet/dry, wind force and temperature).

Take the maximum count of individuals from any one visit, divide it by two and report this as the number of pairs for the whole colony for the season. If a colony is spread over several 'sites', sum the counts for each site to give a total for each visit and again use the maximum from any one visit (ie do not add counts from different dates). Strictly speaking, this method only provides a population index for a colony, but at some sites this index may approximate the number of pairs present.

Make a note of the expected hatching and fledging dates if you intend to monitor productivity. Clutches hatch 23 days after the start of continuous incubation and chicks fledge 33–42 days from hatching.

Breeding season survey – productivity

Information required

- estimated number of fledged young.

Number and timing of visits

Weekly from 10 June to mid-August or whenever the last chicks disappear.

Sites/areas to visit, Safety reminders, Disturbance

As for population survey (above).

Equipment

As for population survey (above) plus a recording form (see Figure 1).

Methods

You should know from previous visits roughly when young are due to fledge. Return to the colony at this time and, during weekly visits, note the status of pairs and chicks on the map using standardised symbols (see below). Transfer these records to a recording form (Figure 1). It may help to number nests and young on the map to allow you to cross-reference these to the recording form. You do not need to be able to relocate individual pairs/chicks from one visit to the next, but you should visit the colony often enough to be able to estimate the number of pairs still present and the number of chicks near fledging age, and to be able to assess whether chicks that disappear have fledged or have been predated or washed out.

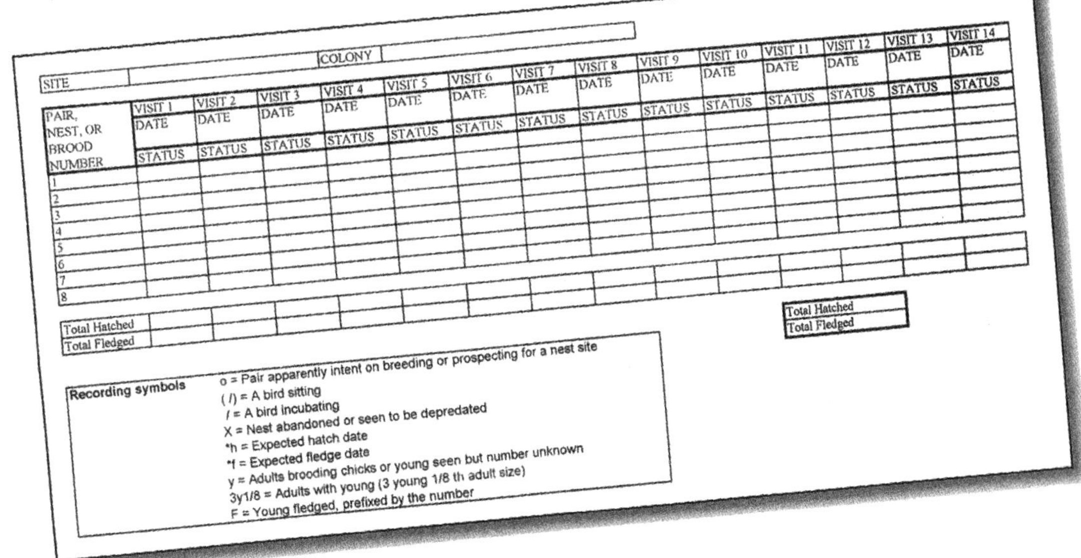

SITE | COLONY

PAIR, NEST, OR BROOD NUMBER	VISIT 1	VISIT 2	VISIT 3	VISIT 4	VISIT 5	VISIT 6	VISIT 7	VISIT 8	VISIT 9	VISIT 10	VISIT 11	VISIT 12	VISIT 13	VISIT 14
	DATE	DATE	DATE	DATE	DATE	DATE	DATE	DATE	DATE	DATE	DATE	DATE	DATE	DATE
	STATUS	STATUS	STATUS	STATUS	STATUS	STATUS	STATUS	STATUS	STATUS	STATUS	STATUS	STATUS	STATUS	STATUS
1														
2														
3														
4														
5														
6														
7														
8														
Total Hatched														
Total Fledged														

Total Hatched
Total Fledged

Recording symbols
o = Pair apparently intent on breeding or prospecting for a nest site
(/) = A bird sitting
/ = A bird incubating
X = Nest abandoned or seen to be depredated
*h = Expected hatch date
*f = Expected fledge date
y = Adults brooding chicks or young seen but number unknown
3y1/8 = Adults with young (3 young 1/8 th adult size)
F = Young fledged, prefixed by the number

Figure 1
An example avocet productivity recording form.

At some sites (eg saltmarsh) avocets move newly hatched young from the breeding site to feeding areas (eg mudflats) which are difficult to observe. In these situations, it may be difficult (or even impossible) to estimate productivity.

Use the following symbols on the map to help you to determine the fate of nests and young:

o	Pairs showing signs of attempting to nest (nest prospecting), ie birds nest-scraping, picking up vegetation or showing prolonged interest in one area (perhaps dropping onto their chest and scraping/kicking out behind with their feet).
(✓)	Bird sitting on the ground for short periods, or eggs seen but clutch not yet complete; bird sitting on the ground looking around alertly; sitting bird picking up vegetation and arranging it around itself; bird picking up and flicking vegetation towards a sitting bird; bird chasing off intruders and repeatedly returning to sit in the same place.
✓	Bird sitting tight, assumed to be incubating; changeover at nest observed. Other indications of birds with eggs are: birds squatting with their legs half bent and using their bills to rearrange the eggs before settling back down; birds shaking their feet as they leave the water; birds lowering themselves carefully to the ground and shuffling into position before settling.
X	Nest abandoned, no adults showing any interest in established nest, or nest seen to be predated (note cause/culprit).
y	Adults brooding chicks, or young seen but number unknown (prefix with a number when young can be counted). Indicate the size of the young in relation to an adult if possible, eg ⅛, ½, etc.
*h	Expected hatching date (23 days after the start of continuous incubation).
*f	Expected fledging date (33 days from the first signs of newly hatched young).
3y ⅛	Adults with young (where there are three young, about ⅛ the size of an adult).
F	Young fledged and the number.

A rough indication of chick age can be gained as follows:

Under 10 days old	Unstable when walking (first couple of days only). Looks like grey ball of fluff. Legs look stocky. Bill straight, feeds by pecking at surface.
Over 10 days old	Very independent, often ignores adults. Head starts to look 'angular'. Bill has distinct upturn. Legs growing long, bird looks 'leggy'. Feeds with adult sweeping motion.

Young are considered to have fledged when they are capable of flying a distance of several metres. Noting the approximate size and/or age of young and their expected fledging dates will help explain any unexpected disappearances of young.

At the end of the season, estimate the number of fledged young. Divide by the number of pairs (from population survey, as above) to report the overall number of chicks per pair.

Avocet breeding behaviour – contributed by Hilary Welch

The following may help in assessing productivity (and to a lesser extent population size) with greater certainty.

- When the eggs are due to hatch the adults become very restless. The sitting bird regularly stands to turn the eggs, or shuffles on the nest, fluffing up its feathers and adjusting its position on the eggs. The second bird in the pair is usually in close attendance, flicking extra items of material in the direction of the nest and keen to take over incubation. This behaviour can start a day or more before any young are seen and probably means the adults can hear the young calling from inside the eggs.
- The young usually hatch one at a time over a period of 24–48 hours, sometimes longer. During this period, one adult may look after the young which have hatched and need to feed, while the other continues to incubate the remaining eggs. You may therefore see a bird still sitting on the nest, but another adult close by with young or brooding. When the young are very small, adults can brood young while sitting flat on the ground, as if on eggs; this can be confusing! More often, brooding adults sit on their 'elbows', and the legs of the young can be seen beneath them. To be 100% sure of the number of young, you need to wait for the adult to stand up; counting the number of legs is not reliable as chicks sometimes stand on one leg.
- Once all the chicks in a brood have hatched, the adults will establish and defend a brood-feeding territory in which to rear them. The feeding territory may be some distance from the nest-site, and birds may later move the brood to an alternative area if they are disturbed (by predators or other avocets), or if better feeding is available elsewhere. As a rule of thumb, however, each family of chicks is likely to be feeding in a discrete area during any one visit.

- Early in the season there are frequent territorial disputes as pairs establish nesting territories, but then, during incubation, there is a period of comparative quiet. Late on in the season, territorial behaviour is most usual between pairs with young or pairs where the eggs are about to hatch. Such behaviour is always worth following up as there may be previously unobserved young present. Territorial behaviour includes:
  - Two birds walking a line, shoulder to shoulder, apparently feeding.
  - A group of birds, usually three or more, in a tight group with heads down making a lot of noise (low bubbling sound), often with fighting skirmishes and attacks on chicks (this most often happens when the chicks are very small and being moved from nesting to feeding territory; the chicks lie flat on the ground while the adults fight, and can be very hard to see).
  - 'Butterfly' flight: adult gliding with wings in a V, legs dangling and jinking its body from side to side as it lands. This is done when there are intruders in a feeding territory, and has only been seen in birds which have young.
- Birds which have fledged seem to prefer to stay on at the breeding site as long as they and the site are not disturbed.

Winter survey

WeBS.

See *Generic wintering bird monitoring methods* in the generic survey methods section.

Stone-curlew
Burhinus oedicnemus

Status

Red listed: BD, BR, SPEC 3 (V)
Schedule 1 of WCA 1981
Annex I of EC Wild Birds Directive

National monitoring

Key areas surveyed annually (RSPB/EN).
Rare Breeding Birds Panel.

Population and distribution

The stone-curlew is a rare breeding species in Britain. The current population is mainly confined to Norfolk, east Suffolk, east Cambridgeshire, Cambridgeshire/Essex, Berkshire/Oxfordshire, and the South Downs. The population and range have declined since the mid-19th century (*Red Data Birds*) because of habitat loss and changes in farming practices. There are currently 166–180 breeding pairs of stone-curlew in the UK (*Population Estimates*).

Ecology

Stone-curlews are only found on habitats with light, free-draining soil types, ie chalk, gravel, sand or limestone, and sparse, short vegetation, including arable land, heathland, young forestry. Breeding stone-curlews can be present from the end of March until the end of October. They nest on open bare ground, lay a clutch of usually two eggs and two broods are occasionally reared within a season.

Breeding season survey – population

A survey method has been written for stone-curlew, adapted from Green and Elliott (1995). However, as this text contains sensitive information, it is available only on written request from the Conservation Science Department of the RSPB.

Reference

Green, R E and Elliott, G (1995) *The RSPB Stone Curlew Wardening and Management Handbook.* RSPB unpubl.

Ringed plover *Charadrius hiaticula*

Status

Amber listed: WI
Non-SPEC

National monitoring

National breeding surveys: 1974, 1984 (BTO).
WeBS.

Population and distribution

In the UK, ringed plover breeding sites are widely distributed around the coast, except for south-west England and south Wales, with the highest densities in the Hebrides and Northern Isles (*88–91 Atlas*). The current UK breeding population is estimated at 8,600 pairs (Prater 1989). Ringed plover are found around almost all of the UK coastline in the winter and occasionally at inland sites. The UK winter population is estimated at 31,000 individuals (*Population Estimates*).

Ecology

In the breeding season most birds are found on sandy or shingle beaches, with high densities breeding on machair. The ringed plover has an extended breeding season from late March to late August, being apparently double-brooded in the south but single-brooded in the north. Many nests are lost and repeat clutches are frequent (*Red Data Birds*).

Breeding season survey – population

Method 1 – Censusing high densities on machair

See *Waders* in the generic survey methods section.

Method 2 – Censusing in other situations

The following is based on Prater (1976) and Prater (1989).

Information required

- maximum number of territorial pairs
- map of the survey boundary.

Number and timing of visits

Three, at least 10 days apart, during May and June.

Time of day

Between 0830 and 1800 BST.

Weather constraints

Avoid poor weather.

Sites/areas to visit

Coastal sites: beach, sand-dune, saltmarsh, machair, farmland, industrial sites. Inland sites including: reservoirs/lake shores, gravel pit, riverbed, moorland, heathland or farmland.

Equipment

- 1:10:000 OS map.

Safety reminders

When near water, work in teams of two wherever possible.

Disturbance

These birds are sensitive to disturbance. No attempt should be made to prove breeding by finding nests or broods.

Methods

Mark the boundary of the survey area on the map. Sub-divide the site into manageable areas. On each visit walk transect lines 150 m apart; if two observers are present, they should walk parallel transects 150 m apart. The locations and behaviour of all ringed plovers seen should be recorded on 1:10,000 field maps using BTO symbols (see Appendix 1). On subsequent visits follow the same transect lines but walk them in the opposite direction to the previous visit. On linear beach habitats walk a few metres below high-tide mark.

Walk at a steady pace and stop to scan ahead with binoculars at regular intervals; incubating birds will walk off nests at 80–100 m range and usually are easily seen. If more than one observer is involved, they should meet at the end of each pair of transect lines to cross-check records and reduce the likelihood of double-counting the same individuals. Make a single summary map of the visit as soon as possible after fieldwork and give weather details, date, times and the directions in which the transects were walked.

If several individual ringed plovers are seen in an area, it is more difficult to determine the number of territorial pairs present. Most territorial birds remain in their territory, watching the observer and often calling anxiously. Males and females can be distinguished with practice (males have larger, blacker breast bands), which helps with interpretation.

Record the following as territorial pair(s):

- One adult recorded alone 50 m or more from other adults = one pair.
- Two individual adults within 50 m of other adults = one pair.
- Two adults together, or two birds recorded as a pair = one pair.
- Three or four adults together = two pairs.
- One to four adults flying into, out of or through the area, or into the site = one to two pairs.
- Five or more adults remaining in the area, either on the ground or circling around (vocal birds only) = three+ pairs.

Exclude the following from estimates of breeding ringed plover populations:

- Five or more birds in a flock on the ground without vocal registrations (assumed non-breeding).
- Five or more birds in a flock flying into, out of or through the area or site.
- Any bird(s) which fly out of or through the site in one direction for more than 150 m without landing.
- Do not include nests in the estimate of breeding numbers. Some nests discovered lack adults in the immediate area; inclusion of the nest as an additional pair may lead to an inflated population estimate if the nesting pair has already been recorded elsewhere.

Report the maximum number of breeding pairs from any one visit.

Winter survey

WeBS.

See *Generic wintering bird monitoring methods* in the generic survey methods section.

References

Prater, A J (1976) Breeding population of the ringed plover in Britain. *Bird Study* 23: 155–161.

Prater, A J (1989) Ringed plover *Charadrius hiaticula* breeding population in the United Kingdom in 1984. *Bird Study* 36: 154–159.

Dotterel *Charadrius morinellus*

Status

Amber listed: BL
Non-SPEC
Schedule 1 of WCA 1981
Annex I of EC Wild Birds Directive

National monitoring

National surveys: 1987–88 and 1998 (SNH).
Rare Breeding Birds Panel.

Population and distribution

In Britain, the dotterel breeds in the montane zone, mainly on plateaux above 800 m. Its stronghold in the UK is the central Scottish Highlands, although a handful of pairs breeds in the north of England and the southern and northern Highlands. Since the *68–72 Atlas*, the number of dotterel estimated to be breeding in Britain has increased, partly owing to a more intensive survey effort and partly to a genuine increase (Galbraith et al 1993b). The most recent population estimate is 840–950 pairs (*Population Estimates*).

Ecology

The dotterel breeds in montane areas (above the former treeline) characterised by heaths of prostrate dwarf-shrubs, small herbs, mosses and lichens. Within this broad habitat type, it prefers areas of 25 ha or more, at or above the altitude where heather *Calluna vulgaris* becomes stunted or prostrate due to exposure (about 800 m above sea-level); heaths of *Racomitrium lanuginosum*, reindeer moss *Cladonia* and *Cetraria* lichens; open fell-fields especially with *Juncus trifidus*; and flat or gently sloping ground. It avoids steeply sloping ground (Galbraith et al 1993a). The dotterel is a solitary or loosely social bird. Birds display and scrape nests in May, but will not necessarily nest at those sites (Galbraith et al 1993b). The breeding season is not synchronised, and nests with eggs may be found from mid-May to the second half of July. Birds are mobile between sites in a single season; males may feed close to the nest or hundreds of metres away and chicks may also move long distances, up to several kilometres (*BWP*). The female lays a single clutch of 2–4 eggs from mid-May to late July, and incubation is almost entirely by the male. Most of the young fledge by early or mid-August (*Red Data Book*).

Breeding season survey – population

Dotterels are difficult birds to survey, partly because of the inaccessibility of their mountain breeding habitat and partly because of their unusual mating system, which involves male parental care with occasional polyandry (Kålås and Byrkjedal 1984).

The survey method given here was devised by the SNH Montane Ecology Project (Whitfield 1994). This method involves a single visit. Based on SNH's studies, the number of single males detected on a single visit represents about 42% of dotterel breeding attempts, thus allowing counts from a single visit to be corrected to give an estimate of the actual number of breeding attempts. The correction factor accounts for any failed breeding attempts that may have been missed; extra visits add little further information. Obtaining a more accurate count of the breeding population, especially at higher altitudes, would require intensive studies (up to three days a week throughout the season to cover an area of about 5 km^2).

Information required

- number of single males (regardless of whether breeding was proven)
- map showing the site boundary and all dotterel registrations.

Number and timing of visits

One visit, between 23 June and 14 July (20 June to 18 July if the timing of site visits is adjusted according to likely differences in timing of breeding).

Time of day

From 1000 BST until dusk.

Weather constraints

Avoid poor weather such as persistent precipitation, low cloud and wind speeds greater than Beaufort force 5. Expect that about 50% of days will be lost to poor weather.

Sites/areas to visit

High (>800 m) level or gently sloping montane plateaux heaths of prostrate dwarf shrubs, small herbs, mosses and lichens.

Equipment

- 1:25 000 OS maps of the area
- prepared recording forms
- Schedule 1 licence.

Safety reminders

A reliable person should know where you are and when you are due back. Carry a compass and know how to use it. If possible, work in teams. Take spare warm clothing, a plastic survival bag, a whistle, food supplies and a first-aid kit.

Disturbance

Observers may have to get quite close to birds to find and sex them, but do not stay too close for too long.

Methods

Mark on the map the area of possible breeding habitat according to the above criteria. If a site is particularly large, it is more efficient for two or more surveyors to walk abreast, particularly if they are experienced at surveying dotterel. Walk 50–100 m apart. This 'team' approach has the benefit of being safer in difficult terrain.

It may take time to reach the site and to determine whether the weather will be suitable. Two surveyors can cover about 3 km^2 a day, including walk-in and walk-off time.

Before carrying out the survey, map out a prearranged route to save time and increase accuracy. Although there are no rules about how much time should be spent surveying, an individual surveyor should spend 2–3.5 hours covering an area of 1 km^2 (about 2–3 minutes for a 100×100 m quadrat). Take as long as is necessary to cover the site adequately (as described below).

Walk at a comfortable pace, with periodic scanning using binoculars about every 50–100 m. About half the time should be spent walking and the other half scanning. All ground within 50–100 m of the surveyor should be covered by each scan, including areas behind you and those already covered. Follow up any signs of dotterels, such as a bird calling out of sight or a bird on the skyline behind you. This will undoubtedly mean deviating from the planned route. Always return to the route, which should be easy to pick up again.

Sexing any birds seen is very important. Since it is the male that incubates and cares for the young, this method is essentially a survey of males, although all dotterel registrations should be recorded. Record the sex and behaviour of any adult seen and the presence of nests or young. Sexing the birds can take time but the confirmation of a nest/brood is not necessary as almost all single males present between late May and mid-July will have a nest or brood (Whitfield 1994, SNH unpubl).

The male and female are very similar with only subtle plumage characteristics separating them. As a generality, the female tends be brighter, more clearly marked than the male. An adult breeding female has a purer black and less streaked crown; the supercilium is wider and purer white; the forehead is less streaked with white; cheeks and throat usually lack narrow dull black streaks; the chest is purer grey and only faintly barred and dusky, not slightly brownish as in the male (*BWP*). The 'back' feathers of the females have less distinct buff edging and so they appear less 'scaly' than the males. Observers unfamiliar with sexing dotterels may find it useful to visit a breeding site in May, when pairs are present, to familiarise themselves with these plumage characteristics. Once the bulk of males is incubating, most females leave Britain (SNH unpubl).

With experience, the sexes can also be separated on vocalisations and behaviour as follows:

Song	A series of peeps, delivered at about 2 per second. Given by male and female, but more often by the female.
Twitter	A song can develop into a twitter when the notes are much faster. Again both male and female, but more often the female.
Distraction-displays	The male will try to lure you away from the nest/chicks and often give a soft squeaking call. This should be noted as a distraction-display. Only males perform this display.
Head-bobbing	A male will start head-bobbing when he has chicks and occasionally does so when he has a nest. The rate of head-bobbing increases with the observer's proximity to the bird and with the chicks' age. A male with a brood also usually alarm-calls (a short 'wheet' call) but the call is soft and may not be heard because of the wind. Again, only males give this display.

Surveyors working together should attract each others' attention (without causing alarm!) whenever they spot or hear a dotterel, as this will help avoid double registrations of the same bird. It may still help for all those involved in the survey to compare notes immediately afterwards to check that different registrations are of different birds. Record all registrations on a map using standard BTO codes (Appendix 1) and cross reference these to a notebook which should also include details of observers' names, times and weather conditions. Compile a single summary map after each visit to a site.

The number of single males detected on a single visit between the prescribed dates will represent approximately 42% of all dotterel breeding attempts. Multiply the number of males detected by 2.38 (1/0.42) to determine the number of dotterel breeding attempts (Whitfield, SNH unpubl). Females, pairs and flocks of birds should be reported separately.

References

Galbraith, H, Murray, S, Duncan, K, Smith, R, Whitfield, D P and Thompson, D B A (1993a) Diet and habitat use of the dotterel (*Charadrius morinellus*) in Scotland. *Ibis* 135: 148–155.

Galbraith, H, Murray, S, Rae, S, Whitfield, D P and Thompson, D B A (1993b) Numbers and distribution of dotterel *Charadrius morinellus* breeding in Great Britain. *Bird Study* 40: 161–169.

Kålås, J-A and Byrkjedal, I (1984) Breeding chronology and mating system of the Eurasian dotterel *Charadrius morinellus*. *Auk* 101: 838–847.

Whitfield, P (1994) *Instructions to Montane Ecology Project Surveyors.* Scottish Natural Heritage, unpubl.

Golden plover *Pluvialis apricaria*

Status

Amber listed: WI
SPEC 4 (S)
Annex I of EC Wild Birds Directive

National monitoring

WeBS.

Population and distribution

Golden plovers are upland breeders, their distribution reflecting that of the UK's montane regions and upland heather moor. Afforestation in some Scottish areas such as Caithness and Sutherland has displaced up to 19% of the population in these areas (Thompson et al 1988). The UK breeding population is estimated to be 22,600 pairs and there are an estimated 310,000 wintering birds in the UK (*Population Estimates*).

Ecology

Golden plovers typically nest on moorland 300–610 m above sea-level, with vegetation varying from complete dominance by ling heather *Calluna vulgaris* on dry ground to co-dominance with cotton-grass *Eriophorum vaginatum* on wet ground where blanket bog has developed. The birds will occasionally breed in barer patches in areas of full moor vegetation (above 15 cm tall). A clutch of usually four eggs is laid from April to mid-May, the timing varying with altitude and season. The winter distribution of birds is markedly different, comprising a range of lowland agricultural habitats, permanent pasture, ploughed fields and winter cereals (*Red Data Birds*).

Breeding season survey – population

See *Waders* in the generic survey methods section.

Winter survey

WeBS.

See *Generic wintering bird monitoring methods* in the generic survey methods section.

Reference

Thompson, D B A, Stroud, D A and Pienkowski, M W (1988) Effects of afforestation on upland birds: consequences for population ecology. Pp. 237-259 in Usher, M B and Thompson, D B A *Ecological Change in the Uplands*. Blackwell Scientific, Oxford.

Grey plover
Pluvialis squatarola

Status

Amber listed: WI, WL
Non-SPEC
Annex II/2 of EC Wild Birds Directive

National monitoring

WeBS.

Population and distribution

The winter distribution of grey plover is almost entirely coastal, on larger, muddy estuaries, particularly in south-east England from the Wash to the Solent and in north-west England. Birds are widespread but less abundant elsewhere in Britain and Ireland and absent from large areas of north and west Scotland, Orkney and Shetland (*Winter Atlas*). Seasonal movements are complex and it is difficult to assess total numbers of grey plover visiting Britain and Ireland during the whole non-breeding period (*Winter Atlas*). An estimated 43,400 individuals winter in the UK and 70,000 visit the UK on spring migration (*Population Estimates*).

Ecology

Birds are often seen roosting at high tide within flocks of knot and dunlin. They feed on estuaries, on the middle and upper shores. Usually individuals are well dispersed and may establish low-tide feeding territories (*Red Data Birds*).

Winter survey

WeBS.

See *Generic wintering bird monitoring methods* in the generic survey methods section.

Lapwing
Vanellus vanellus

Status

Amber listed: WI
Non-SPEC
Annex II/2 of EC Wild Birds Directive

National monitoring

CBC, BBS.
BTO/RSPB breeding survey of England and Wales in 1987 was repeated in 1998.
WeBS.

Population and distribution

Lapwing breed throughout the UK, but are less abundant in the far north-west of Scotland and south-west of England and Wales (*88–91 Atlas*). Over 90% of lapwing breed on agricultural land, and the long-term decline in the breeding population has been partly attributed to the declining availability of mixed-crop farms (*88–91 Atlas*). The lapwing is the most widespread wintering wader in Britain and Ireland. There are an estimated 200,000–250,000 pairs breeding and 43,400 birds wintering in the UK (*Population Estimates*).

Ecology

Lapwings are primarily attracted by suitable ground conditions, requiring ready access to moist soil carrying an appreciable biomass of surface or subsurface organisms. A clutch of usually 3–4 eggs is laid in March–May. Incubation lasts about 28 days, and there is one brood, although a replacement clutch may be laid after egg loss. Young are cared for by both parents, the fledging period is 35–40 days and young are independent soon after fledging (*BWP*).

Breeding season survey – population

A lowland breeding wader population survey method is outlined in the generic survey methods section. The following is based on the BTO/RSPB national lapwing survey (Shrubb and Lack 1991, BTO/RSPB 1998).

Information required
- number of pairs.

Number and timing of visits

One visit in April (first half of the month for southern sites, second half for northern sites).

Time of day

Any time.

Weather constraints

Try to choose a clear, sunny day.

Sites/areas to visit

A mixture of habitats including autumn- and spring-sown cereals, sugar beet, bare ploughed or tilled land, stubble, ley grass, moor grass, permanent grass, saltmarsh, dunes, gravel pits, waste ground, reservoirs, sewage farms and airfields.

Equipment

- recording forms
- 1:25,000 OS map
- telescope.

Safety reminders

No specific advice; general guidelines are given in the *Introduction*.

Disturbance

Keep disturbance to a minimum. Extensive open areas should be checked from roads, tracks and footpaths without disturbing nesting birds. If birds are flushed, retreat as quickly as possible. Always get permission for access to farmland.

Methods

Map the boundary of the survey area. Ensure that every field or open area within the boundary is checked. Proceed around the site recording the approximate location of each breeding pair on the map. All fields should be scanned for territorial males, pairs or birds standing guard near nests, which are easily counted, as are incubating birds on short vegetation. Use the mapped locations of these to estimate the total number of breeding pairs.

Counts are most accurate when birds are undisturbed, so they should be done from roads, tracks and pathways where possible. Even in apparently flat fields, undulations can hide birds, so ensure that the whole field has been scanned. An observer walking through a field (with permission), or a crow flying over, will usually cause all the lapwings to fly up, including any incubating birds. In this case, the total seen may be halved to give the number of pairs. To record the number of pairs associated with each habitat (as in Figure 1), map the habitat types/crop types in each field and allocate each pair to a habitat type. Alternatively, just report the total number of breeding pairs.

Winter survey

WeBS.

See *Generic wintering bird monitoring methods* in the generic survey methods section.

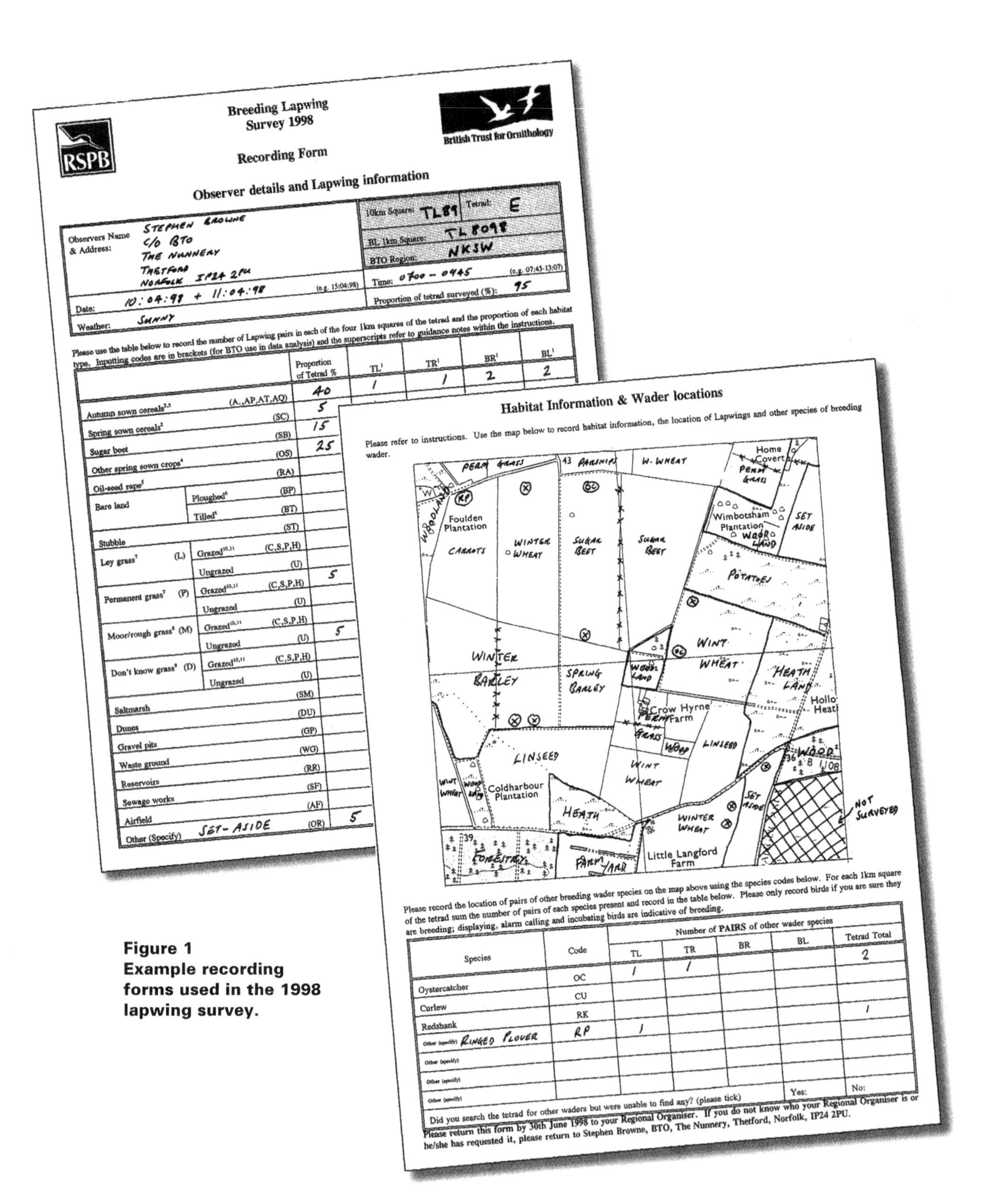

RSPB

Breeding Lapwing Survey 1998

Recording Form

British Trust for Ornithology

Observer details and Lapwing information

Observers Name & Address: STEPHEN BROWNE c/o BTO THE NUNNERY THETFORD NORFOLK IP24 2PU	10km Square: TL89 Tetrad: E
	BL 1km Square: TL8098
	BTO Region: NKSW
	Time: 0700 – 0945 (e.g. 07:45-13:07)
Date: 10:04:98 + 11:04:98 (e.g. 15:04:98)	Proportion of tetrad surveyed (%): 95
Weather: SUNNY	

Please use the table below to record the number of Lapwing pairs in each of the four 1km squares of the tetrad and the proportion of each habitat type. Inputting codes are in brackets (for BTO use in data analysis) and the superscripts refer to guidance notes within the instructions.

			Proportion of Tetrad %	TL[1]	TR[1]	BR[1]	BL[1]
				1	1	2	2
Autumn sown cereals[2,3]		(A.,AP,AT,AQ)	40				
Spring sown cereals[2]		(SC)	5				
Sugar beet		(SB)	15				
Other spring sown crops[4]		(OS)	25				
Oil-seed rape[5]		(RA)					
Bare land	Ploughed[6]	(BP)					
	Tilled[6]	(BT)					
Stubble		(ST)					
Ley grass[7] (L)	Grazed[10,11]	(C,S,P,H)					
	Ungrazed	(U)	5				
Permanent grass[7] (P)	Grazed[10,11]	(C,S,P,H)					
	Ungrazed	(U)					
Moor/rough grass[8] (M)	Grazed[10,11]	(C,S,P,H)					
	Ungrazed	(U)	5				
Don't know grass[9] (D)	Grazed[10,11]	(C,S,P,H)					
	Ungrazed	(U)					
Saltmarsh		(SM)					
Dunes		(DU)					
Gravel pits		(GP)					
Waste ground		(WG)					
Reservoirs		(RR)					
Sewage works		(SF)					
Airfield		(AF)					
Other (Specify) SET-ASIDE		(OR)	5				

Habitat Information & Wader locations

Please refer to instructions. Use the map below to record habitat information, the location of Lapwings and other species of breeding wader.

Please record the location of pairs of other breeding wader species on the map above using the species codes below. For each 1km square of the tetrad sum the number of pairs of each species present and record in the table below. Please only record birds if you are sure they are breeding; displaying, alarm calling and incubating birds are indicative of breeding.

Species	Code	Number of PAIRS of other wader species TL	TR	BR	BL	Tetrad Total
Oystercatcher	OC	1	1			2
Curlew	CU					
Redshank	RK					
Other (specify) RINGED PLOVER	RP	1				1
Other (specify)						
Other (specify)						
Other (specify)						

Did you search the tetrad for other waders but were unable to find any? (please tick) Yes: No:

Please return this form by 30th June 1998 to your Regional Organiser. If you do not know who your Regional Organiser is or he/she has requested it, please return to Stephen Browne, BTO, The Nunnery, Thetford, Norfolk, IP24 2PU.

Figure 1
Example recording forms used in the 1998 lapwing survey.

References

Shrubb, M and Lack, P C (1991) The numbers and distribution of lapwings *V. vanellus* nesting in England and Wales in 1987. *Bird Study* 38: 20–38.

BTO/RSPB (1998) Breeding lapwing survey 1998: fieldwork instructions and example recording form England and Wales.

Knot
Calidris canutus

Status

Amber listed: WI, WL, SPEC 3^{W} $(L)^{W}$
Annex II/2 of EC Wild Birds Directive

National monitoring

WeBS.

Population and distribution

In Britain and Ireland, wintering knot concentrate on the major estuaries, and about a dozen sites support over half the winter population (*Winter Atlas*). These sites include the Wash, Ribble, Humber, Thames, Alt, Morecambe Bay, Solway, Forth and Strangford Lough (*WeBS 1992–93*). An estimated 298,000 individuals winter in the UK (*Population Estimates*).

Ecology

Knot appear to require large, open mudflats, where at high tide they can form large tightly packed flocks. They feed on marine bivalve molluscs and remain gregarious when feeding (*Winter Atlas*).

Winter survey

WeBS.

See *Generic wintering bird monitoring methods* in the generic survey methods section.

Purple sandpiper *Calidris maritima*

Status

Amber listed: BR, WI
SPEC 4 (S)
Schedule 1 of WCA 1981

National monitoring

Rare Breeding Birds Panel.
WeBS.

Population and distribution

This species has been known to breed in the UK since 1978 (Dennis 1983), though only in Scotland and in very low numbers. The preferred breeding habitat is open ground on mountain tops. There are an estimated 21,300 individuals wintering along the UK coastline (*Population Estimates*).

Ecology

Purple sandpipers breed on open ground on hillsides, mountains and arctic tundra. A clutch of 3–4 eggs is laid during May–July. There is one brood (*Red Data Birds*), of which the male takes sole charge. The attending male is, as a rule, very alert, meeting any intruder at a distance of 50 m or more from the young, calling anxiously (Bengtson 1970). The young fledge by July to mid-August.

Breeding season survey – population

Information required

- estimated number of pairs
- maps showing boundaries of the areas covered.

Number and timing of visits

A minimum of two, mid-May to mid-August. At least one visit should be in late May or early June.

Time of day

Any time.

Weather constraints

If possible survey on calm, clear days.

Sites/areas to visit

With so few breeding birds in such a vast area of suitable habitat, the only practical way of surveying this species is to survey those areas of Scotland known or suspected to have held breeding purple sandpipers since 1978. The habitat used by nesting birds in Scotland is similar to that used in Scandinavia: alpine vegetation on high plateaux. This short vegetation is dominated by sedges, mosses and lichens interspersed with rock screes and gravel.

Equipment

- 1:25,000 OS map of the survey area
- 1:10,000 field maps of the survey area
- Schedule 1 licence.

Safety reminders

A reliable person should know where you are and when you are due back. Take a compass, survival bag, waterproofs, whistle, extra clothing, first-aid kit and food.

Disturbance

There is no need to locate nests, which are in any case very hard to find. The whereabouts of any breeding birds should be kept confidential to help prevent disturbance by birdwatchers or egg-collectors.

Methods

Mark the site and all surrounding suitable habitat on the map. Survey the whole site by walking parallel transects 200 m apart. At large sites, it may be necessary to have more than one person surveying adjacent areas, to cover the ground more quickly.

At regular intervals (at least every 100 m) scan in every direction with binoculars as far as the terrain or weather allows, and listen for calls and songs. Use vantage points (eg rocks or hillocks) wherever possible. When individuals or pairs are encountered, determine whether or not these are new birds. It may be necessary to retrace your steps to check on the continued presence of any birds located previously.

The behaviour of the male can provide information on territorial/ breeding status: song-flights are part of territorial behaviour, 'rodent-running' is used to distract observers away from the nest or chicks, and alarm-calls are given when guarding chicks (Bengtson 1970). Pairs may stand together and give single-wing lifts at the approach of the observer (Bengtson 1970). At the end of incubation the female generally departs, so single birds seen in July are usually males with chicks, or failed breeders.

The location and behaviour of any purple sandpipers should be marked on the field maps using standard BTO symbols (see Appendix 1). Note also the time of the observation, behaviour and flight line (in relation to the boundaries of the survey area) of the birds.

After each visit transfer all records to a master visit map. Remove duplicate registrations of birds made in adjacent survey areas by other observers.

Using information on the master visit map, estimate the number of pairs by summing the number of displaying males, pairs seen together, and any additional broods/nests. Also report the number of visits made.

Breeding season survey – breeding success

To estimate population size and breeding success simultaneously, combine the second visit of the population survey with the first visit of the breeding success survey.

Information required
- number and size of any broods seen.

Number and timing of visits

Two visits, one in late June or early July (when young broods should be around), the other in late July (when broods are older).

Time of day, Weather constraints, Sites/areas to visit, Equipment, Safety reminders

As for the population survey (above).

Disturbance

As for the population survey. Take extra care to avoid drawing the attention of predators to the chicks.

Methods

This method only provides crude estimates of breeding success; at best the number of broods and their sizes, at worst the number of broods with at least one chick.

Cover the whole survey area by walking transects 200 m apart, as for the population survey. Record any birds seen on field maps using standard BTO symbols (see Appendix 1) and cross reference any additional notes with the mapped symbols. On the late June/early July visit, note the number and size of any broods seen. Whenever possible, estimate the size of the chicks in relation to an adult and describe the amount of down/feathering on the chicks, this helps to discriminate between broods. On the last visit in late July, the chicks are likely to be quite widely dispersed and the only way to count the whole brood is to sit and watch for a period.

Knowing the number of pairs from the population survey, breeding success can simply be estimated as the proportion that produced broods of at least one chick. If the sizes of all broods are known, success can be estimated as the mean number of chicks per pair.

Winter survey

WeBS.

See *Generic wintering bird monitoring methods* in the generic survey methods section.

References

Bengtson, S A (1970) Breeding behaviour of the purple sandpiper *Calidris maritima* in West Spitsbergen. *Ornis Scand.* 1: 17–25.

Dennis, R H (1983) Purple sandpipers breeding in Scotland. *British Birds* 76: 563–566.

Dunlin
Calidris alpina

Status

Amber listed: WI, WL, SPEC 3^{W} (V^{W})

National monitoring

WeBS.

Population and distribution

The main concentrations of breeding dunlins are in the Flow Country, the high peaty bogs of the Grampians and Pennines, the peatlands of the Northern and Western Isles and the machair of the Western Isles (*88–91 Atlas*). A decline in numbers of breeding dunlins in Scotland has been partly attributed to afforestation (*88–91 Atlas*). The dunlin is one of the most abundant waders wintering on the UK coast, occurring almost anywhere where mud is present. There are estimated to be 9,150–9,900 pairs breeding and 549,000 individuals wintering in the UK (*Population Estimates*).

Ecology

Breeding birds are found in two main habitats: wet moorland, where there are pools and patches of very short vegetation, and coastal zones, particularly the wet areas in the machair of the Western Isles. They lay a clutch of 3–4 eggs between late May and mid-July; there is one brood. As the smallest of Britain's wintering waders, the dunlin needs to feed for nearly all of the winter daylight hours (*Red Data Birds*).

Breeding season survey – population

See *Waders* in the generic survey methods section.

Winter survey

WeBS.

See *Generic wintering bird monitoring methods* in the generic survey methods section.

Ruff
Philomachus pugnax

Status

Amber listed: BR, WL
SPEC 4 (S)
Schedule 1 of WCA 1981
Annex I of EC Wild Birds Directive

National monitoring

Rare Breeding Birds Panel.
WeBS.

Population and distribution

Only a few pairs breed regularly in Britain, mainly in East Anglia, Kent and northern England. Nesting occurs on inland wet meadows, coastal grazing marsh and upland saltmarsh. Ruffs were once widespread in Britain, but land drainage and human persecution led to major declines in the 19th century. The wintering population is found inland and on the coast. There are currently estimated to be 2–24 ruff nests in the UK annually, and 700 ruff wintering (*Population Estimates*).

Ecology

Breeding habitat includes inland wet meadows, coastal grazing marsh and upland saltmarsh. Ruff wintering in Britain and Ireland are usually found singly or in small groups. They feed both nocturnally and during the day, and are often seen feeding in or near the muddy margins of lakes and pools, on seashores and tidal mudflats (*Winter Atlas*).

Breeding season survey – population

No method is given here for this rare breeding species.

Winter survey

WeBS.

See *Generic wintering bird monitoring methods* in the generic survey methods section.

Jack snipe *Lymnocryptes minimus*

Status

Amber listed: SPEC 3^W (V^W)
Annex II/1 of EC Wild Birds Directive

National monitoring

WeBS.

Population and distribution

Jack snipe are not thought to breed in Britain (*88–91 Atlas*), but they are quite widely distributed in the UK in winter (*Winter Atlas*). There is a concentration of jack snipe in Strathclyde, but otherwise the species is most commonly found south of southern Cumbria and away from high ground. There are estimated to be between 10,000 and 100,000 jack snipe wintering in the UK (*Population Estimates*).

Ecology

During the non-breeding period, jack snipe are associated with shallow, wet and muddy habitats, preferring more cover than common snipe (*Winter Atlas*). They feed in dense grass, where their diet comprises insects, molluscs, worms and a variety of plant material, particularly seeds. Birds are mainly nocturnal or crepuscular and roost solitarily, usually near their feeding areas. They are highly cryptic and will lie prone to avoid detection, only flushing if nearly trodden on. Once flushed, a bird makes a typical short, direct escape flight before returning to the ground, only rarely calling.

Winter survey

1. WeBS

A simple record of the birds seen. A suboptimal method for such a cryptic species as it is dependent on accidental flushing and is thus not systematic. See the generic survey methods section.

2. Transect counts

This method has not yet been tested but is repeatable and thus is preferable to WeBS counts for this species. It is based on selecting suitable habitat from maps and then estimating the numbers found in each area of suitable habitat. Linear habitats, such as streams and rivers, are sampled and the sample length is multiplied up by the total available length of the habitat. Non-linear habitats, such as marshes and water meadows, are sampled by walking transects of a known width, and multiplying the number of birds detected by the available area.

Information required

- total number and density of jack snipe seen, by habitat type
- field maps (if necessary).

Number and timing of visits

Once a month, from October to March. If this is not possible, the critical months are December, January and February.

Time of day

0900–1600 BST.

Weather constraints

Avoid counting in poor weather or in windy conditions of Beaufort force 5 or above.

Sites/areas to visit

Shallow, wet and muddy habitat with some vegetation cover, even if not very extensive: swamp, fen, marsh, overgrown floodland, sewage farms, irrigated land, flooded arable fields, water meadows, streams or river banks.

Equipment

- 1:25,000 map
- recording form
- 25 m rope
- lightweight cane or stick about 2 m long.

Safety reminders

No specific advice. See general information in the *Introduction*.

Disturbance

Unavoidable.

Methods

Visit and get to know the site before counting. Clearly define the boundaries of each habitat to be surveyed. Details of count transect positions (start and end locations and direction of observation) should be kept and used each year.

In the case of a linear habitat, decide whether it is to be completely or partially covered. As jack snipe are difficult to flush, both sides of streams or ditches should be covered during complete censuses. Decide on the start and end points for a partial survey and select the stretches to be sampled at random. Do not select the best stretches for coverage otherwise the final population estimate will be too high. Walk the selected stretches. If these are wider than a couple of metres it may be useful to gently tap the ground with a very light stick to ensure that any jack snipe sitting tight are flushed. Count all birds and follow the movements of flushed individuals by eye to avoid double-counting jack snipe returning to the ground shortly after take-off. If a colleague is present, it may be possible to map the positions of the jack snipe as they take off. Such information helps to determine habitat preferences and is a useful habitat management tool. Calculate the number of jack snipe present in a linear habitat as:

$$\frac{\text{No. of jack snipe recorded} \times \text{Total length (m)}}{\text{Length sampled (m)}}$$

In the case of a non-linear habitat, decide on the start and end points of a series of parallel transects. As two people will be walking in parallel dragging a 25-m-long rope between them, the transects should either be every 25 m for a complete survey or at intervals divisible by 25 m (eg 50 m, 75 m, 100 m, etc) for a sample survey. When the survey is finished, transfer the mapped information to a recording form (see Figure 1). Do not include jack snipe flying overhead from one unknown point to another in the recorded total. Calculate the number of jack snipe present in the habitat as:

$$\frac{\text{No. of jack snipe recorded} \times \text{Total area (m}^2\text{)}}{\text{Area sampled (m}^2\text{)}}$$

where the area sampled is 25 × total transect length in m. For both habitat types give confidence intervals around the estimates from the sample surveys wherever possible.

	Linear or non-linear	Total length or area of habitat	Length or area sampled	No. of jack snipe recorded	Total no. of jack snipe in each habitat
Habitat 1	L	TOTL1	SAML1	NOS1	NOS1 x TOTL1 / SAML1
Habitat 2	L	TOTL2	SAML2	NOS2	NOS2 x TOTL2 / SAML2
Habitat 3	NL	TOTAREA1	SAMAREA1[1]	NOS3	NOS3 x TOTAREA1 / SAMAREA1
Habitat 4	NL	TOTAREA2	SAMAREA2[1]	NOS4	NOS4 x TOTAREA2 / SAMAREA2
Overall total	Both habitats	–	–	–	Sum of above

[1] SAMAREA = Total length of transect (in m) x width (25 m)

Figure 1
Jack snipe recording form for a single site made up of two linear habitats (1, 2) and two non-linear habitats (3, 4).

Use BTO codes and protocol (see Appendix 1) when mapping, taking care to record the landing positions of any birds taking flight which have already been counted; if more than one team of two counters is involved, record the times at which birds take flight and land to avoid double-counting. Include site details, the date and time of the visits and count conditions on the recording form. If a single maximum figure is required for each site, quote the peak number of individuals recorded (or estimated if a sample survey) on any one visit.

Contributed by Mark Rehfisch and Steve Holloway

Snipe
Gallinago gallinago

Status

Amber listed: BDM
Non-SPEC
Annex II/1 of EC Wild Birds Directive

National monitoring

BBS.
WeBS.

Population and distribution

Snipe are widespread on the moorland bogs and marshy rough pastures of northern England, Wales, Scotland and Ireland, and patchily distributed in lowland wet habitats (*88–91 Atlas*). There has been a contraction in the breeding range, particularly in the lowlands, partly due to drainage and agricultural intensification. There are an estimated 61,000 breeding pairs of snipe in the UK (*Population Estimates*). Snipe have a widespread winter distribution in the UK. They frequent inland and coastal marshes, bogs and streamsides (*Winter Atlas*). More than 100,000 individuals winter in Great Britain (*Population Estimates*).

Ecology

Snipe require access to shallow water, which can be fresh or brackish. They nest in any tall or dense vegetation separated by more open ground with low tussocks or clumps of sedge *Carex*, rushes *Juncus* and coarse grasses. A clutch of usually four eggs is laid in April–May and there is usually one brood. Incubation lasts 18–20 days, the young are precocial and nidifugous, brooded when small, and the fledging period is 19–20 days (*BWP*).

Breeding season survey – population

See *Waders* in the generic survey methods section.

Winter survey

WeBS.

See *Generic wintering bird monitoring methods* in the generic survey methods section.

Woodcock
Scolopax rusticola

Status

Amber listed: BDM, SPEC 3^W (V^W)
Annex II/1 of EC Wild Birds Directive

National monitoring

WeBS.

Population and distribution

As breeding birds, woodcock favour moist woodland habitat and are widely distributed in the UK. The woodcock's range and abundance expanded during the nineteenth century to its present level (Holloway 1996). However, there are problems with estimating breeding numbers of this secretive species, so recent trends are difficult to measure. Although covered by the BTO/IWC/SOC breeding bird atlases in 1968–72 and 1988–91, woodcock numbers and distribution are thought to have been underestimated because the survey method was designed to encompass all bird species (*88–91 Atlas*) and not the woodcock alone. The woodcock ceased to be included in the BTO's CBC trends in 1995 because the sample size was too small. Annual records of woodcock bags collected by the Game Conservancy Trust provide an index of numbers in winter (Tapper 1992) when the resident woodcock population is swelled by a huge influx of migrant birds from Fennoscandia and continental Europe. At this time they are more widely distributed and will use scrub and hedgerow habitat as well as a wider range of woodland types (*Winter Atlas*). There are an estimated 9,100–23,000 'pairs' of breeding woodcock in the UK (*Population Estimates*), while the wintering population is probably about 800,000 birds (Hirons and Linsley 1989).

Ecology

Woodcock nests are notoriously difficult to locate, even with trained dogs. The survey method must therefore depend on counting males as an index. Male woodcock perform a peculiar roding display-flight at dawn and dusk from late February until July. A dominance hierarchy is thought to be established whereby few first-year males rode and older males are more active (Hirons 1980, 1983). There is some evidence that the frequency of observations of roding males correlates well with the total number of males in a wood (Ferrand 1988).

Breeding season survey – population

This method is based on studies conducted by Hirons (1983), Ferrand (1988) and Hoodless (1994). It provides an index of population size.

Information required

- maximum 'density' of roding males
- a map showing the location of observation points and the limit of visibility in each direction.

Number and timing of visits

Three visits, during May and June.

Time of day

Evening. Be in position from one hour before sunset until one hour after sunset or until it is too dark to see.

Weather constraints

Avoid very cold, wet or windy weather.

Sites/areas to visit

Woodland of all types throughout the UK and extensive patches of bracken in upland areas.

Equipment

- 1:10,000 map and photocopied field maps
- torch.

Safety reminders

Familiarise yourself with the site during the day prior to the first count to ensure that you can find your way home in the dark!

Disturbance

Not relevant if the survey method is followed. Nest searching should be avoided as woodcock are particularly prone to desert the nest if disturbed.

Methods

Take a new field map on each visit to the site and clearly mark the limits to visibility on each occasion. For density calculations, the survey area is that encompassed within the limits of visibility. Slowly walk through the survey area and count all observations of woodcock seen roding during the dusk period. Even if the same individual appears to make several circuits, count each occasion that it comes into view as a new observation.

Woodcock are probably breeding if roding males are seen.

Counts are converted to density on the basis of the area of woodland surveyed. The highest of the three densities of roding males provides an index of the peak density of breeding 'pairs'.

Winter survey

Method 1 – Pinewood bird survey

See the generic survey methods section. Can been used to estimate density of woodcock diurnally.

Method 2 – Nocturnal survey

Based on a technique developed by Hirons and Linsley (1989) and applied by Hoodless (1994).

Information required

- a map of the survey area, showing the fields searched, the crop types and the number of woodcock seen in each field.

Number and timing of visits

At least one visit to each site between the November and January full moons.

Time of day

2000–0300 BST on dark, new moon nights.

Weather constraints

Any weather conditions except foggy/misty nights.

Sites/areas to visit

Agricultural fields selected to be representative of the farming practice of the region.

Equipment

- 1:25,000 map and photocopied field maps
- 200,000–300,000 candle-power spotlamp
- torch and compass.

Safety reminders

If working on foot, ensure that you are familiar with the site, eg the location of ditches and other potential obstacles. Close liaison with the landowner is essential, especially if night shooting of rabbits or foxes takes place on the farm. In some areas, it is also advisable to inform the police of your activities to avoid being stopped on suspicion of poaching. If lamping from a vehicle from public roads, display a sign in the rear window such as 'Wildlife survey – slow-moving vehicle, please pass'.

Disturbance

Do not repeatedly survey the same fields, and when working on foot try to avoid flushing birds (do not approach closer than 20 m).

Methods

Scan fields with the spotlight either from a vehicle or by walking through them. Mark each field surveyed, and if the whole field is not visible, mark the area within each field that was searched. Record the number of woodcock seen in each field. When working from a vehicle, use binoculars to distinguish woodcock from rabbits, lapwings and golden plovers. Ensure that a total area of at least 150 ha is searched to provide a representative sample.

Record the crop type of each surveyed field, eg permanent pasture, ley grass, winter cereal, root crop, stubble/set-aside, bare plough.

Convert counts to a density of wintering woodcock on the basis of the total area searched.

Contributed by Andrew Hoodless

References

Ferrand, Y (1988) Contribution à l'Étude du Comportement du Mâle de Bécasse des Bois *Scolopax rusticola* L. en Période de Reproduction: Méthode de Dénombrement. These de Doctorat, Université Montpellier.

Hirons, G (1980) The significance of roding by woodcock *Scolopax rusticola*: an alternative explanation based on observation of marked birds. *Ibis* 122: 350–354.

Hirons, G (1983) A five-year study of the breeding behaviour and biology of the woodcock in England – a first report. Pp. 51–67 in Kalchreuter, H (ed) *Proc. 2nd European Woodcock and Snipe Workshop.* IWRB, Slimbridge, UK.

Hirons, G and Linsley, M (1989) Counting woodcock. *Game Conservancy Annual Review* 20: 47–48.

Holloway, S (1996) *The Historical Atlas of Breeding Birds in Britain and Ireland 1875–1900.* Poyser, London.

Hoodless, A N (1994) Aspects of the Ecology of the European Woodcock *Scolopax rusticola* L. PhD Thesis, University of Durham.

Tapper, S (1992) *Game Heritage.* Game Conservancy, Fordingbridge.

Black-tailed godwit *Limosa limosa*

Status

Red listed: HD, BR, WL, SPEC 2 (V)
Schedule 1 of WCA 1981
Annex II/2 of EC Wild Birds Directive

National monitoring

Rare Breeding Birds Panel.
WeBS.

Population and distribution

Two races of black-tailed godwit breed in the British Isles: *L. l. islandica* breeds in Iceland and northern Scotland and winters in Britain and Ireland; *L. l. limosa* breeds mainly in the Netherlands and at a few localities in England, Wales and southern Scotland but winters in southern Europe and West Africa. The estimated UK breeding population is 29–53 pairs (largely *limosa*). In winter, black-tailed godwits are more widespread in the country but occur mainly in the muddy zones of estuaries in south, east and north-west England and southern Ireland. The UK wintering population has been estimated at 7,800, with up to 12,400 recorded on autumn migration (*Population Estimates*).

Ecology

Breeds in wet meadows, coastal grazing marshes and moorland bogs, and is loosely colonial. A clutch of four eggs is laid, mainly from the beginning of April to late May. In winter, birds feed on wet grasslands inland, as well as on fine sediment inner estuary sites (*Red Data Birds*).

Breeding season survey – population

Information required

- maximum number of displaying males
- maximum number of birds (divided by two).

Number and timing of visits

Three visits, at least one week apart. For southern England: first visit 1–15 April, second visit 16–31 April, third visit 1–15 May. For northern England and Scotland, the survey period will be a week to six weeks later.

Time of day

Between dawn and 1200 BST. Avoid cold, wet and windy conditions (wind speed not exceeding Beaufort force 3).

Sites/areas to visit

Wet meadows, coastal grazing marshes and moorland bogs. Most likely to be found in eastern England (*limosa*) or northern Scotland (*islandica*), but could occur anywhere where there is suitable habitat.

Equipment

- 1:10,000 OS map
- Schedule 1 licence.

Safety reminders

No specific advice. See the *Introduction* for general information.

Disturbance

The method may involve some disturbance to breeding birds, depending on the site (eg sites where parallel transects are used).

Methods

Mark the boundary of the survey area clearly on a map. Where the birds are nesting in fields (eg wet grassland sites), number all the fields. Large sites such as moorland should be treated as a single field. Ensure that this map is kept with the data to allow the same fields to be re-surveyed in subsequent years.

On each visit, slowly walk along a route which takes you just close enough to each field in the study site to allow you to see any birds that are present. Scan each field carefully from a distance and record the location and behaviour of all birds present using standard BTO codes (Appendix 1). Try to identify whether they are males or females wherever possible. Males are smaller and redder on the underparts than females, a difference which is usually quite obvious within pairs. Established pairs are usually not very far from one another except when the male is performing a ceremonial flight or pursuing a rival male.

On large sites where you are unable to scan from a distance, you may need to walk parallel transects about 300 m apart. Record movements between fields but allocate each bird to the field in which it was first recorded. If first observed in display-flight, allocate the bird to the field in the centre of its flight.

Transfer summary information to visit forms at the end of each visit. Calculate the total number of birds (and divide this by two), the total number of displaying males, and (if possible) the total number of males recorded on each visit. Report all three figures and supply copies of your visit maps. There has been little work so far to test which is the best figure for estimating the number of breeding pairs.

Beware of confusing migrant with breeding birds at sites where both occur. Do not include birds occurring in tight flocks of more than two birds (breeding pairs may interact for short periods and males may follow each other when performing display-flights but this behaviour is usually obviously territorial and accompanied by vigorous calling and/or song).

Winter survey

WeBS.

See *Generic wintering bird monitoring methods* in the generic survey methods section.

Bar-tailed godwit
Limosa lapponica

Status

Amber listed: WI, WL, SPEC 3^{W} (L^{W})
Annex II/2 of EC Wild Birds Directive

National monitoring

WeBS.

Population and distribution

The winter distribution is almost entirely coastal, with birds distributed along the North Sea coasts of Britain as well as north-west England, the Outer Hebrides and Ireland. Fewer occur along the coasts of southern England and the cliff-bound coasts of south-west England, west Wales and western Scotland (*Winter Atlas*). Wintering numbers fluctuate from year to year, with an estimated average of 56,100 individuals in the UK (*Population Estimates*).

Ecology

Bar-tailed godwits feed mainly on the middle to low shores of relatively sandy estuaries, where they take a range of larger molluscs and polychaete worms. There is a slight sexual separation in feeding areas: the longer-billed and longer-legged females are able to exploit sites in deeper water, while the smaller males forage on wet sand or in very shallow water (*Red Data Birds*).

Winter survey

WeBS.

See *Generic wintering bird monitoring methods* in the generic survey methods section.

Whimbrel
Numenius phaeopus

Status

Amber listed: BL
SPEC 4 (S)
Schedule 1 of WCA 1981
Annex II/2 of EC Wild Birds Directive

National monitoring

Rare Breeding Birds Panel.
WeBS.

Population and distribution

The entire UK breeding population is confined to Scotland. Most are found on the exposed heathland and moorland of Shetland, the rest on Orkney, the Western Isles and the north coast of the mainland (*88–91 Atlas*). Numbers of breeding whimbrel have increased since the 1950s on Shetland and Orkney, but not elsewhere (*88–91 Atlas*). There are an estimated 530 breeding pairs of whimbrel in the UK (*Population Estimates*).

Ecology

Whimbrel breed mainly on serpentine heaths, wet moorlands and blanket bogs with short vegetation (*88–91 Atlas*, M Grant pers comm). A clutch of usually four eggs is laid in a scrape scantily lined with grass or moss, in mid-May to early June. Incubation lasts about 26 days, the peak of hatching is 13–20 June, and the young fledge about 29 days after hatching; most chicks have fledged by mid-August (*Red Data Birds*). Whimbrel are single-brooded.

Breeding season survey - population

It is possible that this species will be encountered during wader surveys using the field-by-field method described in the generic methods section. The method given below, however, is based on that used by Richardson (1990) and M Grant (pers comm).

Information required

- maximum number of territories (=pairs).

Number and timing of visits

Two visits, between 20 May and 15 June, separated by one week .

Time of day

Any time of day.

Weather constraints

Avoid cold, wet and windy conditions.

Sites/areas to visit

Heath, bog and acid grassland on Shetland, the Western Isles, the north Scottish mainland and Orkney.

Equipment

- 1:25,000 OS map
- Schedule 1 licence.

Safety reminders

Nothing specific. See general guidelines in the *Introduction*.

Disturbance

Keep to a minimum.

Methods

Mark the boundary of the survey area on a map and take the same field map on both visits. Cover all areas of suitable habitat using parallel transects about 200 m apart (the exact distance depends on the terrain). Mark on the field map the position of any whimbrel actively calling or apparently on territory. Prefix all mapped registrations from visit one with an 'A', and those from visit two with a 'B'. Take the position of the nest (and centre of the territory) to be where alarm-calling is at its most intense, often with both birds of the pair calling actively overhead or undertaking distraction displays. This might be difficult to determine as whimbrel may fly up to 500 m to defend their territories. One member of the pair may also adopt a characteristic secretive 'creeping' run. If the observer is very close to the nest, the pair will often alight nearby and continue to call in an extremely agitated fashion. Establish the number of territories by pooling the information recorded on both visits. This might be difficult in areas of high density; the mean distance between territory centres can be as low as 224 m.

A site map should be prepared showing the area covered together with the number of territories found which equals the number of breeding pairs.

Reference

Richardson, M G (1990) The distribution and status of whimbrel *Numenius p. phaeopus* in Shetland and Britain. *Bird Study* 37: 61–68.

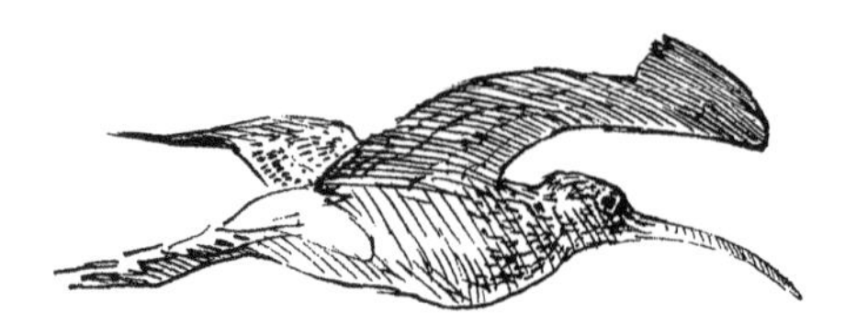

Curlew *Numenius arquata*

Status

Amber listed: BI, WI, SPEC 3^{W} (D^{W})
Annex II/2 of EC Wild Birds Directive

National monitoring

BBS.
WeBS.

Population and distribution

The curlew is a widespread breeding species, occurring throughout much of Britain, but absent from most of south-east England, and sporadic in south-west England, north-west Scotland and parts of Ireland. There have been declines in breeding numbers in Ireland, western Scotland, south-west Wales and southern England which have been attributed to land drainage, re-seeding of moorland and afforestation (*88–91 Atlas*). In winter, the curlew has an essentially coastal distribution in Britain, the preferred habitat being large estuarine mudflats, although birds also winter inland. In Ireland, a higher proportion winters inland, with the Shannon valley being particularly favoured (*Winter Atlas*). There are an estimated 38,000–43,000 pairs breeding and 123,000 individuals wintering in the UK (*Population Estimates*).

Ecology

Curlews typically breed in upland areas, favouring moist moors and heaths, but they also breed on rough grassland. A clutch of usually four eggs is laid between late April and late June and there is one brood (*Red Data Book*).

Breeding season survey – population

See *Waders* in the generic survey methods section.

Winter survey

WeBS.

See *Generic wintering bird monitoring methods* in the generic survey methods section.

Redshank
Tringa totanus

Status

Amber listed: WI, SPEC 2 (D)
Annex II/2 of EC Wild Birds Directive

National monitoring

Saltmarsh breeding survey 1996 (RSPB).
WeBS.

Population and distribution

Breeding redshank are found throughout Britain and Ireland on wet grassland and coastal saltmarshes. Reductions in numbers have occurred in north-east and central Scotland, inland southern England and in central Ireland. Implicated in the decline is the loss of habitat due to drainage and agricultural intensification (*88–91 Atlas*). In comparison with the breeding season, the redshank's winter distribution is more coastal, and the species is found anywhere with suitable feeding habitat on the coasts of Britain and Ireland (*Winter Atlas*). There are estimated to be 32,000–35,000 breeding pairs and 122,000 wintering redshank in the UK (*Population Estimates*).

Ecology

The redshank breeds mainly on the coast, with the highest densities on the middle and upper parts of saltmarshes, and locally on coastal grazing marshes and damp machair. Eggs are laid from April to the end of June with a peak in May, but replacement clutches are frequent. Three-quarters of those wintering in Britain do so on estuaries and some continue feeding on pastures at high tide (*Red Data Birds*).

Breeding season survey – population

See *Waders* in the generic survey methods section.

Winter survey

WeBS.

See *Generic wintering bird monitoring methods* in the generic survey methods section.

Greenshank
Tringa nebularia

Status

Amber listed: BL
Non-SPEC
Schedule 1 of WCA 1981
Annex II/2 of EC Wild Birds Directive

National monitoring

National surveys: 1995 (RSPB/SNH/JNCC), 2005.
WeBS.

Population and distribution

There are an estimated 1,440 summering pairs of greenshanks in the north and west of Scotland (Hancock et al 1997). The range of this species has contracted slightly and birds have disappeared from some areas in recent years mainly due to afforestation of suitable breeding habitat (*88–91 Atlas*). Breeding occurs primarily in peatland areas, but also in open forest bog. Between 600 and 1,000 birds overwinter. Most of the breeding population is thought to winter around the coast of the British Isles, in the south and west of Britain and Ireland (*Winter Atlas*).

Ecology

Greenshanks breed on the open wet moorland and flow country of the north. A clutch of usually four eggs is laid in May to mid-June. There is one brood and the young fledge from the end of May (*Red Data Birds*).

Breeding season survey – population

This method follows Hancock et al (1997).

Information required
- estimated number of summering pairs
- estimated number of pairs that hatched young
- maps showing the area covered, registrations and survey route.

Number and timing of visits

Two visits, the first in the period 10 April to 25 May (highest detectability 15 April to 8 May), the second in the period 26 May to 11 July (highest detectability 1–23 June). At least one of the two visits should be within a high-detectability period.

Time of day

Any time of the day before 1800 BST.

Weather constraints

Do not survey in conditions of poor visibility, continuous rain or wind speeds of Beaufort force 5 or greater.

Sites/areas to visit

The Scottish Highlands, sub-montane zone. The following areas can be excluded from the survey: areas higher than than 800 m above sea-level; slopes steeper than 13% (>25 25-ft contours per 1 km^2 diagonal at 1:25,000); built-up areas; enclosed farmland; dense native or planted forests at thicket stage or older; and offshore islands smaller than 25 ha. All waterbodies should be visited, including lochs, lochans, pool complexes and rivers. Open forests, young plantations and coastal areas (except sea cliffs) should also be surveyed.

Equipment

- 1:25,000 OS maps of the area
- prepared field recording maps
- Schedule 1 licence.

Safety reminders

A reliable person should know where you are and when you are due back. Carry a compass at all times and know how to use it. When surveying remote upland areas, take spare warm clothing, a plastic survival bag, first-aid kit and food supplies.

Disturbance

Minimal; there is no need to search for nests or to get close to adults. Adults with young chicks are likely to be disturbed when pool systems and lochs are checked in June, but keep disturbance to a minimum.

Methods

Mark the boundary of the survey area (using a solid line) and a predetermined survey route (using a dotted line) clearly on a map. An area of 5 km^2 will take about 2–3 days for one person to complete. Use a new field map on each visit to the survey area.

Visit all waterbodies within the survey area and get to within 500 m of all other areas to be surveyed. Record all greenshank registrations on the map using standard BTO notation (Appendix 1). Record the behaviour of any greenshank seen, noting in particular any adult alarm-calling (chipping), any broods seen or adults exhibiting mobbing behaviour.

If working in a pair or team, record the time of each greenshank registration, then resolve any possible duplications immediately after each field visit.

All the registrations from different observers and from different days should be put on a single summary map for each site and for both visits. Code any registrations from the first visit with an 'A', and those from the second visit with a 'B'. Halve the peak count of individuals from any one visit to give the estimated number of summering pairs.

For the second visit only, sum the number of adults either heard 'chipping' at or mobbing the observer for an extended period, and give this figure as the estimated number of pairs that hatched young.

Winter survey

WeBS.

See *Generic wintering bird monitoring methods* in the generic survey methods section.

Reference

Hancock, M H, Gibbons, D W and Thompson, P S (1997) The status of breeding greenshank *Tringa nebularia* in the United Kingdom in 1995. *Bird Study* 44: 290–302.

Turnstone
Arenaria interpres

Status

Amber listed: WI
Non-SPEC

National monitoring

WeBS (especially NEWS).

Population and distribution

A common coastal wader during the winter in Britain and Ireland, turnstone is only scarce along the north-west coast of mainland Scotland and the Inner Hebrides. Within-winter movements exceeding 10 km are rare, and birds tend also to be site-faithful between winters (*Winter Atlas*). An estimated 69,700 individuals winter in the UK (*Population Estimates*).

Ecology

Wintering turnstones are confined to coastal habitat, frequenting estuaries, sandy beaches and particularly rocky shores. They forage in small groups and favour mussel beds or strand lines, where they can turn over or push aside fronds of seaweed or stones to find shrimps, winkles, barnacles and other invertebrates (*Red Data Birds*).

Winter survey

WeBS.

See *Generic wintering bird monitoring methods* in the generic survey methods section.

Red-necked phalarope *Phalaropus lobatus*

Status

Red listed: HD, BR
Non-SPEC
Schedule 1 of WCA 1981
Annex I of EC Wild Birds Directive

National monitoring

Rare Breeding Birds Panel.

Population and distribution

Red-necked phalaropes are a rare breeding summer visitor to Britain and Ireland. They occur in the far north and west, especially Shetland, Orkney, the Outer Hebrides and County Mayo. The breeding population has declined over a long period, mostly as a consequence of habitat loss. Conservation measures aimed at halting this habitat loss are proving successful (*88–91 Atlas*). There are currently about 36 males in Britain (*Population Estimates*).

Ecology

Red-necked phalaropes breed at sites with open water, emergent swamp, wet and dry mire and old peat workings. Birds arrive in May and mate during late May and early June. A clutch of 3–7 eggs is laid between early June and early July; there is one brood. Incubation lasts 18 days and is by the male alone, with hatching in late June to late July. The chicks are quickly able to fend for themselves, although tended by the male; they fledge between late July and mid-August (*Red Data Birds*).

Breeding season survey – population

This method is based on research undertaken on Fetlar by O'Brien (in prep).

Information required

- average number of males
- maps showing sightings of males.

Number and timing of visits

At least twice a week, but more frequently if possible; 20 June to 20 July.

Time of day

Any time of day.

Weather constraints

Due to the disturbance aspect of the method, do not survey in heavy rain or very cold weather.

Sites/areas to visit

Areas with a mixture of shallow pools and emergent vegetation with wet marsh.

Equipment

- chest waders
- 1:10,000 detailed maps of mires
- Schedule 1 licence.

Safety reminders

Tell someone where you are going and when you are due back.

Disturbance

The method is based on disturbing males, either while incubating, or with newly hatched young, so do not stay in one spot for more than about 10 minutes (the typical period for which a male usually leaves the nest to feed) and minimise disturbance to other breeding species nearby, eg terns and skuas.

Method

On Fetlar, paired birds frequently appear up to 2 km from where the male subsequently breeds. The Fetlar mires, which are 2–5 ha in size, each take 1–1.5 hrs to survey on each visit, depending on the number of males present.

If possible, scan the site from a high vantage point to gain an idea of where most of the visible birds are. Plan a survey route to include the perimeter of all areas of wet marsh. Where the site is large, it may be necessary to walk through the centre of the marsh; waders will be necessary. All sightings of males should be marked on detailed maps of the site, along with the apparent stage of the breeding cycle which the male has reached. This can be determined by the male's behaviour: if he is paired then laying will have started but the clutch will still be incomplete; if he is solitary, sneaky and aggressive toward females then he is likely to be incubating a clutch; if he is flying around and calling he is looking after chicks. Record all males. There is no need to record females; by the end of the survey period most females have left the breeding mires anyway.

To calculate the population size, sum the total number of breeding males recorded across all visits and divide by the number of visits. This provides an estimate of the number of breeding pairs.

Breeding season survey – breeding success

Information required

- total number of males acting as if with young.

Number and timing of visits

As for the population survey (above), a minimum of every three days. Check sites from about 18 June to the end of August.

Time of day, Weather constraints, Safety reminders

As for the population survey (above).

Sites/areas to visit

As above, males with young are more likely to be in the very wet areas.

Disturbance

As for the population survey, although only males will be with young.

Methods

As for the population survey. The first males acting as if with young normally appear about 20 June. Males tend to fly from one pool to another, calling, or continually fly overhead. Often two or three other males will join in. The level of response to disturbance varies with the age of chicks and number of males present, although it is usually greatest 4–5 days after chicks have left the nest. As the chicks get older, so the male's response becomes less intense, covers a wider area and is less frequent. Males continue to respond to disturbance for up to 20 days after hatching (contrary to previously quoted Finnish studies).

It is useful to attempt to recognise males individually at this stage. Plumage variation combined with the onset of moult means that this is often relatively straightforward. Record the approximate location of displaying males on a map.

A combination of mapping locations of displaying males and individually identifying them through plumage variation builds up a picture of the total number of males that have managed to hatch successfully at least one egg from a clutch.

Breeding season survey – productivity

Although the method outlined above gives a crude measure of breeding success, it does not quantify the total productivity (ie number of chicks fledged) of the site. A method to do this is outlined below.

Information required
- total number of fledged juveniles.

Number and timing of visits

Every three days, 15 July to 25 August. Do not visit the site any more or less frequently, otherwise the final estimate of the total number of juveniles fledged from that site will not be reliable.

Time of day, Weather constraints, Sites/areas to visit, Safety Reminders

As for the population survey (above).

Disturbance

The method is based on disturbing fledged juveniles. Consequently, do not to stay in one spot for longer than is absolutely necessary.

Methods

The first fledged juveniles are seen from about 21 days after the first eggs hatch. Juveniles hide in emergent vegetation around the edge of open water, so they tend to be more secretive than males with young and require considerably more effort to locate. Plan a route that ensures coverage of all areas of open water. Follow the same route on each visit, as this minimises damage to the vegetation and provides a corridor which fledged juveniles use, thus creating additional good habitat. Map the location of fledged juveniles, as this gives a good idea of preferred areas within the mires and minimises the chances of double-counting.

From detailed observations of individually colour-ringed birds on Fetlar, it is known that fledged juveniles remain on site for about six days and that there is about a 50% chance of seeing a fledged juvenile on any one visit. Thus, if a site is visited once every three days, the total number of fledglings recorded on the site (ie the sum of all individual visit counts) is the best estimate of the total number of juveniles fledged.

Information on fledged juveniles seen on other areas may also be of use. During the work undertaken on Fetlar, very few fledged juveniles were seen away from the breeding mires, though this was not always the case.

Contributed by Mark O'Brien

Reference

O'Brien, M (in prep) *Ecology and Habitat Requirements of Red-necked Phalaropes on Fetlar*. RSPB unpubl.

Arctic skua *Stercorarius parasiticus*

Status

Non-SPEC

National monitoring

Seabird Colony Register (SCR).
Seabird Monitoring Programme (SMP).
Orkney and Shetland were surveyed as a whole in 1992, Hoy in 1996, other colonies 1985–1987, but 1996 figures are available for Handa and St Kilda.
Seabird 2000 (1999–2001).

Population and distribution

Britain holds 1–3% of the world's population of breeding arctic skuas (Lloyd et al 1991). The species breeds on coastal moorlands in the extreme north and west of Scotland. The population increase in the 1970s and early 1980s probably followed that of other seabird species that arctic skuas kleptoparasitise (*88–91 Atlas*). Over half of British arctic skuas breed in Shetland, where numbers have been declining since the late 1980s (Ewins et al 1986, Weirs et al 1988). Although there are several threats to the species, the main problem on Shetland recently has been a lack of food. The UK breeding population is around 3,200 territories (*Population Estimates*).

Ecology

Compared with great skuas, arctic skuas prefer lower slopes and drier ground, breeding on heathland and mossy vegetation. The clutch, usually of two eggs, is laid from late May, and young hatch from late June. Incubation lasts 25–28 days and young fledge to maturity in 25–30 days. The young can leave the nest within a few days of hatching, and the brood usually separates, though chicks do not normally wander more than 100–200 m from the nest (*BWP*).

Breeding season survey – population

The same methods can be used for monitoring population and productivity of both arctic and great skuas. These methods are taken from the *Seabird Monitoring Handbook*.

Information required

- maximum number of Apparently Occupied Territories
- maps showing their location
- the number of non-breeding birds.

Number and timing of visits

Three visits, late May to mid-July, preferably June.

Time of day

Any time, but note the time of day of the count as this can affect the number of non-breeding birds recorded in the colony. If non-breeding birds are to be counted, this is best done just before sunset as non-breeders form into 'clubs' at this time (Klomp and Furness 1990, 1992). Spend as much time as necessary.

Weather constraints

Throughout the season, to minimise the risk of chilling eggs or chicks, avoid colony visits during heavy rain, strong winds, prolonged wet weather or fog.

Sites/areas to visit

Coastal grassy moors in the extreme north and west of Scotland.

Equipment

- 1:10,000 map of the area
- clipboard
- padded hat and/or bamboo cane
- transect markers.

Safety reminders

Only visit an area at dusk if you have previously visited it during the day. A reliable person should always know where you are and when you are due back. Take care on steep slopes, especially next to cliffs (appropriate gripped footwear should be worn). Skuas are likely to dive at anyone entering a breeding colony. A padded hat will provide some peace of mind (birds may injure themselves on hard hats).

Disturbance

Disturbance in colonies with high densities of nesting birds makes counting difficult. In such situations sit or stand still for a few minutes at periodic intervals, to allow the birds to resettle in their territories. Avoid disturbing colonies in wet weather, especially those with high breeding densities where birds are more likely to be put off their eggs. Great care should be taken not to flush young skuas, as they are very vulnerable to predation by neighbouring adult skuas.

Methods

The whole colony needs to be surveyed, therefore one of the first tasks is to find and map its extent. If the colony is very large, with birds nesting at a low density, it may be necessary to check different areas over successive days. Once the extent of the colony is mapped, mark parallel transects on the map. Allow about 500 m between each transect, but make them closer together according to the density of the colony and the nature of the terrain (eg if the ground is undulating and some areas are not visible from the transects). Mark the beginning and end of each transect on the ground to make them more accurate.

Walk each transect, stopping at regular intervals (eg every 200–300 m) and thoroughly scan all round with binoculars and telescope. Record all Apparently Occupied Territories (AOTs).

Score an AOT if any of the following are seen:

- nest, eggs or chicks
- apparently incubating or brooding adult
- adults distracting or alarm-calling
- a pair or single bird in potential breeding habitat.

The following should not be scored as AOTs:

- bird(s) flying past en route to somewhere else

- feeding individual(s)
- single bird (or pair) which is flushed from an area, and which then flies completely out of sight
- three or more skuas of the same species regularly together but not showing signs of territoriality.

Record all evidence of territorial skuas by plotting sightings on large-scale maps using different codes for (eg) nests, eggs and adults giving alarm-calls. Take care not to count the same AOT more than once from the same or different transects.

Territorial birds may utilise prominent mounds in their territories and these can be a useful indicator of an AOT (though territorial birds may use more than one mound). In dense colonies it is worth taking extra time to observe territorial behaviour to avoid assigning to different territories members of the same pair which are standing apart.

The breeding population should be reported as the maximum number of AOTs recorded on a single visit.

Wherever possible, census the number of non-breeding skuas in the colony. These birds tend to gather in groups or 'clubs' that occupy small areas about 50–100 m wide. The number of these birds changes throughout the day; counts just before dusk are best, but always note the time of the count. Report the number of non-breeding birds seen in clubs on each visit.

For arctic skuas, record the relative proportions of light and dark colour phases of territorial birds.

Breeding season survey – productivity

Method 1

This method is the least intensive of the two.

Information required

- number of young fledged per AOT.

Number and timing of visits

Two visits, a week or two apart, first visit about one week after first fledging 20–25 July. (This is ignoring the need to visit in May/June to assess numbers of AOTs.)

Time of day

Any time.

Weather constraints, Sites/areas to visit, Equipment, Safety reminders, Disturbance

As for the population survey (above).

Methods (*Method 1: Young fledged per Apparently Occupied Territory*)

Productivity can be assessed for whole colonies (especially smaller ones, of say <100 AOTs) or for sample areas of larger colonies. If sample areas are to be used, these should be selected randomly. In large colonies, try to follow 50–100 AOTs in total, preferably in two or more areas of similar population size. Further details on sampling techniques are given in the *Introduction* and in the *Seabird Monitoring Handbook*.

Well-grown chicks are those which have lost more than half of their down feathers on the mantle/scapulars/upperwing coverts. Fledged chicks stand up and are easy to see, whereas chicks that have not yet flown crouch and hide. Recent fledglings appear more round-winged than adults and fly rather poorly. Fledgling arctic skuas have a distinctive dark 'scaly' plumage because their dark brown feathers are edged light brown, while fledgling great skuas appear more uniformly dark than adults.

Assess the numbers of AOTs in late May/June using methods given for population monitoring. The visit dates (above) can be further refined by noting the approximate ages of chicks seen in June, thus allowing calculation of when first chicks are likely to fledge. Using similar transect and scanning methods as for population monitoring (see above), map and count fledglings and well-grown chicks. Arctic skuas first fly when about four weeks old and great skuas at about six weeks. Fledglings of both species tend to remain in their natal area for one to three weeks after fledging.

If possible, try to relate each fledgling or chick to a territory. Adults generally defend chicks that cannot fly, but once chicks have fledged the adults are much less inclined to swoop at people and tend to fly off with their fledglings, giving aerial protection. In a large or dense colony, large numbers of flying young can cause confusion; it is important to retreat to a good vantage point to let the birds settle down before continuing the survey. Be aware that some large chicks may be difficult to see, even if vegetation cover is limited. Some prior experience or practice at locating chicks is advisable. If individual fledglings cannot be related to particular territories, use the maximum total count of fledglings and near-fledglings as your estimate.

Keep a separate note of any smaller chicks present (ie ones that are still mainly downy above). If possible, re-check territories containing these small chicks a week or two later, especially if about 20% of the total of young birds are small on the first date. This will improve the accuracy of the final estimate of productivity.

When calculating productivity for the colony as a whole, or for each sample plot, divide the number of large chicks or fledglings produced by the number of AOTs. For two or more sample areas, express the colony's productivity as the mean ± standard deviation of the individual plot figures.

It is useful to record the numbers of dead (uneaten) and dead (eaten) chicks/fledglings seen in the colony.

Method 2

This more intensive method follows the fate of individually marked nests, in either the whole colony or sample areas of a colony.

Information required

- number of young fledged per individually marked nest.

Number and timing of visits

At a minimum, visit every 5–7 days from around the date of first hatching throughout the chick-rearing period. If possible, also visit two or three times at intervals of 10–15 days during the main incubation period. The main incubation period is late May to mid-June for arctic skuas and mid-May to mid-June for great skuas.

Time of day

0900–1600 BST; visit at different times on different days.

Weather constraints, Sites/areas to visit

As for the population survey (above).

Equipment

- marked stakes or bamboo poles
- indelible pen
- ringing/processing equipment.

Safety reminders, Disturbance

As for the population survey.

Methods (*Method 2: young fledged per marked nest*)

On the first visit, during the main incubation period, scan from suitable vantage points to locate incubating adults or other birds at nest-scrapes. Mark each nest with a bamboo pole or wooden stake. To reduce disturbance, try to note the position of several adjacent nests before marking them, rather than pinpointing and marking each nest separately. After each nest or group of nests is marked, locate a suitable vantage point nearby from which to continue searching.

Ideally, place nest-markers a set distance and direction from the nest (say 5 m south-west). Take care to ensure that nest-markers are not too obvious, to avoid attracting the attention of people or predators to nests. The position of marked nests should be noted on 1:10,000 maps, with notes on any useful landmarks nearby, to aid locating nests later.

At the next visit (10–15 days later), and on each subsequent visit, note the contents of each nest already marked and search for and mark any other nests that have appeared since the previous visit.

Once hatching begins, visit every 5–7 days. As a minimum, keep a note of the approximate size/age of each chick. Note also any dead chicks and any evidence of cause of death. Chicks should be rung (by a licensed ringer only) as soon as they are old enough (usually about 10 days) to allow individual identification. Chicks of 10 days and older wander considerably, so ringing is essential in obtaining accurate data on production from individual marked nests. If time allows, on each date weigh each chick to the nearest gram, and measure the wing (flattened, straightened chord of outer wing, excluding down) using a stopped wing rule. This will provide information on growth rates, which may give an indication of food availability. Measuring the wings of dead chicks may indicate age at death. Take care that handled chicks remain crouched and do not wander off (loose vegetation placed gently over their heads may ensure this). Record such information on a standard form (Figure 1).

Assess numbers of chicks fledging by visiting nests every 5–7 days from around the date of first fledging onwards (roughly over the period 10

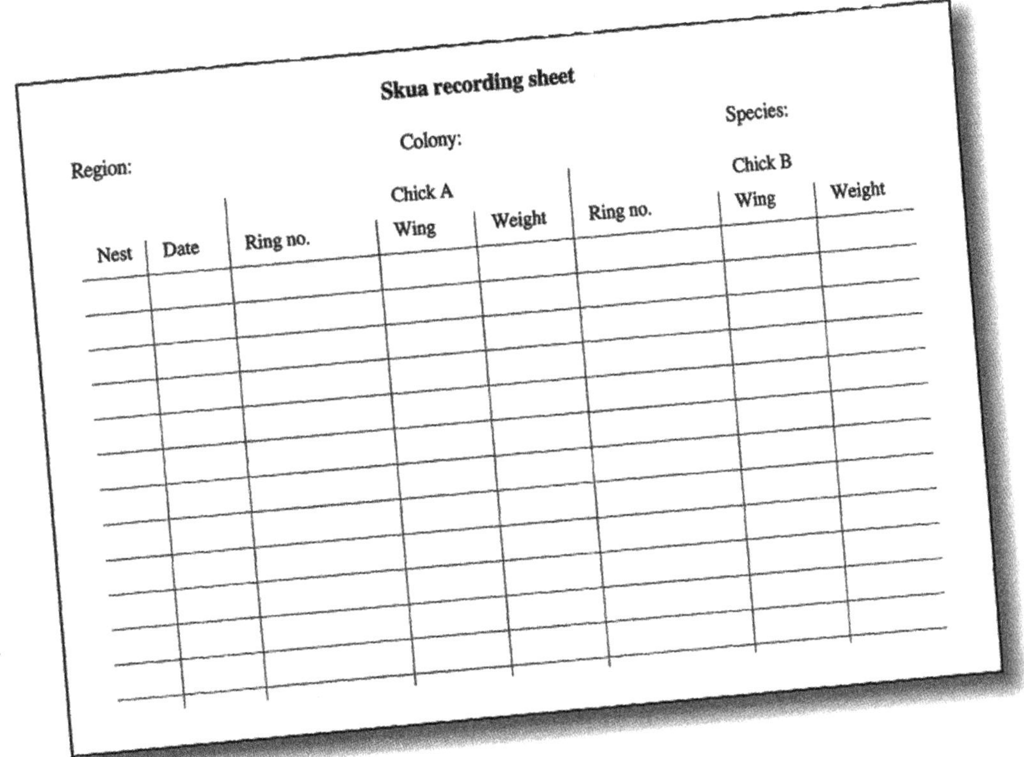

Skua recording sheet

Region: Colony: Species:

Nest	Date	Chick A Ring no.	Wing	Weight	Chick B Ring no.	Wing	Weight

Figure 1
A sheet for recording skua productivity, as recommended by the *Seabird Monitoring Handbook*.

July to 10 August). Record numbers and approximate ages/sizes of chicks associated with each nest. Visits should continue until the outcome of each nest is known. Assume that any chicks surviving to four weeks old (arctic skua) or six weeks (great skua) will fledge successfully.

Express productivity as number of chicks fledged per nest found with eggs. Where two or more sample plots of a colony are studied, use the mean ± standard error of the individual plot figures.

Keep a separate note of any known post-fledging mortality, although quantitative assessment of this is difficult and it is not incorporated in the productivity assessment.

References

Ewins, P G, Wynde, R M and Richardson, M G (1986) *The 1986 Census of Arctic and Great Skuas on Foula, Shetland*. Nature Conservancy Council NE Scotland Region report.

Klomp, N I and Furness, R W (1990) Variations in numbers of nonbreeding great skuas attending a colony. *Ornis Scandinavica* 21: 270–276.

Klomp, N I and Furness, R W (1992) Non-breeders as a buffer against environmental stress: declines in numbers of great skuas on Foula, Shetland, and predictions of future recruitment. *J. Applied Ecology* 29: 341–348.

Lloyd, C, Tasker, M L and Partridge, K (1991) *The Status of Seabirds in Britain and Ireland*. Poyser, London.

Weirs, P G, Ellis, P M, Bird, D B and Prior, A (1988) The distribution and status of arctic and great skuas in Shetland 1985–86. *Scottish Birds* 15: 9–20.

Great skua
Catharacta skua

Status

Amber listed: BI, BL
SPEC 4 (S)

National monitoring

Seabird Colony Register (SCR).
Seabird Monitoring Programme (SMP).
Seabird 2000 (1999–2001).

Population and distribution

The British breeding population of great skuas constitutes 57% of the world population (*Red Data Birds*). They breed in the far north of Scotland, mostly on the Northern and Western Isles. Their numbers have increased steadily since the start of the 20th century, but more recently numbers have fallen in the main breeding area in Shetland. The estimated UK breeding population of great skuas is around 8,500 territories (*Population Estimates*).

Ecology

Great skuas breed in loose colonies on coastal grassy moors, showing preference for wetter areas. A clutch of two eggs is laid in May–June and the young fledge from early July (*Red Data Birds*).

Breeding season survey – population and productivity

See arctic skua for a description of methods.

Common gull
Larus canus

Status

Amber listed: SPEC 2 (D)
Annex II/2 of EC Wild Birds Directive

National monitoring

Seabird Colony Register (SCR).
Seabird Monitoring Programme (SMP).
Seabird 2000 (1999–2001).

Population and distribution

Of the whole European breeding population of common gulls, 15% breed in Britain and Ireland (*Birds in Europe*). Common gulls nest on the ground and require sites which are fairly free of predators, such as small coastal or inland islands, but breeding also occurs inland on moor and bog. The breeding distribution is almost entirely restricted to the north and west of Britain and Ireland (*88–91 Atlas*). There are estimated to be 68,500 pairs of common gull breeding in the UK (*Population Estimates*).

Ecology

Unlike most *Larus* species, common gulls are equally adapted to exposed marine coasts and to inland sites. They are mostly colonial. A clutch of usually three eggs is laid. In northern Britain laying occurs mostly during late May and June. Incubation lasts 22–28 days; the young are cared for by both parents, and fledging takes 35 days (*BWP*).

Breeding season survey – population and productivity

Gull population and productivity monitoring methods (taken from the *Seabird Monitoring Handbook*) are outlined in the generic survey methods section.

Lesser black-backed gull *Larus fuscus*

Status

Amber listed: BI, BL
SPEC 4 (S)
Annex II/2 of EC Wild Birds Directive

National monitoring

JNCC Seabird Colony Register (SCR).
Seabird Monitoring Programme (SMP).
Seabird 2000 (1999–2001).

Population and distribution

The lesser black-backed gull breeds around the coasts of Britain and Ireland, and increasingly colonies are found inland and in urban areas. This species has a more southerly distribution than the herring gull. The population increased between 1940 and 1970 but, unlike herring gulls, their population has continued to increase (Lloyd et al 1991). There are however, marked differences in success between colonies with large declines in some areas (*88–91 Atlas*). The UK breeding population is estimated at 85,000 pairs (*Population Estimates*).

Ecology

Lesser black-backed gulls breed on open flat or sloping sites, sometimes partly sheltered by vegetation, and increasingly on the flat roofs and ledges of buildings. They lay a clutch usually of three eggs from mid-May, incubation lasts 24–27 days and the young fledge 30–40 days after hatching. The young are cared for by both parents; they can leave the nest after a few days, but do not wander far (*BWP*).

Breeding season survey – population and productivity

Gull population and productivity monitoring methods (taken from the *Seabird Monitoring Handbook*) are outlined in the generic survey methods section.

Reference

Lloyd, C, Tasker, M L and Partridge, K (1991) *The Status of Seabirds in Britain and Ireland*. Poyser, London.

Herring gull
Larus argentatus

Status

Amber listed: BDM
Non-SPEC

National monitoring

Seabird Colony Register (SCR).
Seabird Monitoring Programme (SMP).
Seabird 2000 (1999–2001).

Population and distribution

The herring gull breeds mainly around the coasts of Britain and Ireland, but there is an increasing number of colonies inland (*88–91 Atlas*). After a dramatic increase in population levels between 1940 and 1970 the population started to be controlled by culling. It is probably partly as a result of this that the herring gull population halved between 1970 and 1987 (Lloyd et al 1991). There are an estimated 180,000 pairs of herring gull breeding in the UK (*Population Estimates*).

Ecology

Herring gulls breed on open, often sloping sites on the ground, but also on cliff-ledges and increasingly on flat roofs of buildings. They are colonial and lay a clutch usually of three eggs from late April. Incubation lasts 28–30 days, with young fledging 35–40 days after hatching and becoming independent soon after (*BWP*).

Breeding season survey – population and productivity

Gull population and productivity monitoring methods (taken from the *Seabird Monitoring Handbook*) are outlined in the generic survey methods section.

Reference

Lloyd, C, Tasker, M L and Partridge, K (1991) *The Status of Seabirds in Britain and Ireland*. Poyser, London.

Great black-backed gull *Larus marinus*

Status

SPEC 4 (S)
Annex II/2 of EC Wild Birds Directive

National monitoring

Seabird Colony Register (SCR).
Seabird Monitoring Programme (SMP).
Seabird 2000 (1999–2001).

Population and distribution

Britain and Ireland hold about 20% of the European population of breeding great black-backed gulls (*Birds in Europe*). The breeding distribution is concentrated around the coast, with much of the population breeding in Scotland (*88–91 Atlas*). Since 1970, there has been little change in the breeding numbers and range of this species, and there are estimated to be around 20,000 pairs breeding in the UK (*Population Estimates*).

Ecology

Great black-backed gulls are colonial or solitary breeders, preferring more oceanic to land-locked coasts and inlets. They breed on rocky outcrops, moorland, grassy islands and occasionally on buildings. A clutch of 2–3 eggs is laid from May and incubation lasts 27–28 days. Young can wander from the nest a few days after hatching, but stay in the vicinity and fledge to maturity in 7–8 weeks (*BWP*).

Breeding season survey – population and productivity

Gull population and productivity monitoring methods (taken from the *Seabird Monitoring Handbook*) are outlined in the generic survey methods section.

Kittiwake
Rissa tridactyla

Status

Non-SPEC

National monitoring

Seabird Colony Register (SCR).
Seabird Monitoring Programme (SMP).
Seabird 2000 (1999–2001).

Population and distribution

Britain and Ireland hold about 8% of the world breeding population of kittiwakes (*Birds in Europe*). Breeding takes place on steep cliff-faces around the British and Irish coasts, although the species is largely absent from the coastline of south-east England (*88–91 Atlas*). The population growth of some colonies, most noticeably on Shetland, has declined, and this has been linked to a reduction in breeding success as a result of changes in food availability (Heubeck 1989). An estimated 500,000 pairs breed in the UK (*Population Estimates*).

Ecology

Kittiwakes breed in colonies on sheer, usually high cliffs with narrow ledges, close to the sea. The clutch of 2–3 eggs is laid from late May, incubation lasts 25–32 days and the young fledge after about 43 days (*BWP*).

Breeding season survey – population

Information required

- maximum number of Apparently Occupied Nests
- number of unattended empty nests, nests with unattended eggs or dead chicks, occupied trace nests, and adults – if time allows
- maps of the boundaries of the survey area and approximate colony extent.

Number and timing of visits

At least one, but preferably three, visits during the latter half of incubation (when numbers of nests are most stable), usually late May to mid-June. In either case, counts of each section of cliff should be repeated, where possible, as a check on accuracy. If the season appears to be unusually late (indicated by a high proportion of 'trace' nests or unoccupied well-built nests in June), a count in late June is a useful further check.

Time of day

Any time of day.

Weather constraints

Avoid colony visits during heavy rain, strong winds or prolonged wet weather.

Sites/areas to visit

Any seabird cliffs not already covered.

Equipment

- 1:10,000 map of the area
- photographs of the colony (optional)
- telescope (optional)
- boat (optional) and life-jackets.

Safety reminders

When working alone in remote areas ensure someone knows where you are going and when you will return. A second person should be present whenever boats are used, and experience of the sea in the area and boat-handling skills are essential. Rocky shores are always hazardous because of their uneven, slippery surfaces, weed-covered rocks and fissures. On cliffs and crags beware of slippery vegetation at cliff edges and of undercut or loose strata.

Disturbance

There is no need to disturb the kittiwake colony being counted: counting can be done inconspicuously from above (eg adjacent cliff top) or below.

Methods

The procedure outlined here follows Heubeck et al (1986), with modifications.

Clearly define the boundaries of the census area, whether a stretch of coast (recommended) or a colony. This area should be consistent from year to year. Divide colonies into subsections of coastline, using natural features definable on a 1:10,000 map. Check all suitable habitat along the whole census area before or during the main counting period to detect newly established colonies. In particular, look for and map any kittiwake roosts on cliffs, as these can develop into colonies. For coastlines with small or sparse colonies, counts from a boat in calm conditions can provide accurate figures and allow rapid checking for new nesting areas. Large or dense colonies are often very difficult to count accurately from a boat, so as much of the colony as possible should be counted from land.

Subdivide individual cliff-faces, especially where numbers of kittiwakes are high, using obvious ledges, fissures or other features, as this will help avoid under- or double-counting (photographs or rough sketches are helpful).

Count all Apparently Occupied Nests (AONs). Whenever time allows, also count unattended empty nests, unattended eggs or dead chicks, occupied trace nests, and adults.

An AON is defined as a well-built nest capable of containing eggs, with at least one adult present. Poorly built 'trace' nests with adults in attendance are more likely to involve non-breeding birds, but additional counts of these can be useful, as a high proportion of trace nests may indicate a late breeding season or, possibly, a decrease in the proportion of adults breeding.

Keep a note of (and map) any parts of a colony that might not be visible from land. Estimate (minimum–maximum) the number of AONs likely to be hidden, based on numbers on visible sections. However, when reporting these estimates be very clear that they are of unknown reliability, and may not be directly comparable with other counts. If at all possible, check and count hidden sections from a boat on a calm day, especially if you estimate that hidden sections are likely to total more than 10% of the population. Boat counts of nests are quite useful for this species as there is little danger that the nests will be hidden when viewed from below. However, separating trace nests from well-built ones can be difficult if viewing angles are steep.

Report the highest reliable count of AONs from a single count of the whole colony rather than the sum of individual subsection maxima.

Breeding season survey - productivity

For a highly accurate assessment of breeding productivity, visit study plots every two or three days throughout the season, recording the degree of construction of nests, dates of laying, hatching and fledging. This method is probably too labour-intensive for general use and the recommended lower-input method (*Method 1*) requires at least five visits, to map Apparently Occupied Nests (AONs) and record the number of chicks likely to fledge. If nests are not mapped, *Method 2* can still provide useful data with at least two visits, based on comparison of counts of AONs early in the season with counts of chicks around the time of fledging. These two methods are taken from the *Seabird Monitoring Handbook*. Method 1 was developed by Harris (1987, 1989).

Information required

- the number of completed nests and an estimate of the number of chicks fledged (for the whole colony or for each individual plot)
- maps or photographs of the colony and boundaries, showing subcolony sections or selected plots
- maps of individual nests (Method 1 only).

Number and timing of visits

Method 1 – At least five visits. One in May, to check nests, select plots and take photographs. Two or more visits in late May to mid-June to number nests and note chicks which have already hatched. Two or more visits in mid- to late July to count fledged young.

Method 2 – At least two visits. One (preferably more) in early or mid-June to count nests and note the presence of any chicks. One (preferably more) when chicks are estimated to fledge. If fledging date cannot be estimated (chicks are not seen in early June), a visit during 15–20 July is usually suitable.

Time of day

Any time of day.

Weather constraints

Avoid colony visits during heavy rain, strong winds or prolonged wet weather.

Sites/areas to visit

Any colony not already covered.

Equipment

- telescope (optional)
- 1:10,000 map of the area
- *Method 1* – if photographs are not already available for the colony, you will need black-and-white film and a camera
- *Method 1* – transparent overlays and fine-tipped coloured waterproof pens.

Safety reminders

Avoid choosing study plots that cannot be observed from a safe vantage point. Cliff edges may not be as solid as they look and may become more unsafe in wet and windy conditions. If working alone, always ensure someone knows where you have gone and when you intend to return. If working with boats, never work alone. The boat should be operated by an experienced and trained boat handler and life-jackets should be worn at all times. Take necessary equipment to deal with an emergency.

Disturbance

When counting cliff-nesting birds, the count position should not be so close as to cause disturbance interfering with the count and, potentially, with breeding success. Do not climb down a cliff or slope to count birds at closer quarters if it involves disturbing many cliff-nesting birds.

Methods

Breeding success can vary considerably among subgroups in a colony and between colonies on the same stretch of coast. If the colony is small, check as many nests as possible. If it is large, use randomly selected sample plots instead. To do this, identify all potential plots of 50–100 nests that are safely viewable without causing undue disturbance. Then, randomly select as many of these plots as possible to survey. The safety of observers and minimisation of disturbance to birds are paramount in study-plot selection. Where random selection is impractical, disperse the plots throughout the colony as much as possible. Several small plots are more likely to be representative than one or two large plots. Aim for 5–10 plots of 50–100 nests. The initial check and the marking of 50–70 nests on photographs may take an hour or more, but later checks will be both quicker and easier.

Method 1: Mapped nests

Make the first visit in May to check nests, select plots and photograph (using a black-and-white print film) the selected plots (if photographs are not already available). It is important to do this when the birds are present and, preferably, on nests. The same photographs can be used for several years as many of the same nest-sites and ledges will be used each season. Good photographs are essential, with nest-positions clearly distinguishable. Each print should have a maximum of about 70 nests, but prints may be overlapped if necessary to encompass slightly larger plots.

Make large prints (A4 size is ideal) and tape transparent overlays on them. Write on these using a fine-tipped waterproof pen. (Alternatively, mount the negatives as slides, project onto clean white paper, mark the nests, sites and prominent cliff features, and make photocopies for field use.) Sketches showing nest positions can also be used if photographs are not available early enough in the season, but even greater care is required to avoid confusion between nests.

Visit the colony in late May and mid-June and mark the following on the overlays or on sketch-maps:

- nests with birds apparently incubating
- other complete attended nests
- other site-holding birds with even a trace of a nest
- any unattended well-built nests (empty or otherwise).

Do not try to estimate clutch size or to confirm nest contents for standing birds (but keep a note of any clutches or empty nests that are immediately apparent); the basic unit is the well-built nest regardless of its contents.

Number the nests or traces sequentially, and note the state of each on a check-sheet. For an example data sheet which can be used to record details of individual nests, see Figure 1 in shag. Alternatively, to avoid using check-sheets in the field, use different symbols on the photo overlays to indicate different categories of nest. Suitable codes include the following:

I	Apparently incubating adult
c/1	Clutch of one egg
c/0	Empty well-built nest with adult in attendance
c/x	Well-built nest with adult standing, contents unknown

Using different symbols to mark nests or traces on photos can speed recording.

On early visits, keep a note of any chicks which have hatched and their approximate age (see Tables 1–2 for an ageing guide).

On second or later visits, ensure that all nests checked on the first visit are re-checked and their contents noted. Add any additional nests or traces that have appeared since the first visit.

Estimate when first fledging should occur (the incubation period is about 27 days, and chicks fledge from 35 days after hatching). Visit as close to this time as possible, check each nest marked previously and add any new nests. Note the numbers of young at each nest on your check-sheet, with an indication of their age or size. As a minimum, mark on the overlay or check-sheet the number of 'large' young (wing tips reaching or extending beyond tip of tail, little or no down) present alongside each nest. (Use a different coloured pen for each visit.) Also note any young that are not near to fledging, ie with wing tips obviously shorter than the tail, and split them into 'medium' (well-developed black-and-grey upperwing pattern) or 'small' (largely downy). More detailed notes on age are worth taking if time permits (see Tables 1–2 for guidance).

Do not waste time trying to determine the numbers of very small young in late broods. Try to return 5–7 days later and check these late nests. The more checks made, the better the result.

Assess how many young you think may fledge. When doing this, remember that large young sometimes move between nests, that young in broods of two or three sometimes fledge several days apart, and that fledged young may return for several days to their own or other nests. Keep a note of any chicks noticeably smaller than their siblings and check for their presence on the next visit. Assume that any large young (wing tips > tail) noted on the previous check have fledged if they disappear between visits. On the final visit, assume all large and medium chicks remaining will fledge. Keep a separate record of the

Table 1
A guide to assessing the age of kittiwake chicks (Maunder and Threlfall 1972, Harris 1987). Taken from the *Seabird Monitoring Handbook*.

Average age (days)	Description
9	Black tips to feathers of neck just visible
10	Tail feathers erupt
11	Black tips to upper wing-coverts visible
16	Black tips to vanes of tail feathers
25–30	Most down lost but still some on top of head and back
30	Wing tips equal length of tail
36	Wing tips 1–2 cm longer than tail
40–45	Wing tips 3–4 cm longer than tail

Table 2
Alternative suggestions for assessing size- or age-categories of kittiwake chicks. Taken from the *Seabird Monitoring Handbook*.

Size-category	Recording code	Description
a	S (small)	Chick completely downy
b	S	Downy chick, but black tips to upperwing coverts just visible
c	M (medium) or M/S	Clear grey/black pattern visible on upperside of wing, but still some down on upperwing and mainly downy elsewhere
d	L or M/L	No down on upperside of wings, some down elsewhere
e	L (large)	No down visible, wing tips at least equal to length of tail
f	F (fledgeable)	Wing tips 1–2 cm longer than tail
ff	FF (fully fledged)	Wing tips 3–4 cm longer than tail

number of small young. Where many broods (>20%) are still small and downy, please attempt a further visit. If there are fewer small chicks, assume half of them will fledge, as a very rough approximation.

For the whole colony or, where plots are used, for each plot, report productivity as the number of chicks fledged divided by the number of completed nests. Where plots are used, express colony productivity as the mean ± standard error of the figures for the individual plots.

If you wish to follow population changes in your study-plots (or changes in use of particular nest-sites or ledges) in subsequent years, label the transparent overlays with the year and plot boundaries and retain for future reference.

Method 2: Comparison of nest and chick counts

This method provides a useful indication of breeding productivity if few visits are possible in a season, or if insufficient time is available for Method 1 to be used. It is based on a comparison of nest and chick counts. Individual nests are not mapped.

Visit the colony late in the incubation period (ideally in early or mid-June) and count AONs as described under the population census method, above. If you cannot count the whole colony, select plots in the usual way, preferably randomly (see guidelines in the *Introduction*). If other visits are made in June, repeat your counts of AONs.

Note any chicks that are visible during your early visit(s), and their approximate age (see Tables 1–2), as a guide to timing of the season, then revisit the colony when you estimate that the first chicks are capable of fledging (35 days after hatching). A visit around 15–20 July is usually suitable if chicks were not seen in early June.

Count chicks, and split into categories: large, medium and small. If the number of chicks in a specific nest is unclear, use the average brood size from other nests. Record the number of medium and large chicks in July for the whole colony, or for each plot separately, where plots have been used. If fewer than 20% of chicks are still small, assume half of these will fledge. If more than 20% are small, re-do the chick count a week or more later.

If the whole colony has been surveyed, express productivity as the sum of all chicks considered to have fledged, divided by the maximum number of AONs from any one reliable count. If only a sample of the colony has been covered, divide the number of chicks considered to have fledged in each plot by the peak AON count for that plot. Then calculate the mean ± standard error of individual plot figures to give an estimate of colony productivity.

The figures obtained will usually overestimate actual productivity by 10–20%, partly because of subsequent chick loss. Overestimation may exceed 20% in some seasons and will be greatest if the initial count of AONs was made too early in the season or if many of the chicks counted in July do not survive to fledge.

References

Harris, M P (1987) A low input method of monitoring kittiwakes' *Rissa tridactyla* breeding success. *Biological Conservation* 41: 1–10.

Harris, M P (1989) *Development of Monitoring of Seabird Populations and Performance: final report to NCC* (Contractor: Institute of Terrestrial Ecology). Nature Conservancy Council, CSD rep. 941.

Heubeck, M, Richardson, M G and Dore, C P (1986) Monitoring numbers of kittiwakes *Rissa tridactyla* in Shetland. *Seabird* 9: 32–42.

Heubeck, M (ed) (1989) *Seabirds and Sandeels: proceedings of a seminar held in Lerwick Shetland, 15th–16th October 1988*. Shetland Bird Club, Shetland.

Maunder, S P and Threlfall, W (1972) The breeding biology of the black-legged kittiwake in Newfoundland. *Auk* 89: 789–816.

Sandwich tern *Sterna sandvicensis*

Status

Amber listed: BL, SPEC 2 (D)
Annex I of EC Wild Birds Directive

National monitoring

Seabird Colony Register (SCR).
Seabird Monitoring Programme (SMP).
Seabird 2000 (1999–2001).

Population and distribution

The coastal distribution of sandwich terns follows no obvious pattern. The colonies are quite sparse and widely distributed in Britain and Ireland, although most of the largest are on nature reserves (*88–91 Atlas*). The population in the UK has been generally stable since the late 1980s, following an increase (*88–91 Atlas*). There are estimated to be 17,000 breeding pairs of sandwich tern in the UK (*Population Estimates*).

Ecology

Sandwich terns nest in colonies, often associated with other terns or gulls, preferably on sand or shingle beaches, and rarely inland. A clutch of 1–3 eggs is laid during late April or May, and young fledge in late June or July (*Red Data Birds*).

Breeding season survey – population and productivity

Tern population and productivity monitoring methods (taken from the *Seabird Monitoring Handbook*) are outlined in the generic survey methods section.

Roseate tern
Sterna dougallii

Status

Red listed: BD, BR, SPEC 3 (E)
Schedule 1 of WCA 1981
Annex I of EC Wild Birds Directive

National monitoring

Rare Breeding Birds Panel.
Seabird Colony Register (SCR).
Seabird Monitoring Programme (SMP).
Seabird 2000 (1999-2001).

Population and distribution

Roseate terns breed at scattered colonies on small islands around the British and Irish coasts, and always with other tern species. The population since 1969 has declined by 80% and other European colonies have also declined. One possible reason for the decline is human persecution on the wintering grounds (*88–91 Atlas*). There were an estimated 65 pairs of roseate tern breeding in the UK in 1996 (Thompson et al 1997).

Ecology

Roseate tern nest-sites are usually sheltered, often overhung by rock or vegetation, and entrances to puffin burrows are also used. A clutch of 1–3 eggs is laid from May and the young fledge during late July to late August (*Red Data Birds*).

Breeding season survey – population and productivity

Tern population and productivity monitoring methods (taken from the *Seabird Monitoring Handbook*) are outlined in the generic survey methods section.

Reference

Thompson, K R, Brindley, E and Heubeck, M (1997) *Seabird Numbers and Breeding Success in Britain and Ireland, 1996.* UK Nature Conservation no. 21. JNCC, Peterborough.

Arctic tern
Sterna paradisaea

Status

Amber listed: BDM
Non-SPEC
Annex I of EC Wild Birds Directive

National monitoring

Seabird Colony Register (SCR).
Seabird Monitoring Programme (SMP).
RSPB 1989 survey of Orkney and Shetland.
Seabird 2000 (1999–2001).

Population and distribution

Arctic terns are concentrated in the north and west of Britain and Ireland and patchily distributed around the remaining coastline. There have been recent dramatic declines in breeding numbers, associated with food shortages (*88–91 Atlas*). There are estimated to be 44,000 pairs breeding in the UK (*Population Estimates*).

Ecology

Arctic terns nest in colonies, generally along the coastline but also at inland lochs or on open grazed areas. The clutch, usually of two eggs, is laid from mid-May, and the young fledge in July.

Breeding season survey – population and productivity

Tern population and productivity monitoring methods (taken from the *Seabird Monitoring Handbook*) are outlined in the generic survey methods section.

Little tern
Sterna albifrons

Status

Amber listed: SPEC 3 (D)
Schedule 1 of WCA 1981
Annex I of EC Wild Birds Directive

National monitoring

Seabird Colony Register (SCR).
Seabird Monitoring Programme (SMP).
Seabird 2000 (1999–2001).

Population and distribution

Little tern colonies are scattered around the coasts of Britain and Ireland, with many concentrated around the south-east of England (*88–91 Atlas*). Colonies are susceptible to human disturbance, tidal flooding and predation, so numbers fluctuate (*88–91 Atlas*). There are estimated to be 2,400 pairs of little tern breeding in the UK (*Population Estimates*).

Ecology

Loose colonies are formed on sand or shingle beaches. These are normally single-species and consist of fewer than 100 pairs. A clutch of 2–3 eggs is laid from mid-May, and the young fledge in July (*Red Data Birds*).

Breeding season survey – population and productivity

Tern population and productivity monitoring methods (taken from the *Seabird Monitoring Handbook*) are outlined in the generic survey methods section.

Guillemot
Uria aalge

Status

Amber listed: BI, BL
Non-SPEC

National monitoring

Seabird Colony Register (SCR).
Seabird Monitoring Programme (SMP).
Seabird 2000 (1999–2001).

Population and distribution

Colonies are distributed around the coasts of Britain and Ireland except for the south-east of England (*88–91 Atlas*). Colonies occur where suitable feeding and nesting habitats combine, and the largest are in Shetland, Orkney and northern Scotland (*88–91 Atlas*). Numbers are fairly stable and there are estimated to be 1,100,000 adults breeding in the UK (*Population Estimates*).

Ecology

Colonies are on coastal cliffs and rock stacks. A single egg is laid from May, and the young leave the colony about 20 days after hatching. The young are only one-third grown when they leave the colony, and remain in the care of the adult male on the sea from the end of June to about mid-September (*Red Data Birds*).

Breeding season survey – population

Guillemot populations are very difficult to census. The species breeds at high densities, and builds no nest. Two methods, both taken from the *Seabird Monitoring Handbook*, are outlined here. *Method 1* is a whole-colony census, while *Method 2* is a census of sample plots. With this species, plot counts cannot be used if criteria for number and timing of visits, time of day and weather constraints are not met.

Information required

- *Method 1* – the total number of individual adults on land in the entire colony
- *Method 2* – the number of individual adults on land in each sample plot.

Number and timing of visits

Preferably during the first three weeks of June.
Method 1 – one to five visits (the more the better)
Method 2 – five to ten visits.

Time of day

0800–1600 BST.

Weather constraints

Avoid foggy or wet and windy conditions (winds stronger than Beaufort force 4). Note the weather at the time of each count.

Sites/areas to visit

Any colonies not currently being monitored.

Equipment

- camera with black-and-white film.

Safety reminders

Avoid choosing study plots that cannot be observed from a safe vantage point. Cliff edges may not be as solid as they look and may become more unsafe in wet and windy conditions. If working alone, always ensure someone knows where you have gone and when you intend to return. If cliffs or steep slopes are to be climbed to gain access to beaches, use the correct safety equipment and do not work alone. If working with boats, never work alone. The boat should be operated by an experienced and trained boat handler and life jackets should be worn at all times. Take necessary equipment in case of an emergency. Be aware of the signs of Lyme's disease, which can be caught from seabird ticks.

Disturbance

When counting cliff-nesting birds, ensure that the count position is not so close as to disturb nesting birds. Do not climb down a cliff or slope to count birds at close quarters if it disturbs cliff-nesting birds. Try to minimise the time spent inside each subcolony.

Methods

Both methods count individual adult birds on land. Counts of breeding pairs are virtually impossible without highly intensive observations of mapped study plots. There can be considerable variation in the numbers of adults present at any one time, and this is the main reason that repeat counts are necessary.

Method 1: Whole-colony counts

Define the colony's limits and divide into subcolonies using physical features such as cliff faces, ledges and fissures. Mark these boundaries clearly on a map of the colony, and keep count data separate for each subcolony. Count individuals on potential breeding ledges, but do not count birds only loosely associated with the colony, ie at the base of the cliffs or in the sea. Do not attempt to exclude obvious non-breeding areas on the cliff face itself, as such judgements can be subjective.

Flat-topped subcolonies, for example on stacks, can be extremely difficult to count where there is no good vantage point to view from above. In such cases aerial photography might be the most accurate method. Boulder beaches, ledges, gullies and caves are also difficult to count; in most cases an estimate will have to suffice.

Keep a note of (and map) any parts of the colony that are not visible from land. Estimate (minimum–maximum) the number of hidden birds based on the numbers in visible sections, but bear in mind that these may not necessarily show similar densities to hidden sections. Make these estimates clear when reporting results. If at all possible, count hidden sections from a boat on a calm day, especially if you estimate that hidden sections are likely to total more than 10% of the total population.

Count all sections visible from land on five separate dates spread throughout the first three weeks of June. Report the mean number of adults counted (± standard deviation) as well as the number on each individual date. Do not attempt to convert from individual adults to pairs using correction factors. For remote colonies, or where time is limited, a single count may be all that can be made.

If possible, photograph the entire length of the colony to produce a permanent record of its precise limits. This will allow future expansions or contractions of the colony to be more readily detected.

Method 2: Plot counts

This method can only be used to assess overall colony size if the percentage of the total colony area is known and if sample plots are located at random within the colony. Otherwise the method only allows an estimate of between-year comparisons in population size. Mark the boundaries of the colony on maps or photographs. Divide those parts of the colony that can be viewed easily and safely from land into potential plots of about 100–300 birds.

Number all potential plots and randomly select those to be surveyed from among them. Select as many plots as can be counted in the time available, though five is the absolute minimum. Photograph all selected plots and mark on their boundaries. Make sure that the boundaries of the plots are unambiguously defined. Count all adults on plots on 5–10 separate dates spread throughout the period early to mid-June.

Do not count birds on tidal rocks at the base of the cliff or just above the high-tide mark. Do not hurry the plot counts; count as accurately as possible. While counting birds along a ledge, ignore birds which land behind or which take off ahead of the position you have reached in your count. Report the number of individuals on land in each sample plot. Once established, the same study plots should be re-surveyed each year.

Breeding season survey – productivity

Judging which guillemots bred and which chicks fledged is not simple; therefore, productivity monitoring in this species requires some effort. This method, adapted from Drury et al (1981), Murphy et al (1986) and Harris (1989a, b), was taken from the *Seabird Monitoring Handbook*.

Information required

- the number of chicks of at least 15 days old which disappear (assumed fledged).

Number and timing of visits

At least three visits in late incubation/early hatching period, and then visits every 1–2 days once most chicks are hatched.

Time of day

Vary the time of day.

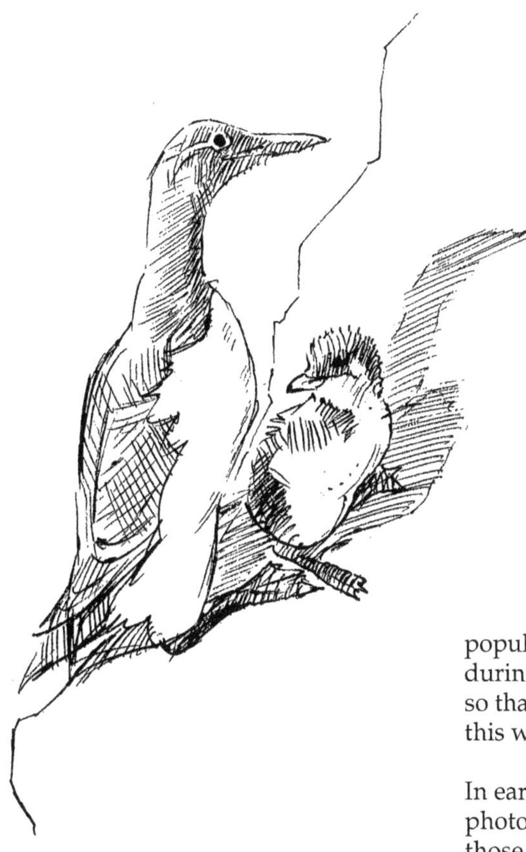

Sites/areas to visit

Any colonies not already monitored.

Equipment

- camera with black-and-white film
- transparent overlays and marker pens.

Safety reminders, Disturbance

As for population survey (above).

Methods

Select a minimum of five plots per colony, each containing about 50 breeding pairs. Do not simply select areas that are easier to count (such as those with a lower density of birds). If it is not possible to locate plots randomly, disperse them throughout the colony. See the *Introduction* or the *Seabird Monitoring Handbook* for information on sampling methods.

Productivity plots can be located within those used for population monitoring, but do not have to be. Photograph each plot during incubation and tape a transparent overlay on each photograph so that it can be annotated. Large, clear prints are very important for this work.

In early June, view the plot from the same point from which it was photographed and mark the position of all active sites. Active sites are those containing:

(a) birds with an egg,
(b) birds with a chick, or
(c) birds that appear to be incubating, and
(d) any other pairs attending a site which appears capable of supporting an egg (bearing in mind that some eggs are laid on highly unsuitable sites).

The last category will include pairs that, having failed at an earlier stage in the season, usually continue to occupy the nest-site. Unpaired guillemots will often preen each other, so care is needed, but actual pairs will usually be more obvious.

Initially, make at least three visits to each plot at different times of day, including some when large numbers of guillemots are present, until you are satisfied that you have found most of the occupied sites within each plot. Record any chicks without an adult in attendance. Keep a note of whether each sitting bird and each member of a pair is of the bridled or unbridled form. This will help reduce mistakes in identifying particular pairs.

Number the active sites, and record their contents every one or two days. If this is not possible, then start visiting the colony when the first chicks are close to fledging. Add any new active sites as necessary. Any young disappearing when aged 15 days or more (known from regular checks or estimated on the basis of size and plumage development; see Table 1) can be considered to have been reared successfully. The more visits made during the chick period, and the longer their duration on each date, the more accurate will be your assessments of chicks' ages. In particular, if you detect most chicks when recently hatched, it will be much easier to confirm whether or not they fledge.

Fledging periods for guillemots are very variable and range from 15 to 30 days, with the mean fledging time being 22 days after hatching. The

Table 1
Guide to ageing guillemot chicks (taken from the *Seabird Monitoring Handbook*).

Age (days)	Description of typical chick
<2	Obviously small and weak-looking, unable to move around ledge and barely able to sit upright. Egg-tooth very prominent. Completely covered in down feathers (which look spiky on guillemot chicks).
4–5	Feather quills emerge.
6–7	Contour feathers begin to erupt from quills, with little down remaining on wings. No white on head, which is still dark and down covered. Egg tooth faded but still present.
c 12	Contour feathers well developed everywhere, except on the head and neck. Beginnings of dark mask through eye (about 10–12 days) with some white on chin/below mask but little or no white above rear extension of mask.
c 15	Plumage well developed, with white of belly almost continuous with the more recently developed white of throat and cheek regions (though usually a sparse of dark down feathers remaining in throat region). Obvious black mask from base of bill through eye, with thinner black line extending behind eye, contrasting with white cheek and with greyer plumage above (usually some white above rear extension of mask also). Egg tooth sometimes still present, but flakes off easily.
>15	Major plumage development completed by now, except for further growth of feathers (especially on wings) and loss of most of the remaining spiky down (producing cleaner looking black and white plumage). Body size and weight continue to increase, and oldest birds can look strikingly large and big-billed (though smaller chicks may be quite capable of leaving the ledge and swimming out to sea).

survival of older chicks is high, so it is safe to assume that any chicks which disappear after 15 days and/or when they are well feathered will have fledged.

For each plot report productivity as:

no. of young fledged from all active sites = (a) + (b) + (c)

and

no. of young fledged from active and regular sites
= (a) + (b) + (c) + (d)

Express the productivity of the colony as a whole as the mean (± standard error) of the productivity of all the plots.

Make notes if you have any reason to suppose the season, or the results, may have been atypical. It is usually convenient, and statistically valid, to use the same study plots each year but it is not necessary to do so.

References

Drury, W H, Ramsdell, C and French, J B, Jr (1981) Ecological studies in the Bering Straits region: final report Research Unit 237. *US Department of Commerce, National Oceanic and Atmospheric Administration Outer Continental Shelf Environmental Assessment Program, Boulder* 11: 175–488.

Harris, M P (1989a) *Development of Monitoring of seabird populations and performance: final report to NCC*. Nature Conservancy Council CSD rep. 941.

Harris, M P (1989b) Variation in the correction factor used for converting counts of individual guillemots *Uria aalge* into breeding pairs. *Ibis* 131: 85–93.

Murphy, E C, Springer, A M and Roseneau, D G (1986) Population status of common guillemots *Uria aalge* at a colony in western Alaska: results and simulations. *Ibis* 128: 348–363.

Razorbill *Alca torda*

Status

Amber listed: BL
SPEC 4 (S)

National monitoring

Seabird Colony Register (SCR).
Seabird Monitoring Programme (SMP).
Seabird 2000 (1999–2001).

Population and distribution

Colonies are distributed around the coasts of Britain and Ireland except for the south-east of England (*88–91 Atlas*). Razorbills are less dependent on sheer cliffs than are guillemots. Colonies occur where suitable feeding and nesting habitats combine, the largest being in Shetland, Orkney and northern Scotland (*88–91 Atlas*). Numbers are fairly stable and there are estimated to be 160,000 adults breeding in the UK (*Population Estimates*).

Ecology

Colonies are on coastal cliffs and rock stacks. A single egg is laid in a crevice, on a ledge or in a hole in the cliff, from May onwards. The young leave the colony about 20 days after hatching, at which time they are unfledged and only one-third grown; they remain in the care of the adult male on the sea from the end of June to about mid-September (*Red Data Birds*).

Breeding season survey – population

Counts of razorbills and guillemots use similar methods but differ in detail. The three methods given here are taken from the *Seabird Monitoring Handbook.* As for guillemot, *Method 1* is a whole-colony census and *Method 2* is a census of sample plots; *Method 3* is specific to boulder beaches.

Information required

- number of individual adults on each visit in the entire colony (*Method 1*) or in each sample plot (*Method 2*)
- mean number of individual adults across all visits in the entire colony or in each sample plot
- *Method 3* – ratio of birds visible from the vantage point: total birds present for each subcolony
- *Method 3* – mean ratio from values for each subcolony
- *Method 3* – estimate of likely number of hidden birds in the remaining boulder subcolonies.

Number and timing of visits

Methods 1 and 3 – Up to five visits.
Method 2 – 5–10 visits.
Preferably during the first three weeks of June.

Time of day

0800–1600 BST.

Weather constraints

Avoid foggy or wet and windy conditions (winds stronger than Beaufort force 4).

Sites/areas to visit

Any colonies not currently being monitored.

Equipment

- camera
- black-and-white film
- 1:10,000 OS map.

Safety reminders

Avoid choosing study plots that cannot be observed from a safe vantage point. Cliff edges may not be as solid as they look and may become more unsafe in wet and windy conditions. If working alone always ensure someone knows where you have gone and when you intend to return. If cliffs or steep slopes are to be climbed to gain access to beaches, use the correct safety equipment and do not work alone. If working with boats, never work alone. The boat should be operated by an experienced and trained boat handler and life-jackets should be worn at all times. Take necessary equipment in case of an emergency. Be aware of the signs of Lyme's disease, which can be caught from seabird ticks.

Disturbance

When counting cliff-nesting birds, ensure that the count position is not so close as to disturb nesting birds. Do not climb down a cliff or slope to count birds at close quarters if it is likely to disturb cliff-nesting birds. Try to minimise the time spent inside each subcolony – no more than 30 minutes among any breeding group. A quick search by several observers causes less disturbance than a prolonged search by one. For routine monitoring do not visit more often than once a week.

Methods

Counts of razorbills can vary greatly in their difficulty depending on the type of site. Where birds nest on cliffs, counting is quite straightforward; the main problems are that some birds may be difficult to pick up among guillemots, while others may be scattered on small recesses or ledges. Groups nesting in crevices or among boulders are more difficult to count. Razorbill methods are based on those originally devised for guillemots, so both whole-colony and plot-monitoring counts of the two species can be made at the same time if the species are well mixed. Ideally, however, razorbill and guillemot plots should be kept separate, particularly if more than about 10% of the razorbill population nests in areas not used by guillemots.

Method 1: Whole-colony counts

Mark the limits of the colony, and the boundaries of the subdivisions to be used for counting, on an OS map. If possible, photograph the whole colony to provide a precise record of its extent. This makes it easier to record any contractions or expansions in colony area.

Count *all* visible birds on cliffs except for those only loosely associated with the colony (ie on intertidal rocks or on the sea). Do not attempt to judge the breeding status of individual birds – even if they appear unattached or are not obviously associated with a potential crevice, they should be counted.

Problems will be encountered with ledges, gullies and caves as it is not advisable to climb on cliffs during the breeding season. Where there are any parts of the colony that cannot be counted safely from land or from a boat, estimate the number of hidden birds (minimum–maximum) based on the number of birds on the visible sections. If more than 10% of the colony is not visible from land, it becomes more important to try to view hidden areas from a boat. Make it very clear whether or not your counts are estimates of this sort.

If possible, count on up to five different days spread throughout the first three weeks of June. Report the mean number of adults counted (± standard deviation), as well as those for the individual dates. *Do not* attempt to convert from individual adults to pairs using correction factors.

Method 2: Plot counts

This method can only be used to assess overall colony size if the percentage of the total colony area covered by all plots combined is known, and if plots are located at random within the colony. Otherwise, the method only allows between-year comparisons of estimated population size.

Divide those parts of the colony where razorbills are present, and which can be viewed safely and without difficulty from land, into potential plots of about 20–100 birds. Number all potential plots and select at least five to be counted using random number tables, decimal dice or a pack of cards (excluding face cards). Further advice on sampling is in the *Introduction*.

Photograph all selected plots and mark the plot boundaries on each photograph. Make sure that the boundaries of the plots can be unambiguously defined. Count all adults on plots on 5–10 separate dates spread throughout the first three weeks of June. Do not count birds on the sea or on tidal rocks. Report the number of adults in each plot on each visit and the mean (± standard deviation) across all visits. Once established, the same study plots should be resurveyed each year.

Method 3: Boulder colonies

Divide into discrete subcolonies or breeding groups and select at least five accessible subcolony groups at random. Count visible individuals from a suitable vantage point, then move carefully into each subcolony and count the actual number of individuals present by direct observation and by flushing from crevices. Try to minimise the time spent in each subcolony, especially when many eggs or small chicks are present. Calculate the ratio of the number of birds visible from the vantage point to the total number present for each of the five subcolonies/groups. Calculate a mean ratio from the five individual values and use this to estimate the likely number of hidden birds in the remaining boulder subcolonies. Although repeat counts of birds visible from vantage points in boulder subcolonies will be useful, do not actually enter boulder colonies more than once to count hidden birds because of the disturbance this will cause.

Breeding season survey – productivity

Three methods are given here, all taken from the *Seabird Monitoring Handbook*; each is for a different sort of nest-site. *Method 1* is for open nest-sites, visible from a distance (eg cliffs); *Method 2* is for enclosed but accessible nest-sites (eg boulders or deep crevices); *Method 3* is for enclosed and inaccessible nest-sites.

The method used depends on the type of site. Where the types of site used are mixed within a colony, it may be necessary to combine methods to obtain representative results. Monitor all the site types (containing more than 10% of the colony) that occur within a colony, as productivity may vary between them. To combine the different methods, determine the proportion of birds in each type and ensure that the number of random plots in each type reflects this. For example, if there are twice as many birds in boulder subcolonies as on cliffs then allocate twice as many plots in the boulder area. For each type, select plots (preferably randomly, see the *Introduction* for advice) until you have as many as required, but a minimum of five. Report productivity as the mean (± standard error) across all plots.

Information required

- *Method 1* – number of young assumed fledged divided by the number of active sites
- *Method 1* –number of young assumed fledged divided by the number of active and regular sites
- *Method 2* – number of large chicks divided by the number of active nest-sites relocated and checked
- *Method 3* – the number of apparently successful sites divided by the number occupied early in the season.

Number and timing of visits

Method 1 – At least six visits.
Method 2 – At least two visits.
Method 3 – Many visits.
Mid-May to late June/early July. In general, any counts which do not meet these conditions cannot be used.

Time of day

Vary the time of day.

Weather constraints

Avoid wet and windy conditions to avoid damage to eggs and chicks.

Sites/areas to visit

Any colonies not already surveyed.

Equipment

As for the population counts (above), plus:

- transparent overlays and marker pens
- bamboo canes/stakes/markers
- torch (Method 2).

Safety reminders and disturbance

As for the population survey (above).

Methods

Method 1: Open, accessible nest-sites

Divide the whole area into clearly definable plots of 10–50 nest-sites that can be viewed easily and map the boundaries of each on photographs. Be sure that all safely observable areas of cliff are included.

Select at least five plots per colony to count. In small colonies it might be possible to count all plots. Do not simply select areas that are easier to count, eg areas with a lower bird density. Whenever possible, select plots at random or otherwise dispersed throughout the colony. Productivity plots could be located within those used for population monitoring but do not have to be.

Tape a transparent overlay to the photograph(s) of each of the sample plots; this allows them to be annotated. Large, clear photographic prints are required for this work. If the colony holds few or sparsely distributed razorbills on large areas of cliff, covered by several photographs, these may need to be treated as individual subplots. The same photographs may be used year after year so long as there have been no major changes to cliffs.

Make at least three visits to each plot during the late incubation/early chick-rearing period (mid-May to early June) and view each plot from the same point from which it was photographed. Mark the positions of all active sites. These are:

(a) birds with an egg,
(b) birds with a chick, and
(c) birds sitting tight.

Also mark any regular but inactive sites; these are:

(d) any other pairs attending a site that appears capable of supporting an egg.

Record any chicks without an adult in attendance.

Number the active sites on each plot and record their contents, ideally every one or two days. If this is not possible then start visiting the colony when the first chicks are close to fledging. The survival of older chicks is high, so assume that any chicks which disappear after 15 days and/or when they are well-feathered have fledged. Add any new active sites as necessary. The more visits made during the chick-rearing period, and the longer their duration on each date, the more accurate will be the assessment of chick age. In particular, if you detect most chicks when they have just hatched, it will be easier to confirm that they have reached 15 days of age.

For each plot report productivity as:

no. of young fledged from all active sites = (a) + (b) + (c)

and

no. of young fledged from active and regular sites = (a) + (b) + (c) + (d)

Calculate the average productivity of cliff-breeding razorbills in a colony as the mean ± standard error of the productivity of individual plots. Plots containing fewer than 10 active and regular sites should be combined with the nearest plot before calculating the mean. If all plots have fewer than 10 sites (or all except for one or two), combine all plots.

Make notes if you have any reason to suppose the season, or the results, may have been atypical.

Method 2: Enclosed but accessible nest-sites (eg boulders or deep crevices)

It is not necessary to measure productivity of razorbills nesting among boulders or in burrows where these form <10% of the whole colony.

Divide the whole area into plots containing 10–50 razorbill pairs and sketch their approximate position on maps or photographs. Randomly select a minimum of five of these that are safely accessible and mark their boundaries unambiguously on large-scale photographs or maps.

Visit each plot after the peak of laying, which is usually in mid-May. Try to find as many active nests (with egg, chick or tightly sitting adult) as possible. A bird sitting tight can be taken as evidence of an active nest. Avoid flushing incubating birds if possible, and try to spend no more than 30 minutes among any breeding group. Many birds will be nesting in shallow recesses that can be observed easily but in some cases a torch will be necessary to check deeper crevices. Feel with your arm into holes where you believe a nest may be. Some might only be checkable by touch and it might even be necessary to use a bamboo cane, but be careful not to break the egg. Avoid putting birds off their egg if at all possible.

Mark the entrance to each located site with a stake or other marker. Some nests will be missed and others will be inaccessible but this is relatively unimportant as long as a reasonable sample size of nests is found (10 or more). Re-check the marked sites in the pre-fledging period, usually 15–20 June in northern Britain. The chick can normally be seen or felt, although some chicks will hide in crevices off the main nest-chamber. Fortunately, hidden chicks will call. Pull out any chicks that are not visible (use a piece of blunt, bent wire to do so if necessary) to check age. Note any replacement eggs laid and, if possible, check these in late June. Otherwise assume replacements fail.

Assume that large chicks more than 10 days old (outer wing longer than 60 mm) are likely to fledge. Keep a note of smaller chicks and check them again a week or so later.

More frequent checks during the incubation and chick-rearing periods will improve the accuracy of the productivity estimate, but this should be balanced against the possibility of causing excessive disturbance. For routine monitoring it is not recommended that checks are carried out more frequently than weekly. Limit visits to no more than 30 minutes.

For each plot, productivity is expressed as the number of large chicks present divided by the total number of active sites re-located where presence or absence of a large chick was determined. Sites where a replacement egg was laid and a chick subsequently found count as successful sites. If only one visit can be made to check fledging, calculate productivity on the assumption that all chicks fledged and all replacement eggs failed.

Method 3: Enclosed and inaccessible nest-sites

This method measures the productivity of razorbills breeding among boulders or crevices where most sites are not observable. It is labour intensive, and unnecessary in colonies where less than 10% of the population breeds in such areas. Where similar but accessible sites occur, it is usually preferable to monitor those sites instead using Method 2.

Divide the area into plots likely to contain 10–30 pairs and sketch the plots on rough maps or photographs. Randomly select a minimum of five plots. Mark the boundaries of these plots unambiguously on large-scale maps or photographs.

Mark the position of breeding sites clearly on the photograph or map early in the season and during the incubation period by observing birds attending sites and visiting hidden sites during incubation changeovers. If a bird is seen entering or leaving a site on at least two separate days it can be considered an active site.

When birds are feeding large chicks (around mid- to late June), make a few watches to determine which sites have fish taken to them; these can be considered successful sites. Feeding watches are best done in the early morning when feeding frequency is highest.

Report the number of apparently successful sites divided by the number of sites occupied early in the season for each plot separately and the mean (± standard error) of the individual plot productivities. This method is likely to overestimate success compared with the two other methods but can still provide useful information if other methods cannot be used.

Black guillemot *Cepphus grylle*

Status

Amber listed: SPEC 2 (D)

National monitoring

Seabird Colony Register (SCR).
Seabird Monitoring Programme (SMP).
There was a full survey between 1982 and 1990, JNCC.
Seabird 2000 (1999–2001).

Population and distribution

Approximately 20% of the European population of black guillemots breeds in Britain and Ireland (Lloyd et al 1991). In Britain they mainly breed around the north-west coast of Scotland, Orkney and Shetland, but are quite evenly scattered around the Irish coastline (*88–91 Atlas*). Black guillemots are particularly vulnerable to oil spills and to predation by rats and mink (Walsh et al 1994). The UK breeding population is currently estimated to be around 37,000 adults (*Population Estimates*).

Ecology

Black guillemots breed on the coast, preferentially near shallow water. Their nest-sites are typically in natural holes, crevices, caves and boulder beaches. The clutch of 1–2 eggs is laid from mid-May, incubation lasts 23–40 days, and the young fledge 31–51 days after hatching (*BWP*).

Breeding season survey – population

Black guillemots usually nest in pairs or small groups scattered along the coast. Surveys should therefore aim to cover sections of coastline rather than discrete 'colonies'. Nest-sites are difficult to count with any accuracy because of their scattered distribution and inaccessibility. Carefully timed counts of individual adults provide the most accurate method.

Information required
- maximum number of individual adults
- plumage of all adult birds.

Number and timing of visits

Two visits, a week or more apart during the first three weeks of April, though counts later in April or early May are acceptable.

Time of day

From first light to about two hours later (about 0600–0800 BST in the north, but as late as 0900 further south).

Weather constraints

Winds should not be stronger than Beaufort force 4, and preferably not blowing onshore. Sea-swell should be slight to moderate. Make a note of weather conditions at the time of the count.

Sites/areas to visit

Any colony not currently monitored. Breeding habitat includes mainly rocky shore coastline, cliffs and offshore islands.

Equipment

- 1:10,000 OS maps of the coastline
- boat (optional) and life-jackets.

Safety reminders

When working alone in remote areas ensure someone knows where you are going and when you will return. When working near water always work in pairs. Rocky shores are always hazardous because of their uneven, slippery surfaces, weed-covered rocks and fissures. Beware of slippery vegetation at cliff edges and of undercut or loose strata.

Disturbance

It may be necessary to flush adults into the water to count them accurately; this is more successful and least disruptive to the birds if carried out in the pre-laying period.

Methods

This method, which is taken from the *Seabird Monitoring Handbook*, is based on techniques described in Tasker and Reynolds (1983) and Ewins (1985a, b). It is the recommended method for censusing and monitoring black guillemots. The basic count unit is 'adults associated with a colony'. A recording form is shown in Figure 1.

Count accuracy is influenced by the familiarity of the observer with the study area and with their experience in counting black guillemots. Ideally novices should accompany experienced fieldworkers to learn the method; failing this they should make several 'test' visits to the areas to be surveyed.

Counts may be made from sea or land. Land-based counts are usually slightly more accurate (unless cliffs are high, say >70 m), but a much longer stretch of coast (up to 35 km) can be covered each morning in a suitable small boat. The use of boats is preferred for high cliffs, offshore islands, coasts with many islets where birds might be hidden on the blind sides, or long stretches of coast with little or scattered suitable breeding habitat.

On large islands and the mainland, birds normally only nest on cliffs over 10 m high as those are inaccessible to mammalian predators. Thus counting can concentrate on such areas. However, on uninhabited offshore islands birds may be much less restricted in nesting sites.

Move along the coast counting all birds on the sea within about 300 m of the shore and any on land. Birds ashore are often difficult to see but a cliff-top observer can flush them onto the water by clapping, shouting or other noises. Scan the sea frequently, particularly in bad light.

JNCC / Seabird Group black guillemot survey

Summary of instructions (see *Seabird monitoring handbook* for further details):
Surveys must be done between 0600 and 0900 BST, late March to early May. Avoid days with winds > Beaufort Force 4. Give priority to coastline with possible black guillemot breeding habitat, and cover on foot or by boat. Count all adult-plumaged black guillemots within 300 m of shore (including birds ashore, or disturbed from shore). Mark each group of adults encountered on a map of the coastline covered, and provide details for each group on a separate form.

Site name (refer to Ordnance Survey map) ....

Grid reference: Start .... **Finish** ....

Date: .... **Time (BST):** ....

Counted from (circle one): land, sea

Shore slope: gentle <30°, med. <70°, steep ≤90°

Cliff height (metres): <1, 1-10, 11-25, 26-50, 51-100, 101-150, 151-200, >200

Wind speed (Beaufort scale): 0, 1, 2, 3, 4

State of sea: calm, medium, rough

Sea-swell: none, slight, moderate, heavy

Precipitation type: snow, rain, none

Precipitation amount: heavy, light, showers Cloud cover (in eighths): ....

Light conditions: good, moderate, poor (any specific problems, e.g. sun-glare? ....)

Shore substrate (circle one or more): solid rock, loose rock, boulders, scree, sand/shingle, marsh, artificial, other

Birds seen onshore or within 300 m of the shore:

No. of birds in full adult summer plumage: .... Other plumage: ....

Total birds within 300 m of shore: ....

Number of additional birds seen >300 m from shore: ....

Notes on birds' activity (on land, calling, courtship, etc.):

Situation of probable nesting area: 'open' cliff, geo, cave, grassy, boulders, artificial, other

Prospecting/breeding gulls present: Herring, GBB, none Proximity (m): ....

Other notes (PTO if required): ....

Figure 1
The JNCC/Seabird Group black guillemot survey recording form, from the *Seabird Monitoring Handbook*.

Separate the count into:

- birds in adult summer plumage
- birds in other plumages (largely or partly grey, or with dark bars visible in white wing-patch)
- birds >300 m offshore (thus less obviously associated with potential breeding habitat).

Record any feeding birds separately; these should not be considered 'associated' with the colony.

For each group of birds, also record:

- their location (grid reference if possible; most easily done if counts are recorded directly on 1:10,000 maps)
- their probable breeding habitat (eg cliff-crevices, talus slopes, burrows, boulder beaches)
- weather
- other notes such as presence of breeding gulls, signs of mammalian predators, behaviour (eg display, whether many birds were ashore, whether they flushed easily or stayed close inshore).

For monitoring purposes count all suitable coastline in a given study area, or randomly select lengths of coast from the entire length of potential breeding habitat. Many short stretches are preferable to a few long stretches. Lengths of coast with 50+ adults are ideal.

Count the whole coastline or the sample stretches twice, a week or more apart. Report the higher of the two counts. For continuous stretches of coast, or adjacent islets, sum all the counts from any single day before calculating the mean of several days' counts. This allows for possible short-range movements of birds within larger areas. If smaller sample areas are used, the maxima for each area can be used provided the count areas are well separated (as a rough guide, by 500 m or more).

For between-year comparisons it is better to compare the mean (rather than the maximum) of two counts in one year with a single count in another year, provided there is no good reason (apart from higher numbers) to believe that one count is better than another. This is because higher numbers are likely to be detected in a year when two counts were made, thus potentially biasing comparisons. Counts should, in any case, be excluded from calculations of means if they are made during poor conditions, for example if the weather deteriorates or sea-swell increases after counts have begun.

Breeding season survey - productivity

Black guillemots use a variety of nest-sites, often inaccessible, under boulders, in burrows and crevices, and occasionally artificial sites in harbour-walls, piers or buildings. The measure of productivity obtained by *Method 1* will not be particularly accurate, but should be sufficient to detect any major changes in breeding output. *Method 2* provides a measure of the number of successful nests in groups nesting among boulders, but not of overall productivity since it cannot distinguish between nests fledging one and two chicks. Both methods are taken from the *Seabird Monitoring Handbook.*

Information required

- maps showing the boundaries of the survey areas and the locations of the birds

Method 1

- number of large chicks divided by the number of active sites relocated and checked
- brood size
- number of failed attempts

Method 2

- the number of successful nest-sites divided by the number occupied early in the season.

Number and timing of visits

Method 1 – At least one in early May and at least one in mid- or late July.

Method 2 – Variable, dependent on the duration of each individual 'watch'.

Time of day

Method 1 – To check nest contents: any time of day. To watch for nest visits: before 0900 BST.

Method 2 – Before 0900 BST.

Weather constraints

Do not visit in wet or windy conditions.

Sites/areas to visit

As for the population survey (above).

Equipment

Method 1
- stakes or other markers
- small torch and mirror (optional)
- small stick (optional)
- bird bag and wing rule (optional).

Method 2
- numbered stakes or other markers, or
- a comprehensive large-size photograph of the area.

Safety reminders

As for the population survey.

Disturbance

Disturb any incubating or brooding adults as little as possible. Chicks should be handled for the minimum time possible and never in wet or extremely cold and windy conditions. Always determine the presence of eggs in crevices with care.

Methods

Method 1: Accessible nest-sites

This method is adapted from that developed by Harris (1989) for puffins and shags. It can be used wherever most nest-sites are accessible enough to permit their contents to be checked.

Identify potential nest-sites by watching adults arriving or leaving. Check these sites after the peak of laying (usually early May). In some cases, the nest may be visible without too much difficulty. A small torch and mirror may be useful to illuminate dark recesses. A tightly sitting adult or a view of egg(s) can be taken as evidence of an 'active' nest-site. Otherwise, attempt to feel into the burrow/crevice with an arm or stick. Any incubating bird will move off, allowing the egg(s) to be felt on the floor of the nest-chamber. Be careful not to break the egg(s).

Aim for a sample of 10–50 accessible nest-sites (the more the better) with eggs or sitting adults. Ideally these should be dispersed throughout the breeding area. Mark all sites where breeding is detected with stakes or other markers.

Re-check the nest-sites when the birds have large chicks and before the first chicks fledge (usually mid- to late July in northern Scotland). Feel inside for the chicks, bearing in mind that there may be more than one; if necessary, remove one temporarily (bring a bag for the purpose) when searching for a second chick. Assume that any chicks more than three weeks old (wing length 90 mm: Ewins 1992) are likely to fledge.

If possible, re-check nest-sites one or more times to follow up smaller chicks. If any chicks previously noted as 'large' are found dead, subtract them from the number of chicks deemed to have fledged.

Productivity estimation is most accurate if regular (say, weekly) checks are made from the late nestling period onwards, and if more stringent criteria are used to assess numbers of fledged chicks. As the first chicks may fledge when 31–32 days old, those which reach 30 days (wing length usually >115 mm: Ewins 1992), or which could have reached that age between the weekly nest-checks, will have a very high probability of fledging. Thus if weekly nest-checks are made, assume that only those which reach 30 days will fledge.

Productivity is expressed as the number of large chicks divided by the total number of marked breeding sites that were refound, where presence or absence of a chick or chicks was determined. Keep a note of the number of failed attempts and of broods of one and two chicks, and of how many visits were made. Results based on a single count of chicks are more likely to overestimate productivity.

Method 2: Inaccessible nest-sites

This method is adapted from Harris (1989) and is suitable for black guillemots nesting among inaccessible boulders or at other sites where nest contents cannot be checked directly.

Find a vantage point from which the colony or breeding group can be watched from a distance. Observe the area during the early incubation period (usually late May: Ewins 1989) for birds changing over at the nest. The morning peak of activity is best for this (ie before 0900 BST), as changeovers occur most frequently in the first two hours of daylight. Mark these sites with numbered stakes or other markers (eg small cairns of stones), or plot them on large-size photographs taken from the observation point.

When birds are feeding large chicks but before fledging begins (usually mid- to late July in northern Britain), make a few watches to determine which nest-sites have fish delivered to them. This is best done in the early morning when feeding frequency is highest. (At regularly observed sites, fledging should begin about 31 days after the first adults are seen bringing fish back to the colony).

Express productivity as the number of successful nest-sites divided by the number occupied early in the season. Brood size will usually be unknown.

References

Ewins, P J (1985a) Colony attendance and censusing of black guillemots *Cepphus grylle* in Shetland. *Bird Study* 32: 176–185.

Ewins, P J (1985b) *Results of Tystie Cepphus grylle Pre-breeding Distribution Surveys in Shetland, 1982–84, with Monitoring Recommendations.* Report to Shetland Oil Terminal Environmental Advisory Group unpubl.

Ewins P J (1989) The breeding biology of black guillemots (*Cepphus grylle*) in Shetland. *Ibis* 131: 507–520.

Ewins, P J (1992) Growth of black guillemot chicks in Shetland in 1983–84. *Seabird* 14: 3–14.
Harris, M P (1989) *Development of Monitoring of Seabird Populations and Performance*. Final report to NCC (Contractor: Institute of Terrestrial Ecology). Nature Conservancy Council, CSD Rep. 941.
Lloyd, C, Tasker, M L and Partridge, K (1991) *The Status of Seabirds in Britain and Ireland*. Poyser, London.
Tasker, M L and Reynolds, P (1983) *A Survey of Tystie (Black Guillemot) Cepphus grylle Distribution in Orkney, April 1983*. Nature Conservancy Council NE Scotland Regional Rep. 1.
Walsh, P M, Brindley, E and Heubeck, M (1994) *Seabird Numbers and Breeding Success in Britain and Ireland, 1994*. RSPB/JNCC report.

Puffin
Fratercula arctica

Status

Amber listed: BL, SPEC 2 (V)

National monitoring

Seabird Colony Register (SCR).
Seabird Monitoring Programme (SMP).
Seabird 2000 (1999–2001).

Population and distribution

Britain and Ireland are estimated to hold 8% of the world's population of puffins (Lloyd et al 1991). They are burrow-nesting birds, preferring to stay away from human disturbance and predators such as rats and cats. Their colonies are situated around the coast of Britain and Ireland with much of the population in the north-west of Scotland, Shetland and Orkney (*88–91 Atlas*). The UK has an estimated 903,000 breeding individuals (*Population Estimates*).

Ecology

Puffins breed along the coast or on islands, preferring grass-covered peaty turf on gently sloping cliffs free from human disturbance and rats. Nests are in burrows in which a single egg is laid from mid-April. Young fledge about 38 days after hatching (*BWP*).

Breeding season survey – population

Breeding puffins are difficult to count accurately because of their large colonies, burrow-nesting habit and the constantly fluctuating number of birds at each colony. Three main methods are suggested here, all taken from the *Seabird Monitoring Handbook*. The choice of method depends on the structure of the colony and the methods used previously at the site.

Method 1a – Counts of Apparently Occupied Burrows/Sites (AOBs) using quadrats, selected with unrestricted random sampling.

Method 1b – Counts of AOBs using quadrats, selected with stratified random sampling.

Method 2 – Counts of AOBs using transects.

Method 3 – Counts of individuals (this method is the least preferred).

Information required

Methods 1–2

- estimated number of Apparently Occupied Burrows (AOBs)
- a map showing the colony boundaries and the location of the sample quadrats/transects.

Method 3

- number of individuals on land
- number of individuals flying over the colony or over the sea within 200 m of shore
- number of individuals on the sea within 200 m of the shore.

Number and timing of visits

Methods 1–2 – One visit per quadrat/transect, late April to mid-May.
Method 3 – As many visits as possible, early in the season during the pre-laying period.

Time of day

Methods 1–2 – Any time during the day.
Method 3 – Evenings or in foggy conditions.

Sites/areas to visit

Any colony.

Equipment

- 1:10,000 OS map of the area

and, for Methods 1–2 only

- a compass
- enough stakes to mark each quadrat
- a rope (2.52 m for 20 m^2 quadrats or 2.96 m for 30 m^2 quadrats).

Safety reminders

If working alone, always ensure someone knows where you are and when you intend to return. If cliffs or steep slopes are to be climbed, use the correct safety equipment (eg ropes and harnesses) and do not work alone. Beware of soft and eroding cliff edges and overhangs; wear footwear with plenty of grip.

Disturbance

Avoid any area of the colony which may have burrows in danger of collapse. Avoid walking in the colony for any longer than is absolutely necessary.

Methods

Sampling is best carried out before or during the laying period, when birds are digging or cleaning burrows and when ground vegetation is short; late April is optimum in south-east Scotland, early to mid-May in north-west Scotland. However, acceptable counts can be made at any time from late April to early August, although assessments of population change should preferably be based on counts made at the same time of year.

Method 1: Quadrat

a) Unrestricted random sampling

This method is for use where colony density is not known to vary markedly, or where there is no objective basis for subdividing the colony.

Map the limits of the colony and identify clearly any inaccessible areas. At large colonies, which are difficult to count completely, superimpose a grid on the map and randomly select about 30 grid intersections (sampling advice is given in the introduction) within accessible parts of the colony.

Place quadrats at the selected points *regardless* of whether or not there are any burrows present, even if the whole area is bare rock. Fix the location of each point by measuring angles (taking bearings) using a compass, and pacing out distances from fixed points, such as rock outcrops or other structures. These points may be permanently staked and used each year, or new points selected annually. Permanent quadrats increase the precision with which change can be estimated. Other forms of marker (eg small cairns of stones) may be needed on some substrates. Ensure that stakes penetrate well into the ground, to reduce the likelihood that they will be displaced by burrowing puffins but be careful to avoid penetrating any burrows. If using permanent quadrats, keep the original map of grid points so that the positions of missing stakes can be relocated.

Circular quadrats of 20–30 m^2 are the most convenient shape and size of quadrat. Rotate a 2.52- or 2.96-m rope around a fixed stake at each randomly selected point and count the number of apparently occupied burrows (AOBs) within the area covered.

Count all apparently occupied burrows where >50% of the burrow entrance lies within the quadrat, regardless of the direction of the tunnel. AOBs are characterised by signs of regular use, eg fresh digging, hatched eggshells or fish in the entrance. Rabbit burrows are usually larger, usually have much soil outside, and often have droppings at the entrance and conspicuous runs leading away through the vegetation. There is no simple way to separate Manx shearwater and puffin burrows, although mixed colonies are relatively uncommon.

If possible, keep a separate note of burrows that seem to be unoccupied, for example having overgrown entrances or no evidence of use.

To obtain an estimate of the total population, calculate the mean number of AOBs in each quadrat (sum of estimates for each quadrat divided by the number of quadrats taken). Divide this by the area of each quadrat (eg 20 m^2). Multiply the result (mean density) by the total area (in m^2) of the colony (derived from a map plotted on graph paper or using a planimeter) to give the final figure. Report this figure ± standard error.

b) Stratified random sampling

This method can be used where areas of the colony are known to differ markedly in density and where these areas can be defined on an objective basis. The method is described in greater detail by Harris and Rothery (1988).

Define the limits of the colony and mark them on a map. Map and determine the relative proportions of the areas of markedly different density ('strata') within the colony. Distinguish between accessible and inaccessible areas. Measure or estimate the area of each stratum.

Many colonies will be too large to count completely. In this case, superimpose a grid on the map and select quadrat points randomly within each stratum. The methods of placing quadrats, the type of quadrat used and the manner of counting occupied burrows are the same as in Method 1a.

Calculate the mean number of AOBs per quadrat (± standard error) as in Method 1a. Obtain an estimate of the total population (± standard error) within each stratum by dividing the mean number of AOBs per quadrat by the area of each quadrat, and multiplying this figure by the total area (in m^2) of the stratum. Add together the figures for each stratum to obtain a total population estimate.

Method 2: Transects

In colonies where there are no marked differences in density between different areas, transects should preferably be placed randomly within the colony as a whole. Previous knowledge of the colony, or a preliminary survey of relative burrow densities, will be needed. To determine positions for transects, divide a map of the colony into bands 3 m wide perpendicular to the long axis of the colony (thus, a colony 300 m long would have 100 potential transects), number them and select several using random methods. Otherwise, position transects so that all distinct subgroups (especially strata of different density) are sampled (again, randomly if possible). At some colonies, there may be few accessible locations to position transects.

Establish the positions of the transects within the colony, including apparently unoccupied areas beyond the boundary of the colony. The number of transects will depend on time available and the shape of the colony (more transects in long, thin colonies). Larger numbers of transects improve the prospects of demonstrating that changes in populations are statistically significant; five or more are suggested. Mark each transect using stakes placed at intervals around their boundaries. Transect lines should be permanently marked so that the same lines can be checked at each count.

Count all apparently occupied burrows along each 3-m-wide transect. Running ropes along the boundaries of the transect will improve accuracy. Include all burrow entrances lying 50% or more within the transect strip, regardless of the direction of the tunnel. Record numbers for each transect separately.

To obtain an estimate of the total population, calculate the mean number of AOBs within in each transect (sum of estimates for each transect divided by the number of transects). Divide this by the area of each transect. Multiply the result (mean density) by the total area (in m^2) of the colony (derived from a map plotted on graph paper or using a planimeter) to give the final figure. Report this figure ± standard error.

If transects were placed within particular subgroups (strata) of the colony, use the same method *but for each stratum separately*, then sum the population estimates from each stratum for the whole colony figure.

Report the raw data (colony area, number and sizes of transects, numbers of AOBs in each transect) and any calculations you have carried out.

Method 3: Counts of individuals

This method is only useful for estimating the general size of small colonies. At best it only provides a broad indication of colony size.

Count the number of individuals present above ground, on as many dates as possible during a season. Count separately the numbers of birds flying over the colony or nearby sea, and birds on the sea within 200 m of shore.

Record the peak number of birds on land early in the season during the pre-laying period (before mid-April on North Sea coasts, late April in north-west Scotland) and any subsidiary counts made at the same time of flying birds and birds on the sea. If counts cannot be made at this stage, try to make them before June, when substantial numbers of immatures begin to attend the colony. Also record the peak numbers seen at any stage of the season.

Note any positive evidence of breeding, for example adults entering burrows with fish. Puffins occasionally attend mixed seabird colonies without breeding there. Relating counts of visible birds to breeding numbers is difficult, but the likely order of magnitude of the population (eg 10–100 or 100–1,000 adults) may be indicated by the counts.

Breeding season survey – productivity

These methods are taken from the *Seabird Monitoring Handbook*. Accurate productivity monitoring is labour-intensive because of the burrow-nesting habits of the species. If confusion with Manx shearwater burrows is not a problem, a reasonably good estimate can be obtained from two visits, each lasting about a day. Puffins do not tolerate much disturbance when nesting, so none of the methods that have been developed involves handling the adult. Studies indicate that, where this is avoided, productivity is not affected (Harris 1980). The recommended method involves direct examination of a sample of nests (*Method 1*). Where confusion with shearwater burrows is a problem, or where burrows are among rocks preventing physical examination, time-consuming observational methods (*Methods 2–3*) are the only practical alternatives (although they may provide less reliable results).

Information required

- *Method 1* – number of chicks located in burrows in which eggs were found earlier in the season (preferred method)
- *Method 2* – as for method 1, except observations are required to check that the burrows are occupied by puffins
- *Method 3* – number of nest-sites occupied early in the season, and the number of these that were successful.

Number and timing of visits

Method 1 – At least two visits: one in early May, one in early to mid-July.
Method 2 – At least three: one in April, one in early to mid-May and one in early to mid-July.
Method 3 – At least four: two between April and mid-May, two between mid-June and mid-July.

Time of day

Any time of day. Observations of chicks being fed are best done in the early morning when feeding frequency is highest.

Sites/areas to visit

Any colony.

Equipment

- 1:10,000 map of the area
- numbered stakes (1 m long)
- thin bamboo sticks 15–50 cm long
- hide (optional).

Safety reminders

As for population monitoring.

Disturbance

As for population monitoring. Do not handle adults. When the burrows are being probed for presence of eggs or chicks, this should be done gently and carefully. When stakes are being put into the ground, take care not to drive them into burrows, or to collapse any nearby ones.

Methods

Method 1: Staked burrows

This method was developed by Harris (1989). It is suitable for colonies where burrows are in soil and where there are no Manx shearwaters present, and involves feeling down a burrow with a short length of bamboo or stick.

Try for a sample of 100+ burrows spread throughout the colony. Select a series of sticks or thin bamboos 15–50 cm long. Take the longest, lie on the ground and push the stick and your arm down the burrow. Any incubating puffin will move off the egg, which can usually be felt with the stick on the floor of the nest-chamber. If the stick is too long to go around a bend in the burrow, try again with a shorter one. Be careful not to break the egg. Any burrow where an egg is felt is then staked. Bear in mind that the vegetation may grow quite tall and you will want to find the burrow again, so stakes may have to be 1 m or more long. Burrows are best checked when dry.

Re-check the burrows when most pairs have very large chicks, but before the first chicks fledge. It is usually easy to determine if the nest has been successful, either by feeling the chick, by finding the chick's latrine at the first bend of the burrow, or by searching for moulted down among the nest-lining. Keep a note of any evidence of predation of chicks or flooding of burrows and any chicks found dead.

Productivity is expressed as the number of chicks found divided by the total number of staked burrows re-found (ie burrows not re-found are not included in calculations).

Method 2: Staked burrows plus observations from a hide

This method is adapted from Harris (1989) and is suitable for colonies where birds' nests are accessible and where they are shared with Manx shearwaters.

Find a vantage point from which burrows can be watched from a distance with binoculars (preferably from a hide). Mark all visible burrows with large numbered stakes and, early in the season, record which burrows are being regularly used by puffins. As a rough guide, burrows entered by a puffin on two separate dates may be included in your sample. Aim for a sample of about 100 puffin burrows if possible.

Check these burrows for the presence of an egg in early to mid-May, as in Method 1. Assume the burrow occupants, if any, to be puffins, based on earlier observations, unless there are indications (eg calls) that Manx shearwaters are present.

Check for chicks as in Method 1. Check the shape of the bill by touch, or remove the chick to confirm it is a puffin.

Report the number of chicks divided by the number of puffin burrows in which eggs were found.

Method 3: Mapped burrows plus observations from a hide

This method is adapted from procedures developed by Harris (1989). It is suitable for assessing productivity in colonies where nest-site entrances cannot be approached and/or the nest-chamber is inaccessible (eg steep slopes, scree, deep crevices among rocks). It can be used in colonies shared with Manx shearwaters.

Find a vantage point where burrows or the entrances to potential nesting crevices can be watched from a distance.

Mark the position of all visible burrows or crevices regularly used by puffins early in the season on a good photograph or, where the entrance can be safely approached, using a numbered stake. As a rough guide, burrows or crevices entered by a puffin on two separate dates may be included in your sample. Aim for a sample of about 100 nest-sites if possible, although many colonies may be small, or only small parts may be visible.

When birds are feeding large chicks, make a few watches to determine into which burrows/crevices adults take fish. Express productivity as the number of successful nest-sites divided by the number occupied early in the season.

References

Harris, M P (1980) Breeding performance of puffins *Fratercula arctica* in relation to nest density, laying date and year. *Ibis* 122: 193–209.

Harris, M P (1989) *Development of Monitoring of Seabird Populations and Performance.* Final report to NCC (Contractor: Institute of Terrestrial Ecology) Nature Conservancy Council, CSD Rep. 941.

Harris, M P and Rothery, D (1988) Monitoring of puffin burrows on Dun, St Kilda, 1977–87. *Bird Study* 35: 97–99.

Lloyd, C, Tasker, M L and Partridge, K (1991) *The Status of Seabirds in Britain and Ireland.* Poyser, London.

Barn owl
Tyto alba

Status

Amber listed: BDM, SPEC 3 (D)
Schedule 1 of WCA 1981

National monitoring

Three-year breeding survey (UK), 1994–1997 (BTO, Hawk and Owl Trust).

Population and distribution

In the UK, barn owls are at the northern limit of their range and appear to be limited by both altitude and latitude. This partly explains their UK distribution, notably in Scotland. Their absence from areas of the Midlands and central England probably reflects the absence of suitable hunting habitat and localised altitudinal features. They occupy areas of farmland and rough grazing. Numbers have decreased during the twentieth century in most European countries (*Birds in Europe*), possibly as a result of the loss of suitable habitat (*88–91 Atlas*). There are estimated to be 4,400 breeding pairs of barn owl in Britain, 600–900 pairs in Ireland and 40 in the Channel Isles (*88–91 Atlas*), although these figures are now considered to be underestimates (M Toms pers comm).

Ecology

Largely nocturnal and crepuscular, barn owls roost in trees as well as buildings, and are sedentary. The main habitat requirement is the presence of rank tussocky grassland containing short-tailed voles *Microtus agrestis*, the barn owl's main prey. Permanent grassland, hay meadows and the grassy edges of fields all provide good habitat for voles, mice and shrews and are therefore good hunting areas for barn owls. Woodland edges are also favoured, as are hedgerows, river banks, sea walls, ditches and dykes. Other sites that may be used are golf courses, parkland, areas of scrub or waste ground, or road verges. A clutch of 4–7 eggs is laid in April, with larger clutches being produced when prey is especially abundant. Incubation lasts 29–34 days. The young leave the nest-site at about 60 days and they disperse up to 20 km from the nest (*Red Data Birds*).

Breeding season survey – population

The following survey method is based on that devised for the BTO/Hawk and Owl Trust survey of breeding barn owls (Toms 1995).

Information required

- number of potential barn owl breeding sites.
- number of these sites found to be occupied by barn owls.
- a map showing potential and occupied nest-sites.

Number and timing of visits

At least two visits: one between November and January, and the other between 1 June and 16 July. Extra visits will probably be required.

Time of day

Winter visit: any time of the day. Summer visit: preferably late afternoon.

Weather constraints

Do not survey if the weather is bad (eg rain, snow or wind). Watching a site for birds emerging or returning is best done on cloudless still nights.

Sites/areas to visit

All potential sites in any areas of suitable habitat. There is no need to search in built-up areas or dense woodland, except at the periphery.

Equipment

- 1:10,000 map of the survey area
- recording forms
- Schedule 1 licence.

Safety reminders

Work in teams at night where possible and always tell someone where you are going and when you expect to return. Keep away from farm machinery and chemicals. It is *essential* that you wear goggles/face protection when inspecting nest-holes. Only climb trees when strictly necessary, always use a ladder and ensure that you are accompanied. Remember, hay bales may move and can be dangerous; do not attempt to move them.

Disturbance

During winter months barn owls may struggle to survive. If you suspect that a barn owl is present at a particular site, do not flush it out, but simply record the site. A barn owl disturbed in daylight may be mobbed by other birds and will be reluctant to return to the site. The winter fieldwork must be completed by the end of January to avoid possible disturbance to tawny owls which may be incubating eggs or even brooding small young during February (Taylor 1991). Do the minimum necessary to prove breeding has occurred or been attempted. At potential nest-sites look for signs of occupancy, then carry out a site watch. Only after these procedures fail to confirm or refute breeding should you inspect the nest-site. Do not visit nest-sites in the breeding season without a Schedule 1 licence

Methods

Mark the boundary of the survey area on the map. Outline and hatch any areas which were deemed unsuitable and not searched.

The survey periods have been carefully chosen to maximise the chances of breeding owls being found and to minimise the risk of disturbance (Percival 1990), so please adhere to them.

Winter visits

Systematically search for all suitable potential nest sites in buildings, trees, bale stacks, etc, in the survey area and mark all those found on the field map with an X. Record every suspect site, ie every tree with a suspected suitable cavity or nest-box; every building with possible ledges, crevices or nest-boxes; every hay-bale stack (even though they may not be there later in the year), and all quarries and cliff-faces. Count each tree as a single potential site. However, several farm buildings grouped around a yard should be classified as a single site.

Potential buildings
Hay lofts and ledges, the tops of walls, the ends of roof beams and gaps in walls are all worth checking for pellets and feathers. Gaps in walls, even half-brick-sized holes leading into cavities may be used, as may ventilation and ducting tubes, storage bins and old water tanks in roofs. In old/derelict houses, the loft space and cavities alongside chimney breasts are favoured, as is space under floorboards or corners of upper rooms. Barn owls may inhabit buildings occupied by people or animals, eg grain stores or milking parlours.

Nest-boxes
These may be on beams inside agricultural buildings, on exterior walls or telegraph poles, or up in trees (any tree is suitable).

Potential trees
Barn owls generally favour isolated trees in fields or parkland, along hedgerows or roadside verges. Dense woodland can be excluded from the search as there are unlikely to be any barn owls present, but check up to three trees in from the edge or along a ride. Large cavities in the main trunks of deciduous trees are favoured. Any tree trunk with a diameter less than 45 cm can normally be disregarded. Entrance holes can be quite small (15 cm) but often lead into cavities with large chambers. Short, squat, dumpy-looking trees with bulbous cavities are ideal candidates. The entrance holes are an average of 4.5m above ground. If any of the trees you check is a potential site, record it and check in June.

Bale stacks
All hay-bale stacks should be recorded as potential nest-sites.

Cliff and quarry sites
A very few barn owls nest on heavily ridged or creviced cliff or quarry faces. Record these as potential sites.

Only record any known roost-sites if they qualify as potential nest-sites. Make extra visits to the survey area if necessary, eg if the area is large and/or complex.

Complete the potential nest-sites recording sheet (Figure 1). There is room to fill in totals for up to four visits to the site, but the overall total number of potential nest-sites of each kind is the most important information to record. Copy the total numbers of potential nest-sites of each kind onto the summer visit site occupancy recording sheet (Figure 2).

Summer visit

At each potential nest-site, look for signs of barn owl presence. These may be pellets, feathers or white splashings. You may see barn owls in the area but not know where they are nesting, or you may hear about sightings by local people. In all these cases, it is necessary to watch for barn owls at dusk. Get into position an hour before sunset and watch until an hour after sunset. At potential nest-sites with signs of barn owls, watch for the birds emerging or entering the site. If barn owls are seen in the area but you do not know where they may be nesting, watch near where they are seen, follow them or plot their direction of travel on a map and locate possible nest-sites along this path.

At any dusk watch, if you see an adult bird carrying prey to the nest (birds may carry prey to both nest- and roost-sites), or hear the hissing and snoring noises of the young owlets (sometimes heard some distance

POTENTIAL NEST SITES RECORD SHEET

PROJECT BARN OWL:

10KM SQUARE: ☐☐☐☐ TETRAD: ☐

NAME: ______________________ TELEPHONE: ______________________

Year: (circle) 94/95 95/96 96/97

ADDRESS: ______________________

PLEASE RECORD THE NUMBER OF POTENTIAL BARN OWL NEST SITES OF EACH TYPE FOUND IN EACH 1KM OF THE TETRAD.
See instructions on reverse. IMPORTANT NOTE: COPY THESE TOTALS ONTO THE SITE OCCUPANCY RECORD SHEET.(summer survey).

(1) Site type		(2) No. of sites in each 1km square of the tetrad NW	NE	SW	SE	(3) Total number of sites
Building						
Agricultural/used	AU					
Agricultural/disused	AD					
Agricultural/box in building	AN					
Bale stack in building	BB					
Industrial/used	IU					
Industrial/disused	ID					
Industrial/box in building	IN					
Domestic/used	DU					
Domestic/disused	DD					
Domestic/box in building	DN					
Church/used	CU					
Church or ruins/disused	CD					
Military/disused	MD					
Box in other building	NO					
Other (state*)						
*						
*						
(4) TOTAL						

(1) Site Type		(2) No. of sites in each 1km square of the tetrad NW	NE	SW	SE	(3) Total number of sites
Tree Cavity						
Elm	TE					
Oak	TO					
Ash	TA					
Willow	TW					
Beech	TB					
Tree nest box	TN					
*						
*						
Other structure						
Bale stack (outside)						
Pole nest box						
*						
*						
(4) TOTAL						

(5) GRAND TOTALS					

IMPORTANT NOTE:

COPY COLUMN (1) DETAILS AND COLUMN (2) TOTALS ONTO THE SITE OCCUPANCY RECORD SHEET READY FOR THE SUMMER SURVEY. THIS WILL HELP YOU AND US TO CHECK THAT THE INFORMATION HAS BEEN CORRECTLY RECORDED.

Figure 1
Barn owl potential nest-sites recording sheet.

SITE OCCUPANCY RECORD SHEET

PROJECT BARN OWL:

10KM SQUARE: ☐☐☐☐ TETRAD: ☐

NAME: ______________________ TELEPHONE: ______________________

YEAR (circle): 94/95 95/96 96/97

ADDRESS: ______________________

Please record the occupancy of potential Barn Owl nest sites of each type found in each 1km of this tetrad.
See instructions on reverse. IMPORTANT NOTE: **Please return this form by 15 September at the latest, to ensure its inclusion in the data analysis.**

(1) Site type		(2) Total sites NW	NE	SW	SE	(3) No. of occupied sites: Barn Owl NW	NE	SW	SE	Tawny Owl NW	NE	SW	SE	Little Owl NW	NE	SW	SE	Kestrel NW	NE	SW	SE	Jackdaw NW	NE	SW	SE	Stock Dove NW	NE	SW	SE	(4) Other Species (state)	No.	1km	(5) No. sites unoccupied NW	NE	SW	SE
Building																																				
Agricultural/used	AU																																			
Agricultural/disused	AD																																			
Agricultural/box in building	AN																																			
Bale stack in building	BB																																			
Industrial/used																																				
Industrial/disused	ID																																			
Industrial/box in building	IN																																			
Domestic/used	DU																																			
Domestic/disused	DD																																			
Domestic/box in building	DN																																			
Church/used	CU																																			
Church or ruins/disused	CD																																			
Military/disused	MD																																			
Box in other building	NO																																			
Other (state*)																																				
*																																				
*																																				
Tree cavity																																				
Elm	TE																																			
Oak	TO																																			
Ash	TA																																			
Willow	TW																																			
Beech	TB																																			
Tree nest box	TN																																			
*																																				
Other structure																																				
Bale stack (outside)	BS																																			
Pole nest box	PB																																			
*																																				
*																																				
(6) TOTAL																																				

Figure 2
Barn owl site occupancy recording sheet.

away), count this as a breeding site and do not approach it any more closely.

If you do not see any signs of barn owls near a potential nest-site, or during a dusk site watch, you will have to inspect the site. Remember, you should have a licence to do this and always wear goggles or face protection.

Buildings and nest-sites
Checking is best done by two people. One person should enter the building by the safest route and the other should stand about 10 m away, watching for any birds to fly out. Do not deliberately flush birds out; if at all possible, watch the site from a distance. If you do have to look inside a confined nest space, always wear goggles or face protection. Stay as quiet as possible; only one person should approach the site and they should be as quick as possible. If barn owls are seen, if the site contains a carpet of barn owl pellets, or if the visible remains of round white eggs or young are found, count it as a breeding site.

The map filled in during the winter visits will show all potential nest sites (X). Mark on this the sites which showed signs of barn owl presence (f), which were breeding barn owl sites (●) and which were sites lost or destroyed during the survey (■). Fill in the barn owl site occupancy recording sheet (Figure 2). Copy over the number of potential nest-sites noted in each category in the winter (from Figure 1) into the 'Total sites' column of Figure 2. Record the number of each type of site occupied by barn owls (and all other bird species too). Calculate how many of each type of site were unoccupied. Sum each column and record the totals in the bottom line.

Report the number of breeding pairs of barn owl as the total number of occupied barn owl nest-sites found.

References

Percival, S M (1990) Population trends in British barn owls *Tyto alba*, and tawny owls *Strix aluco*, in relation to environmental change. *BTO Research Rep.* 57.

Taylor, I R (1991) Effects of nest inspections and radiotagging on barn owl breeding success. *J. Wildlife Management* 55: 312–315.

Toms, M (1995) *Project Barn Owl 1995: fieldwork instructions; information sheets and recording sheets.* A BTO/Hawk and Owl Trust Collaboration. BTO, Norfolk.

Short-eared owl
Asio flammeus

Status

Amber listed: SPEC 3 (V)
Annex I of EC Wild Birds Directive

National monitoring

None.

Population and distribution

The breeding distribution of short-eared owls in Britain is similar to the distribution of heather moorland, with most pairs in Scotland and the north of England. Population size varies greatly from year to year in relation to small mammal prey availability (*88–91 Atlas*). There are estimated to be 1,000-3,500 pairs of short-eared owl breeding in Britain (*Population Estimates*).

Ecology

Short-eared owl sites change from year to year depending on the availability of small mammal prey. Because of this, short-eared owls may be absent from many areas of apparently suitable habitat. Favoured breeding sites are areas away from human disturbance with plenty of small mammal prey, for example moorland heath, newly afforested hillsides, extensive rough grazing, marshes, bogs, sand-dunes and inshore islands. The actual nest-site is usually in long grass, heather or rushes, often on sloping ground. Short-eared owls will not be found in dense woodland, built-up or heavily grazed areas. Woodland fringe is sometimes occupied by pairs adjacent to young forestry or on coastal wooded strips with rough grazing or marsh (Lockie 1955, Glue 1977). A clutch of 4–8 eggs is laid from mid- to late March through to July; a replacement clutch is laid after early egg loss. Incubation lasts about 25 days and the young fledge after 24–27 days, but leave the nest at 12–17 days. Young are to be found in or close to the nest from late April to July. The young are able to fly well after about 35 days.

Breeding season survey – population

This is a difficult species to survey, given its unpredictable nature (particularly the between-year demographic fluctuations) and the difficulty in surveying wide expanses of open moorland. This method is based on information contained in Lawton Roberts and Bowman (1986), though it has not been tested in the field.

Information required

- number of confirmed, probable and possible breeding pairs
- a map of the survey area with the boundary and summarised registrations marked.

Number and timing of visits

At least two, between early April and the end of May. If breeding is not confirmed, a further visit should be made in June.

Time of day

0700–1900 BST.

Weather constraints

Avoid poor weather.

Sites/areas to visit

Any suitable habitat. Short-eared owls breed on heather moorland and a variety of grassland sites, notably young forestry and coastal/island rough grazing and areas of grass/scrub mix (see *Ecology*, above).

Equipment

- 1:25,000 OS map (1:50,000 will do).

Safety reminders

When working in more remote areas, ensure someone knows where you are and when you are due back. Always carry a compass. If possible, surveyors should not work alone in remote areas, and spare clothing, a survival bag, whistle, first-aid kit and food should be carried.

Disturbance

Disturbance should be minimal. No visits to nests are necessary to confirm breeding.

Methods

Take a new field map on each visit to the site, and mark clearly the boundary of the survey area on the map.

On the first visit, walk to within 250 m of each spot. Map all those areas that are unsuitable for hunting and breeding (see above) and exclude them from further searching. Search for short-eared owl activity and signs of short-eared owl presence elsewhere.

On the second visit (before the end of May), search all suitable areas. Record all short-eared owl observations and attempt to confirm breeding at all potential nesting sites. Even if breeding is confirmed at a site on the first visit, a second is still necessary, as other pairs may be breeding nearby.

If breeding is not confirmed at all potential nest-sites on the second visit, undertake a third in June. Watch birds, locate nest-sites and confirm breeding behaviour wherever possible (this can be done from a discreet distance, without a visit to the nest itself).

During each visit look for hunting males, which tend to be more active diurnally in the pre-egg-laying period and when feeding young than during incubation. Males will carry food to the nest, often landing close to the incubating female; they can also be found resting off-duty within 100 m of the nest. Females will often rise from the nest and land beside or close to an incoming male. Parents will also rise and mob observers, giving distinctive barking or quacking calls. The territorial display and call notes, accompanied invariably by wing-clapping, are sound confirmation of breeding. Beware of ranging young from late June/July onwards, as these can easily be recorded as adult birds by an inexperienced observer. During each visit, the presence (or absence) of short-eared owls, their activity, sex and details of any nests located should be recorded on the map using standard BTO codes (Appendix 1) and cross-referenced to any more-detailed comments.

At the end of the survey, determine the number of confirmed, probable and possible breeding pairs in the survey area using the following rules.

Breeding is confirmed if:

- A nest containing eggs or young is found.
- An adult is seen carrying food for the young.
- A used nest or eggshells are found (occupied or laid within the period of the survey).
- Recently fledged young are found.
- Territorial display and call notes accompanied by wing-clapping are seen.
- Agitated behaviour or anxiety calls are given by the adults (see above).

Breeding is probable if:

- A pair of short-eared owls is seen in suitable nesting habitat in the breeding season.
- Birds are seen visiting a probable nest-site.

Breeding is possible if:

- A short-eared owl is observed in the breeding season in possible nesting habitat.

Report the minimum number of breeding pairs of short-eared owls as the number of confirmed pairs; and the maximum number as the confirmed + probable + possible breeding pairs.

References

Glue, D E (1977) Feeding ecology of the short-eared owl in Britain and Ireland. *Bird Study* 24: 70–78.

Lawton Roberts, J and Bowman, N (1986) Diet and ecology of short-eared owls *Asio flammeus* breeding on heather moor. *Bird Study* 33: 12–17.

Lockie, J D (1955) The breeding habits and food of short-eared owls after a vole plague. *Bird Study* 2: 53–67.

Nightjar *Caprimulgus europaeus*

Status

Red listed: BD, SPEC 2 (D)
Annex I of EC Wild Birds Directive

National monitoring

National surveys: 1992 (RSPB and BTO, with support from Forestry Commission), 2002.

Population and distribution

In Britain, nightjars breed from south-east England to south-west Scotland, although they are concentrated in south and east England. There have been marked long-term declines in distribution and numbers of British and Irish populations (*88–91 Atlas*), although the most recent (1992) survey has shown a partial recovery (Morris et al 1994). There have also been marked declines in the central, northern and north-west European populations (*Birds in Europe*). The main reason for these long-term declines in nightjar numbers is habitat loss. The 1992 survey estimated the British population at 3,400 males (Morris et al 1994).

Ecology

In Britain, nightjars breed principally on lowland heath and in young conifer plantations. The recent recovery in the British population has been attributed to an increase in suitable habitat as a consequence of clearfell forestry (Morris et al 1994). A clutch of 1–3 eggs is laid from mid-May to mid-July. Second clutches are infrequent, and a second brood is possible only if the first clutch was started before the second week in June. The young begin flying at about 15 days old, but they are dependent on their parents until 30 days old (*Red Data Birds*). In Britain, nightjars churr from mid-May to mid-August, peaking in June (though in the Brecks the peak extends to mid-July: Cadbury 1981).

Breeding season survey – population

Information required
- number of singing (churring) males.

Number and timing of visits

At least two visits, from June until mid-July.

Time of day

At dusk (2200–2330 BST in June) or about an hour before dawn (0200–0400 BST). Pre-dawn counts are preferable.

Weather constraints

Avoid surveying in winds greater than Beaufort force 3. Dry, drizzly and humid conditions are all suitable.

Sites/areas to visit

All regularly occupied sites, formerly occupied sites and potential sites, including clearfell, young forestry plantations and lowland heath.

Equipment

- 1:10,000 OS map (but 1:25,000 will do)
- clipboard
- torch
- compass
- whistle
- mobile phone/radio (desirable but not essential).

Safety reminders

Ensure someone knows where you are and when you are due back. If possible, visit all sites during the day to familiarise yourself with them. Map-reading skills (and possibly a compass) are essential, especially in featureless forest or large expanses of heath. Take a whistle and make sure you can find your way back to your car. Take precautions against Lyme's disease: guard against tick bites (tuck your trousers into your socks) and check for ticks at the first opportunity. Try to work in pairs and, if possible, have a mobile phone/radio.

Disturbance

There should be little disturbance because nightjars are detected by song, not by looking for nests or by sighting birds.

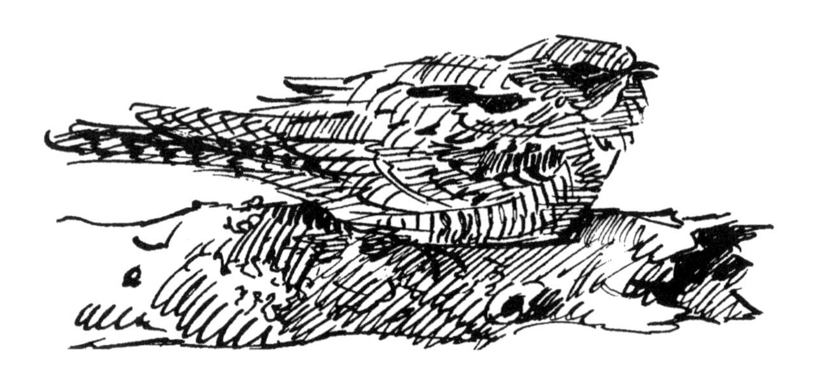

Methods

Large sites should be split into numbered sections and may require coordinated counts. Do not attempt to count more than 80 ha of suitable habitat at a time (equivalent to slightly less than a 1-km square marked on an OS map). If resources are limited, sample larger areas. To do this, stratify plantations into two age-classes (1–10 years and 11–20 years), and consider restocked forestry and new plantation separately. See the *Introduction* for sampling methods.

During daylight hours and using a map, familiarise yourself with any landmarks and work out the most efficient route. For the survey, arrive before dusk (or, for early morning visits, one and a half hours before dawn) and walk your allocated compartment/heathland block from about 20 minutes after sunset. Walk at a steady pace and stop every few minutes to listen for churring. Cup your hands behind your ears and rotate your head to help gauge sound direction. Try to get to within about 100 m of each point by walking the area in a systematic way, either in parallel transects or following field edges or plantation paths and boundaries.

Mark the position of all churring nightjars on a map using a number code for each bird and a letter code for each visit: thus A1, A2, A3, etc, for each male found on the first visit, B1, B2, B3, etc, for each found on the second visit, and so on; over the course of two visits, the same bird may be recorded as A1 and B1.

Mark all other nightjar registrations on the same map, cross-referencing them with other notes to help interpret the maps. Wherever possible, use standard BTO mapping codes (Appendix 1). Note simultaneously churring males by joining their positions on the map with a dotted line; these records will undoubtedly be of two separate males on different territories. If churring is heard from two locations not simultaneously but up to 30 seconds apart, record this as different males if the two are more than 400 m apart, otherwise record this as the same male that has moved. Indicate the direction in which a bird has moved with arrowed lines.

After the second visit, transfer all records to a single master map using the lettering as outlined above. Circle those records thought to be from the same territory and report the number of separate churring males on the site.

In previous national nightjar surveys, habitat information was collected. For advice on this aspect of the survey see Morris et al (1994).

Breeding season survey – breeding success

Finding nests and subsequent monitoring is a very time-consuming and potentially disruptive operation. Nest-finding is not recommended unless undertaken by experienced fieldworkers. Information on breeding success or productivity is collected annually at some sites, but these are detailed studies. Information on nest-finding methods is given in Bowden (1988) and Bowden and Green (1991).

References

Bowden, C G R (1988) *Thetford Forest Nightjar Project: first year progress report to Forestry Commission*. RSPB unpubl.

Bowden, C G R and Green, R E (1991) *The Ecology of Nightjars on Pine Plantations in Thetford Forest*. RSPB unpubl.

Cadbury, C J (1981) Nightjar census methods. *Bird Study* 28:1–4.

Morris, A, Burges, D, Fuller, R J, Evans, A D and Smith, K W (1994) The status and distribution of nightjars (*Caprimulgus europaeus*) in Britain in 1992. *Bird Study* 41: 181–191.

Green woodpecker *Picus viridis*

Status

Amber listed: SPEC 2 (D)

National monitoring

CBC, BBS.

Population and distribution

Green woodpeckers are equally likely to be found in deciduous woodland or feeding out in open farmland, parkland and large gardens. Some of the highest densities are found in western parts of England and South Wales. They are absent from Ireland. In the south-east of England they are absent from the fens and relatively intensive arable areas. In Scotland, green woodpeckers are more common in the south-east but have expanded into central and eastern Scotland in recent years and now occur as far north as Ross-shire (*88–91 Atlas*). The European population of green woodpeckers declined by almost a half between 1970 and 1990 (*Birds in Europe*). There are estimated to be 15,000 pairs breeding in the UK (*Population Estimates*).

Ecology

This woodpecker is reliant on ants (adults and pupae) for food. As a result the egg-laying period is protracted, from early April to mid-July or later, with young in the nest in late July/August (Glue and Boswell 1994).

Breeding season survey – population

This method is based on the BTO Common Birds Census territory-mapping method. Because this species occurs at low densities, the point-count method (used for great spotted woodpecker) is not suitable (Bealey and Sutherland 1982). Where survey plots are small, density may be overestimated, as territories may overlap plot boundaries. Survey plots should thus be as large as practically possible. The standard CBC method requires 8–10 visits to a plot, but here the number of visits has been reduced because: only a single species is being sought; survey plots need to be large; and the survey period is more restricted.

Information required

- estimated number of pairs
- map showing summarised registrations, the survey boundary and survey route.

Number and timing of visits

Make 5–8 visits at roughly fortnightly intervals (when the weather is suitable), from late March to early July. The more visits that are made, the more precise the final estimate.

Time of day

Any time of day, although feeding adults are often at their most active during the maximum heat of the day (1200–1500 BST), seeking ant prey. Parents can also be readily observed carrying food loads from late afternoon through into the evening.

Weather constraints

Avoid poor weather conditions, especially high winds (greater than Beaufort force 4) and heavy rain.

Sites/areas to visit

The preferred habitat combines old deciduous trees for nesting, with nearby feeding grounds that have an abundance of ants. Conifers do not usually provide suitable nesting habitat, but conifer forests can provide plenty of ants. Such combinations of habitat occur in semi-open landscapes with small woodlands, hedgerows, scattered old trees or forest edges; tree-lined riverine areas are commonly occupied. Foraging areas include grassland, heaths, plantations, orchards, golf courses and lawns. Green woodpeckers prefer to feed on the ground, although they may seek insects on the surface of branches. They look for ants' nests by flying systematically along woodland edges, paths or embankments.

Equipment

- 1:10,000 map.

Safety reminders

Nothing specific. See the *Introduction* for general guidelines.

Disturbance

This method involves little or no disturbance.

Methods

Mark the boundary of the survey area on a map. Work out a route which gets to within 200 m of every point. Walk at a comfortable pace, stopping at least every 50 paces to scan the surrounding trees and bushes and particularly to watch for birds feeding on the ground. On each visit mark on the map the position of all sightings of green woodpeckers and any calls that are heard. Use the appropriate BTO symbols (Appendix 1) to annotate the map.

If possible note the sex of individuals seen: adult male and female green woodpeckers are similar, although the broad malar stripe (on side of face) is completely black on the female but red with a black border on the male. It is not really possible to separate the sexes by call. The best-known, laughing or yaffling call is heard from late March to early May and is made by both male and female. Green woodpeckers have several other calls, but very rarely drum (*BWP*).

As young grow they are often noisy, especially during cool spells of weather when they become hungry. As the young get bigger they start to thrust their heads out of nest-holes, often in a conspicuous fashion, and parents may become very demonstrative.

After your final visit, transfer all the mapped records to a single master map. Using the standard CBC procedures (see the generic survey methods section), estimate the number of pairs on the site.

A singing or observed male counts as one pair. If a male is not observed, a female, a group of fledglings or a nest counts as one pair. Beware of records of roaming unmated males, which may confuse interpretations of the maps.

References

Bealey, C E and Sutherland, M P (1982) Woodland birds of the West Sussex Weald. *Sussex Bird Report* 35: 69–73.

Glue, D E and Boswell, T (1994) The comparative nesting ecology of the three British breeding woodpeckers. *British Birds* 87: 253–269.

Great spotted woodpecker *Dendrocopos major*

Status

Non-SPEC

National monitoring

CBC, BBS.

Population and distribution

Great spotted woodpeckers are present throughout much of England and Wales, except for upland and fenland areas that lack trees. In Scotland, they are restricted to lowland and central areas, and none breed in Ireland (*88–91 Atlas*). There are no accurate counts of the number of breeding great spotted woodpeckers in Britain, but an estimate based on densities in different woodland types and the areas of those woodlands in Britain suggests a population of between 25,000 and 30,000 pairs (*88–91 Atlas*).

Ecology

Great spotted woodpeckers pair up in March and April, lay their eggs in early May, and fledge young in June or early July. They breed mainly in mature deciduous woodland (Smith 1987). They have a shorter, louder, lower pitched and firmer drum than lesser spotted woodpeckers (*BWP*).

Breeding season survey – population

This is a point-count method which is considered more suitable for large woodland areas (*Census Techniques*); it is adapted from Bibby et al (1985) and Smith (1987). In the past, Common Birds Census territory-mapping techniques have been used for this species, but territories are so large (10 ha +) that CBC methods may not always be appropriate. Also birds will dispute and display at territory boundaries so that mapped clusters of registrations do not necessarily correspond to territories. Point counts allow a large area to be covered in a short time.

Information required

- number of birds detected in two separate distance bands during two five-minute counts (one early in April and one later) from at least 30 systematically placed counting stations
- map showing the survey boundary and the locations of point counts.

Number and timing of visits

Two visits: one early in April and one two weeks later.

Time of day

From one hour after dawn to midday.

Weather constraints

Poor weather conditions should be avoided, especially high winds (greater than Beaufort force 4) and heavy rain.

Sites/areas to visit

Anywhere with sufficiently large trees to accommodate nest-holes. More isolated or scattered trees are less favoured unless adjoined by larger stands of broadleaved, coniferous or in particular mixed tree species (*BWP*). Woodland with mature and a few dead trees is favoured.

Equipment

- 1:10,000 map of the site and field recording maps
- prepared recording forms.

Safety reminders

No specific advice. See the *Introduction* for general safety reminders.

Disturbance

The method involves little disturbance. Try to walk quietly through the survey area as the final population estimates may be biased if birds are scared away from the observer.

Methods

Mark the boundary of the survey area on a map and work out the area of the site. The point-count method involves the observer standing at a number of fixed points (the count stations) within the area and recording all the great spotted woodpeckers seen or heard within the allotted time.

Placing the counting stations

Place these systematically within the study plot, no closer than 150 m apart to help prevent double-counting. Systematic selection involves placing a grid (of squares not less than 150 × 150 m) over the area and positioning the counting stations at the intersections of the grid. The number of points selected depends on the size and shape of the area to be surveyed, but try to ensure that there are at least 30 counting stations in the study plot. It is a good idea to select a few more points than will actually be needed in case any of the points are unsuitable. In principle, the counting stations could be located at random, though systematic placing is, in practice, often more simple and there is no reason to assume that in a homogeneous habitat like woodland systematic sampling would lead to any biases.

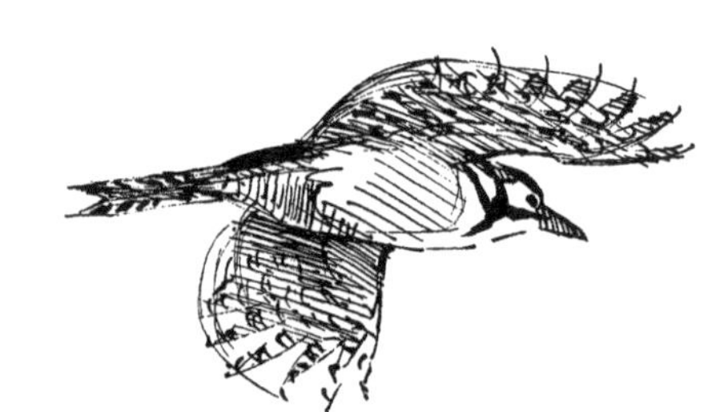

Counts

Work out a route that will take you to all of the points. On the first visit to the site, mark the locations of all of the points (eg with coloured marking tape on trees) to make them easier to find on the return visit. On arrival at a point wait for at least five minutes to let birds recover from any disturbance and then count for exactly five minutes. All great spotted woodpeckers should be counted, whether detected by eye or ear, except any flying right through the area. The birds should be recorded in distance bands, ie those detected within a 30-m radius of the observer or those detected beyond 30 m. An example of the type of recording form that could be used is given in Figure 1. Note whether the birds were seen or heard, give a brief behavioural description and, if seen, note the sex. Male and female great spotted woodpeckers share the same repertoire of non-vocal drumming and a variety of contact-calls.

GREAT SPOTTED WOODPECKER SURVEY

Observer: ____________ Date: Visit 1: ________ Visit 2: ________ Weather: Visit 1: ____________ Visit 2: ____________

6 Figure grid reference: ____________ Total area of survey: _____ ha

VISIT 1 Point no.	No. of birds <30 m	>30 m	Comments	VISIT 2 Point no.	No. of birds <30 m	>30 m	Comments	Max. no. of birds <30m	>30m
1				1					
2				2					
3				3					
4				4					
5				5					
6				6					
7				7					
8				8					
9				9					
10				10					
11				11					
12				12					
13				13					
14				14					
15				15					
16				16					
17				17					
18				18					
19				19					
20				20					
21				21					
22				22					
23				23					
24				24					
25				25					
26				26					
27				27					
28				28					
29				29					
30				30					

Figure 1
An example recording form to be used when surveying great spotted woodpeckers.

Analysis

For each of the distance bands (0–30 m and 30 m +) you will have two counts: one early in April and one later. Use the maximum of these early and late values for each distance band in subsequent calculations. Add up these maxima across all the points so that you have a sum of maxima for the near distance band (n_1) and a sum in the far distance band (n_2). The total number of birds detected (n) is $n_1 + n_2$. Knowing the number of points (m) and the preselected distance radius (r), the density (D) of birds on the study plot is calculated with the following equation:

$$D = \ln(n/n_2) \times n/m\pi r^2$$

where ln is the natural log or $\log_e$ (there is a button to calculate this on most calculators). π is a constant with a value of 3.1416 (also available at the push of a button on most calculators).

If r is in metres then the resulting value will be the density of great spotted woodpeckers per square metre; to convert this to the number per hectare, multiply D by 10,000. To then work out the total number of individual breeding woodpeckers in the survey area, multiply the density per hectare by the total number of hectares in the survey area (see Box 1).

It may seem odd to ask for counts to be split into different distance bands, but this allows a very crude calculation of the manner in which detectability falls with distance from the observer. This in turn allows the number of birds present at a distance, but which were not detected, to be estimated.

The population figure is of individuals presumed to be breeding birds. Because many of the registrations may be of calling birds, there will have been little chance to sex individuals. In April the observer is

Box 1
A worked example of calculating the population of great spotted woodpeckers.

In a wood of 120 ha there were 40 point positions to count at, there were two visits and the prearranged distance for separating registrations was 30 m.

The overall total number of great spotted woodpeckers recorded was one at <30 m and five at >30 m.

$$\text{Density} = \ln(n/n_2) \times n/m(\pi r^2)$$

Where: ln is the equivalent of $\log_e$
n is the total number of birds counted
n_2 is the number beyond the fixed radius (r)
n_1 is the number counted within radius (r) so that $n=n_1+n_2$
m is the total number of point counts
r is the fixed radius

Thus: $D = \ln(6/5) \times 6/40(\pi 30^2)$
$= 0.18 \times 5.3 \times 10^{-5}$
$= 9.55 \times 10^{-6}$ birds per m^2
$= 0.095$ birds per ha
$= 11$ individuals in the survey area (120 ha)

This gives an estimated 5 pairs of great spotted woodpeckers.

equally likely to see or hear male and female great spotted woodpeckers, thus halving the population total gives an estimate of the number of breeding pairs.

Breeding season survey – productivity

Information required
- The number of viable young per pair.

Number and timing of visits

As many as possible from late May to mid-June to locate nest-sites, and at least one visit to count the number of viable young. If the young are 14+ days old when first counted, no further visits are necessary.

Time of day

Locate nests just after dawn to catch the main period of activity. Count young in the nest at around midday.

Equipment

A ladder and a mirror on a pole, with the mirror at an angle of 45°. Have the mirror illuminated if possible.

Safety reminders

When using a ladder, one person should secure the bottom of the ladder while another climbs. Beware of trees that are too rotten to take the weight of a person and a ladder.

Disturbance

Before counting the young, try to wait until both adults are away from the nest and the young have just been fed. Be as quick as possible; always use the mirror to count the young without handling them.

Methods

In order to locate great spotted woodpecker nests, visit any sites that were occupied the year before and/or search for fresh wood chippings on the ground around likely trees. During the initial visits search the whole survey area thoroughly. If you locate trees which are likely to have a woodpecker's nest, watch for a couple of hours for adults feeding young. Feeding activity peaks in the morning and afternoon, with a lull at midday. During the week before fledging, the young become noisier and nests are easier to locate.

The fledging period is 20–24 days, and young of 14+ days are considered to be viable. A 14-day-old great spotted woodpecker will have all its feathers, but the feathers will still be growing and the ring around the eye will be bare. All feather growth, except for the flight feathers and tail, will be complete by day 20 (*BWP*). Count the young in the nest by looking with the mirror, and report the number of viable young per nest.

References

Smith, K W (1987) The ecology of the great spotted woodpecker. *RSPB Conservation Review* 1: 74–77.

Bibby, C J, Phillips, B N and Seddon, J E (1985) Birds of restocked conifer plantations in Wales. *J. Applied Ecology* 22: 619–633.

Lesser spotted woodpecker *Dendrocopos minor*

Status

Non-SPEC

National monitoring

None.

Population and distribution

This is the least common of the three British woodpeckers. Absent from Ireland, it is most common in the south and east of England but sparsely distributed over most of its range and rarely breeds north of Cumbria (*88–91 Atlas*). Numbers rose in the 1970s and fell back again to their former level in the 1980s. Much of this change has been attributed to the presence of elm trees, dead and dying from Dutch elm disease which temporarily increased the birds' food supply (*88–91 Atlas*). BTO CBC figures (Crick et al 1997) show a population continuing to fall, adding to the general cause for concern that surrounds this species. The estimated population of breeding lesser spotted woodpeckers in Britain is 3,000–6,000 pairs (*Population Estimates*).

Ecology

Nesting takes place almost exclusively in broadleaved woodland habitats, in a variety of situations: along wood edges, in orchards, shelterbelts, thick hedges and parkland dotted with established trees. Lesser spotted woodpeckers nest comparatively more frequently in suburban areas than do great spotted or green woodpeckers (Glue and Boswell 1994). They give a quieter, higher-pitched and more brittle-sounding drum (male only) than great spotted woodpecker. They also use an advertising call which is perhaps the most frequent and prominent call: a series of relatively long units with a soft squeaking quality, like 'pee-pee . . .'. A clutch of 4–6 eggs is laid in May–June; incubation is 11–12 days. Young are cared for and fed by both parents and the fledging period is 18–20 days (*BWP*).

Breeding season survey – population

This species is more sparsely distributed and more difficult to locate than great spotted woodpecker, and would be under-represented by the point-count survey method which is recommended for that species (Bealey and Sutherland 1982). The method given here is based on that described by Wiktander et al (1992); it is a more intensive method and may be difficult to achieve in large blocks of woodland. An alternative approach would be to survey the area using a standard CBC (see the generic survey methods section), although this would mean more visits, many of them not at the optimum time of year for this species.

Information required

- maximum number of pairs recorded on any one visit
- a map of the survey area with the boundary clearly marked.

Number and timing of visits

At least three visits, in April.

Time of day

From one hour after dawn to midday.

Weather constraints

Poor weather should be avoided, especially high winds (greater than Beaufort force 4) and heavy rain.

Sites/areas to visit

Any area containing broadleaved trees, including small field copses, open woodland, parkland with plenty of rotting timber, old orchards, shrubberies, gardens, avenues of ancient trees, trees alongside streams, hedgerows with mature trees, golf courses, and gravel pits and reservoirs with scrubby margins.

Equipment

- map of the survey area
- compass (optional).

Safety reminders

Nothing specific. See the *Introduction* for general guidelines.

Disturbance

Disturbance is unlikely using this method, as observation is done from the ground and does not involve any visits to nests.

Methods

Mark the boundary of the survey area on the map. Work out a route which enables you to walk to within 50 m of each point in the survey area, even though this means that you will only be able to cover 0.5 km^2 in a single day. Obviously, a coordinated team of people could cover a larger area.

Walk through the area at a comfortable pace as quietly as possible, paying particular attention to bird movement in the tree canopy, and checking any movement observed using binoculars. At intervals of no more than 50 paces, stop to scan the trees around you to look for birds foraging on the branches and foliage of trees and scrub or catching insects in the air. If disturbed, lesser spotted woodpeckers will typically hide under or behind a branch, mostly high in a tree. Birds recorded drumming in April will be establishing a territory. Nest excavation can be a noisy and busy affair during April and is often accompanied by courtship and display.

Disputes involving more than two birds (usually males) are quite conspicuous with much display and movement. These occur mainly in April with birds chasing one another in the air. Between chases, these birds may pause together on a branch so you cannot always assume that two (unsexed) birds together are the members of a pair.

On each visit, use a different field map with the route marked on it, although you should alternate the direction of the route on each visit. Mark any woodpecker registrations on the map, cross-referencing these to a notebook allowing more room for comments.

Report the maximum number of pairs counted on any one visit to the area. Pair(s) are defined as follows:

Single birds (males or females) = one pair.
Two (or more) birds of the same sex together = two (or more) pairs.
A male and female together = one pair.

The following records, for example, would all constitute two pairs: two males together; two females together; two males and one female; two females and one male; two males and two females.

References

Bealey, C E and Sutherland, M P (1982) Woodland birds of the West Sussex Weald. *Sussex Bird Report* 35: 69–73.

Crick, H Q P, Baillie, S R, Balmer, D E, Bashford, R I, Dudley, C, Glue, D E, Gregory, R D, Marchant, J H, Peach, W J and Wilson, A M (1997) *Breeding Birds in the Wider Countryside: their conservation status (1971–95)*. BTO Research Report 187.

Glue, D E and Boswell, T (1994) The comparative nesting ecology of the three British breeding woodpeckers. *British Birds* 87: 253–269.

Wiktander, U, Nilsson, I N, Nilsson, S G, Olsson, O, Pettersson, B and Stagen, A (1992) Occurrence of the lesser spotted woodpecker *Dendrocopos minor* in relation to area of deciduous forest. *Ornis Fennica* 69: 113–118.

Woodlark
Lullula arborea

Status

Red listed: BD, BL, SPEC 2 (V)
Schedule 1 of WCA 1981
Annex I of EC Wild Birds Directive

National monitoring

National breeding surveys, 1986 (BTO/JNCC) and 1997 (RSPB/BTO/EN/JNCC); to be repeated at 10-year intervals.
Rare Breeding Birds Panel.

Population and distribution

Although the woodlark is still relatively rare in Britain, its population grew rapidly between the mid-1980s and mid-1990s: an estimated population of 250 pairs in 1986 (Sitters et al 1996) had grown to more than 1,500 pairs in 1997 (S Wotton pers comm). Its range, however, is still restricted, and since the 1970s it has been largely confined to the south-west, the New Forest, the Hampshire/Surrey border, Breckland and part of the Suffolk coast, with outlying populations in Nottinghamshire and Lincolnshire. The population is more or less evenly split between heaths and young conifer plantations. Numbers at any site can be very variable from year to year, and the population is expanding into new areas and habitats.

Ecology

The habitats most frequently occupied are burnt or heavily grazed heather or grass heaths, restocked conifer plantations, tree nurseries, cleared woodland, derelict farmland, areas of disturbed ground and set-aside on acidic, sandy soils (S Wotton pers comm). A clutch of 3–4 eggs is laid between March and early August, with replacement nesting after failure at the egg or chick stage; second broods are frequent (*Red Data Birds*).

Breeding season survey – population

The method documented here is the same as that used in the 1997 national woodlark survey (Wotton 1997).

Information required
- number of territories.

Number and timing of visits

Three visits between mid-February and the end of May.

Visit 1: 15 February to 21 March
Visit 2: 22 March to 25 April
Visit 3: 26 April to 1 June

Time of day

Before midday.

Weather constraints

Visit on clear, dry days with little wind.

Sites/areas to visit

Heather and grass heaths which are mown, grazed or burnt, and conifer plantations recently cleared and replanted (up to eight years ago). Other well-drained generally acidic sites on sand, gravel or chalk, with areas of short vegetation (less than 10 cm high) and/or patches of bare ground.

Equipment

- Schedule 1 licence
- 1:10,000 OS map.

Safety reminders

Nothing specific. See general health and safety guidelines in the *Introduction*.

Disturbance

None necessary.

Methods

Mark the boundary of the survey area clearly on a map. Walk to within 100 m of every point within the survey area. On each visit, mark on the map the positions of all singing males (birds which sing are assumed to be male), flying birds, feeding pairs (any pairs observed assumed to be one male and one female), possible nests and adults feeding chicks. It will be easier to position records on the map correctly if landmarks such as isolated trees are also marked. Use BTO (CBC) symbols (see Appendix 1) for all registrations. Record bird movements; indicating flight directions will help to establish territory limits and possible nest locations. Woodlarks will fly quite long distances (up to 200–400 m) between, for example, a foraging area and a song perch. The position of singing birds can be difficult to establish as males sing from the air, from a perch or from the ground. Woodlarks often sit tight until the last possible minute and might even be flushed behind the observer.

Field notes should be checked on the same day as the visit and transcribed if necessary.

After the third visit estimate the number of territories. A singing male constitutes a territory. Strictly, neighbouring birds can only be separated with certainty if they are heard simultaneously. However, singing males can be considered as separate individuals if they are heard at points more than 200 m apart. Although this distance could depend on the quality of the habitat, where birds occur at higher densities they will be more likely to be recorded singing simultaneously anyway. Most territories are likely to be detected through the presence of a singing male but sometimes evidence of territories (eg a bird feeding young) may be obtained even though a singing bird was never recorded. Please take these observations into account when estimating the total number of territories on the site.

Report the estimated number of territories on the site. Also provide a map showing the boundary of the site, a summary map of registrations and a brief description of the habitat of the site (eg grass heath, conifer plantation replanted three years ago, etc).

References

Sitters, H P, Fuller, R J, Hoblyn, R A, Wright, M T, Cowie, N and Bowden, C G R (1996) The woodlark *Lullula arborea* in Britain: population trends, distribution and habitat occupancy. *Bird Study* 43: 172–187.

Wotton, S 1997 (1997) *National Woodlark Survey: Survey Instructions.* RSPB/BTO/EN/JNCC.

Skylark
Alauda arvensis

Status

Red listed: BD, SPEC 3 (V)
Annex II/2 of the EC Wild Birds Directive

National monitoring

CBC, BBS.
National breeding survey, 1997 (BTO).
National winter survey, 1997–98 (JNCC/BTO).

Population and distribution

Ubiquitous in the UK, both in winter and summer, often common. Population has declined markedly over recent decades.

Ecology

The skylark occupies a wide range of open habitats, from saltmarsh and farmland to young regenerating woodland, but avoids tall vegetation such as mature woodland. Its presence is easily detected by the male's characteristic song-flight, although song activity declines as the breeding season progresses. The birds nest on the ground in an excavated cup lined with grass, and the female lays a clutch of 2–5 eggs which are incubated by her alone. Both parents feed the chicks, which leave the nest at around eight days. The chicks are then fed by the parents until fledging at around 16 days, and sometimes for longer. Resident birds winter in small groups or pairs, and large winter flocks are probably mainly continental migrants.

Breeding season survey – population

This method is based on the 1997 BTO survey of breeding skylarks (Browne 1997).

Information required
- number of territories
- map of the site boundary and registrations.

Number and timing of visits

Four visits, evenly spaced (every two weeks) between mid-April and mid-June.

Time of day

If possible try to start fieldwork within two hours of sunrise, but not before.

Weather constraints

Avoid cold, windy or wet days – but try not to let persistently bad weather prevent you from carrying out one of your visits, as it is better to make a relatively inefficient visit than to miss one entirely or to submit data from only a partial visit.

Sites/areas to visit

All suitable habitat, which covers a wide range: lowland, upland and coastal, farmland and heath.

Equipment

- 1:2,500 OS map if possible; see *Maps* in the *Introduction*.

Safety reminders

Nothing specific. See general health and safety guidelines in the *Introduction*.

Disturbance

Keep disturbance to a minimum.

Methods

Map the boundary of the survey area. Use a different photocopied field map for each visit and copy all of the information to a new map at the end. An example recording form is shown in Figure 1. Work out a route that gets you to every part of the survey area. You do not need to survey habitats which are clearly unsuitable for skylarks, such as woodland or urban areas. Walk at a slow pace and use standard BTO codes (see Appendix 1) to record all skylark registrations. It is advisable to alternate the direction of the route walked on each visit. Attempt to locate by sight all singing males; they may be nearer or further away than their song suggests and it is very important to plot their location as accurately as possible. Record the location of singing and non-singing birds and the location of birds carrying food or nesting material. The territorial behaviour of skylarks is such that when they fly up from within their territory they start singing, circle around their territory and then alight within it. The recording of any part of this behaviour gives an indication of the rough location of the territory, but the most accurate pointer to a skylark's territory is the point from where skylarks fly up or alight.

Having completed the field visits, combine information from the field maps onto a summary map and circle each territory. Estimate the number of skylark territories from the summary map. Some further advice on how to estimate number of territories from this type of mapped information is given under Common Birds Census in the generic survey methods section.

Winter survey

This method is based on the JNCC/BTO wintering skylark survey 1997/98 (Gillings 1997).

Information required

- peak number of birds seen on a single visit.

Number and timing of visits

Once a month in November, December and January.

Time of day

From dawn to dusk.

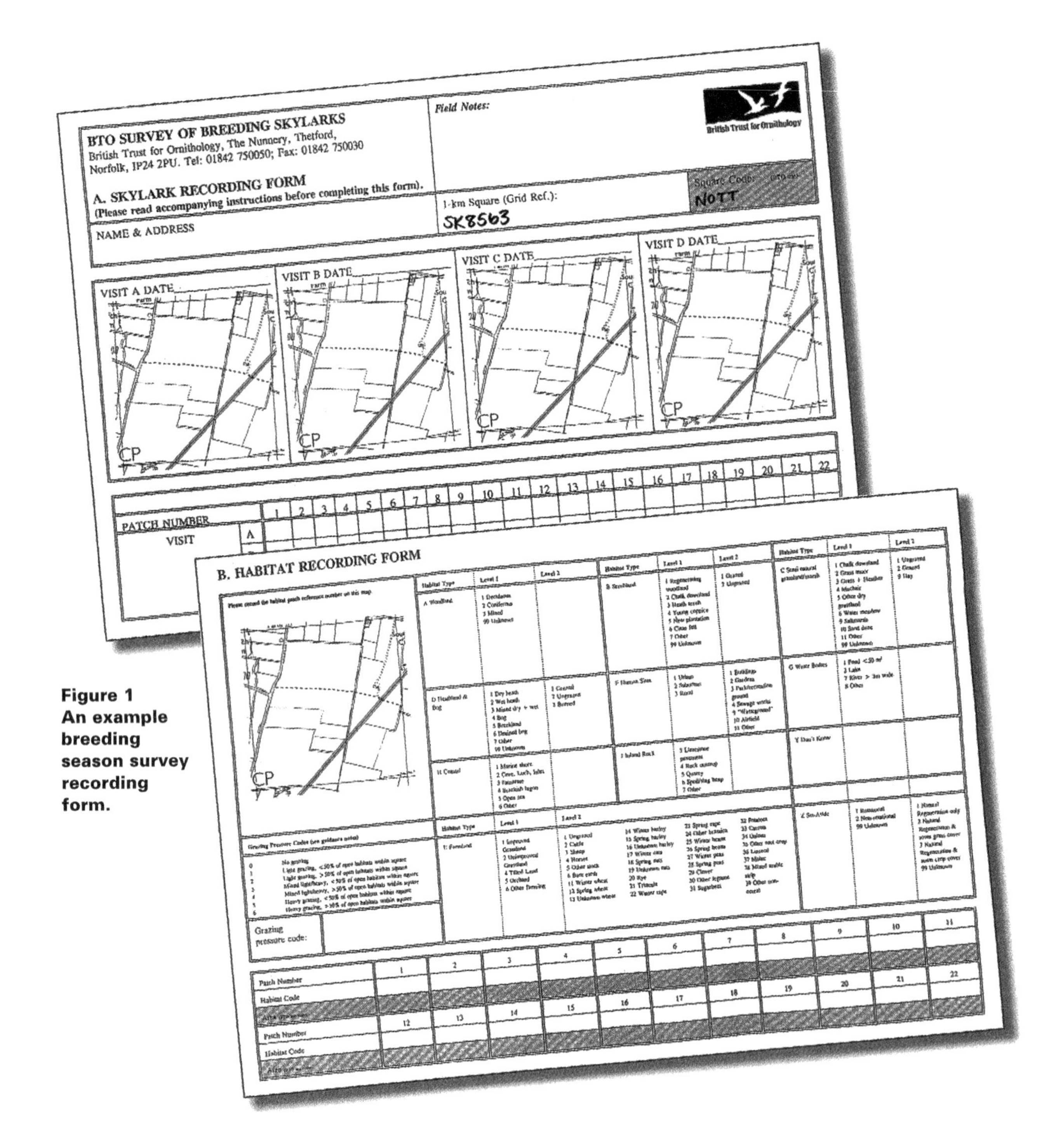

BTO SURVEY OF BREEDING SKYLARKS
British Trust for Ornithology, The Nunnery, Thetford,
Norfolk, IP24 2PU. Tel: 01842 750050; Fax: 01842 750030

A. SKYLARK RECORDING FORM
(Please read accompanying instructions before completing this form).

Field Notes:

British Trust for Ornithology

Square Code: NOTT

1-km Square (Grid Ref.): SK8563

NAME & ADDRESS

VISIT A DATE | VISIT B DATE | VISIT C DATE | VISIT D DATE

PATCH NUMBER 1 2 3 4 5 6 7 8 9 10 11 12 13 14 15 16 17 18 19 20 21 22

VISIT A

B. HABITAT RECORDING FORM

Grazing pressure code:

Patch Number	1	2	3	4	5	6	7	8	9	10	11
Habitat Code											

Patch Number	12	13	14	15	16	17	18	19	20	21	22
Habitat Code											

Figure 1
An example breeding season survey recording form.

Weather constraints, Sites/areas to visit , Equipment

As for the breeding season survey (above).

Safety reminders

Carry a mobile phone in winter. Ensure your clothes are suitable for the conditions. Do not go out in severe weather.

Disturbance

It is necessary to disturb skylarks to find them in winter, otherwise they may be undetectable when feeding on the ground.

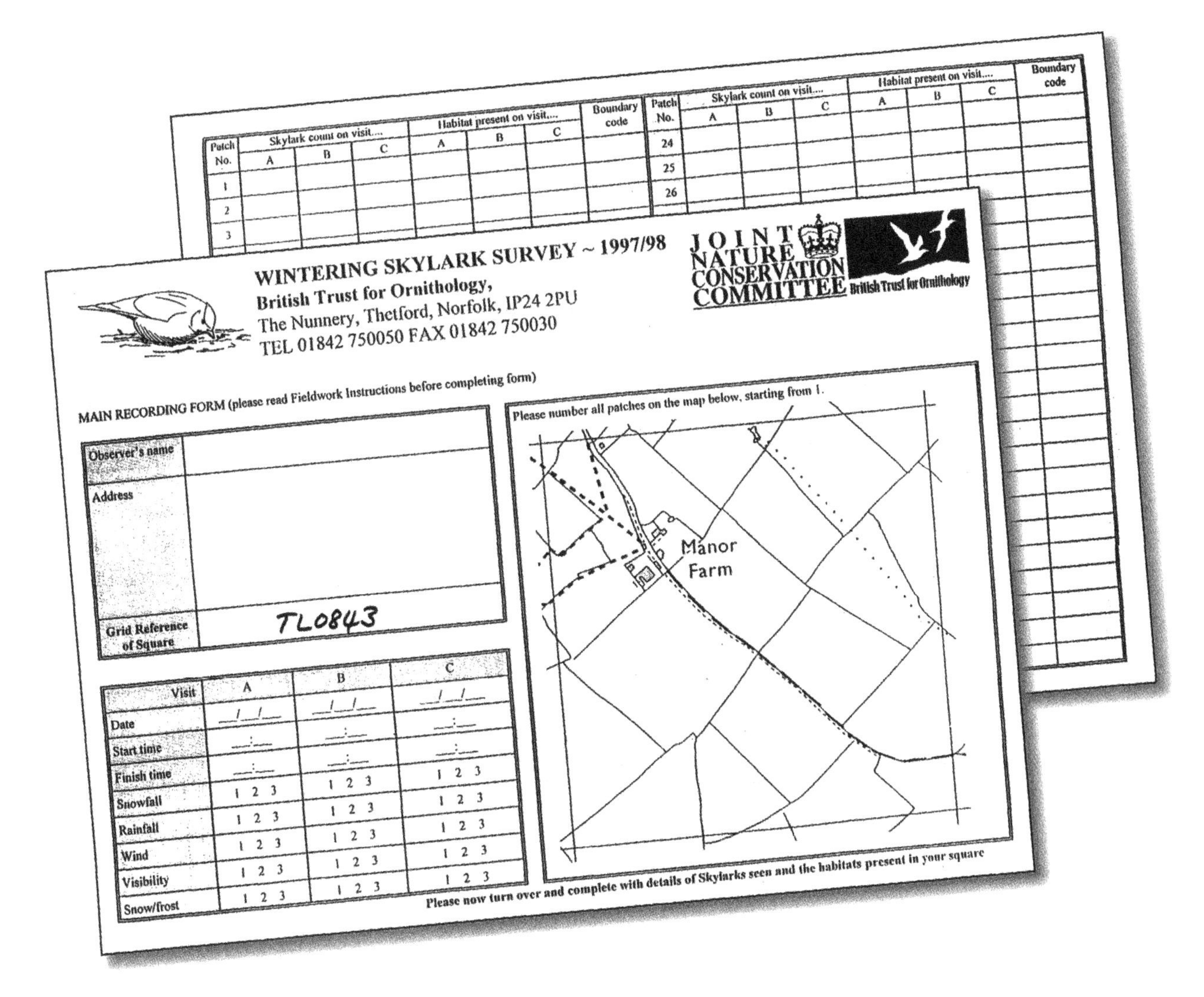

Patch No.	Skylark count on visit.... A	B	C	Habitat present on visit.... A	B	C	Boundary code
1							
2							
3							

Patch No.	Skylark count on visit.... A	B	C	Habitat present on visit.... A	B	C	Boundary code
24							
25							
26							

WINTERING SKYLARK SURVEY ~ 1997/98
British Trust for Ornithology,
The Nunnery, Thetford, Norfolk, IP24 2PU
TEL 01842 750050 FAX 01842 750030

JOINT NATURE CONSERVATION COMMITTEE
British Trust for Ornithology

MAIN RECORDING FORM (please read Fieldwork Instructions before completing form)

Observer's name	
Address	
Grid Reference of Square	TL0843

Visit	A	B	C
Date	__/__/__	__/__/__	__/__/__
Start time	__:__	__:__	__:__
Finish time	__:__	__:__	__:__
Snowfall	1 2 3	1 2 3	1 2 3
Rainfall	1 2 3	1 2 3	1 2 3
Wind	1 2 3	1 2 3	1 2 3
Visibility	1 2 3	1 2 3	1 2 3
Snow/frost	1 2 3	1 2 3	1 2 3

Please number all patches on the map below, starting from 1.

Please now turn over and complete with details of Skylarks seen and the habitats present in your square

Figure 2
An example winter survey recording form.

Methods

Walk each field or area to within 150 m of each point, mapping the location and size of skylark flocks. An example recording form is given in Figure 2. You do not need to survey clearly unsuitable habitats such as woodland or gardens. If the birds leave the area, note where they land and do not count them a second time. Look for colour-ringed birds if possible, and carefully note the leg on which each colour combination was seen.

Report the peak number of skylarks seen on any one visit.

References

Browne, S (1997) *BTO Survey of Breeding Skylarks: instructions and forms.* BTO, Thetford.

Gillings, S (1997) *BTO Wintering Skylark Survey: instructions and forms.* BTO, Thetford.

Sand martin
Riparia riparia

Status

Amber listed: SPEC 3 (D)

National monitoring

BTO Sand Martin Enquiry 1962–1968.

Population and distribution

Sand martins breed throughout the UK, except for the Hebrides and Shetland, and they depend largely on the availability of soft substrate banks for their burrow nesting sites (*88–91 Atlas*). There have been recent crashes in the UK breeding population due to severe droughts in the wintering grounds in Africa (Jones 1987). The breeding population of sand martins in the UK is estimated at 85,000–270,000 nests (*Population Estimates*).

Ecology

Prefers sandy, loamy or other workable banks, excavations, cliffs and earth mounds suitable for tunnelling to form nest-chambers. Such sites often occur naturally along rivers and streams, or by lakes and on sea coasts, or artificially where sand or other materials are extracted in conditions which leave some of the faces undisturbed. Burrows are higher when taller banks are available, eg from a sample of nests from different habitats, nests were higher in quarries (about 4 m) than along river banks (about 2 m) (Morgan 1979). A clutch of 4–6 eggs is laid during May–June (*BWP*).

Breeding season survey – population

The main survey method is adapted from Jones (1986).

Information required

- number of apparently occupied burrows in each colony
- number of unoccupied burrows in each colony
- a map of colony location and the boundary of the survey area.

Number and timing of visits

Two visits, in May and June.
The breeding season is earlier in southern Britain than northern Britain. Approximate dates for the second visit are (Morgan 1979):

Southern England:	mid-May
Northern England:	late May to mid-June
Scotland:	late June.

It is important that the second visit is not much later than that suggested. Sand martins will often dig new holes for second broods, and late counts may therefore overestimate the number of breeding pairs.

Time of day

Any time of the day.

Weather constraints

Avoid poor weather conditions.

Sites/areas to visit

River banks, sand and gravel quarries, coastal cliffs and dunes, sand banks by roads, lake pits, chalk pits.

Equipment

- 1:25,000 OS field map.

Safety reminders

Quarries, sand banks and any kind of pit (clay, chalk, sand) are dangerous places. When you ask for access permission to any such sites, find out about any dangers on the site. Always follow the site manager's advice. Wear a hard hat in quarries. Do not climb up any kind of soft bank unless supported by a ladder. Where possible use binoculars to get a closer look at burrows.

Disturbance

This method does not involve disturbance to birds in burrows.

Methods

Mark the survey boundary on to a map. Although sand martins will often excavate a new burrow on arrival each year, there may still be signs of old burrows from the previous year. The burrows will be in relatively soft substrate, quite high up a bank. Although the entrance holes may only be a few centimetres in diameter, the burrows themselves are over 30 cm deep. Sand martins will usually nest in colonies of up to 50 pairs, but 100 or more is not unusual and they are also known to nest singly or in small groups.

On the first visit, search the whole survey area for suitable habitat and mark all potential sites on to the map, even if there are no signs of excavated burrows.

On the second visit, return to all the potentially suitable habitat within the survey area and count the number of apparently occupied burrows within each colony.

Include the following as apparently occupied burrows.

- Chicks, fledglings or adults are seen at the burrow entrance.
- There are claw marks outside the burrow.
- There are lines of faeces below the burrow.
- The burrow is not obviously out of use, but occupancy cannot be confirmed.

Exclude the following burrows.

- The burrow has vegetation growing out of it.
- There are cobwebs over the entrance.
- It is shallow and visibly blind-ended.

This method may slightly overestimate the population size and will be of most use in detecting large-scale population changes. *Always* make the first visit to locate suitable habitat, as breeding sites vary considerably in distribution and size between years.

Report the number of apparently occupied burrows and the number of unoccupied burrows in each colony (the latter will allow analysis of apparent variation in occupancy rates between observers). Also provide the grid reference for the middle of each colony, or a map of the survey area with the colonies marked.

References

Jones, G (1986) The distribution and abundance of sand martins breeding in central Scotland. *Scottish Birds* 14: 33–38.
Jones, G (1987) Selection against large size in the sand martin *Riparia riparia* during dramatic population crash. *Ibis* 129: 274–280.
Morgan, R A (1979) Sand martin nest record cards. *Bird Study* 26: 129–132.

Yellow wagtail *Motacilla flava*

Status

Currently meets BDM criterion for amber listing (and would also be BI and SPEC 2 should taxonomic opinion favour specific status for the British race *flavissima*)
Non-SPEC

National monitoring

BBS.

Population and distribution

The yellow wagtail is a widely distributed breeding bird concentrated in south-east England. It is rather sparsely distributed in southern Scotland and eastern Wales, areas which are at the edge of its UK range (*88–91 Atlas*). The distribution has contracted over the last 50 years, mostly due to habitat change (*88–91 Atlas*). The CBC index decreased by 35% between 1971 and 1996, and the WeBS index, which is less likely to be representative of the population as a whole, dropped by 75% during 1974–96. There are an estimated 50,000 yellow wagtail territories in the UK (*Population Estimates*).

Ecology

Most yellow wagtails have arrived by 1 May. Males sing to declare a territory and attract a mate, but song activity declines once mated, although males will occasionally sing while females are incubating. After the young have fledged (or the nesting attempt has failed), the male may advertise for another mate, sometimes on a new territory. Feeding areas are often some distance from the nest, sometimes up to 1 km away, although these may vary from day to day depending on weather. Birds nesting in silage fields usually fail due to the early cut, but may re-lay in nearby cereal fields.

Breeding season survey – population

Information required

- number of territories
- map showing the boundary of the survey area and locations of territories.

Number and timing of visits

Three visits at least one week apart between 15 May and 15 June.

Time of day

0900–1400 BST in May, but at any time of day in June.

Weather constraints

Avoid cold or wet weather or winds stronger than Beaufort force 4.

Sites/areas to visit

The lower reaches of slow-flowing rivers, in water meadows, grazed semi-marshland and cattle pastures; sewage farms; reservoir margins or arable fields. Further north, birds will more typically occupy neutral grassland, mainly on sandy soils. They prefer wide river valleys and nest in hay meadows and cereal fields. Sites are frequently traditional, although to the human eye they are often indistinguishable from unoccupied areas; open country is important and wooded areas are avoided.

Equipment

- 1:10,000 OS map
- A4 photocopied map of survey area for use in the field
- A4 photocopied map of survey area for compiling the final visit map.

Safety reminders

Nothing specific. See general health and safety guidelines in the *Introduction*.

Disturbance

None necessary.

Methods

Mark the boundary of the survey area on a map. Work out a route that will take you around each field boundary in enclosed habitat and along parallel transects, 50–70 m apart, in unenclosed habitat. Walk at a slow pace alternating the direction of the route on successive visits. Map the location, number, sex and activity of yellow wagtails seen using standard BTO codes (Appendix 1). Use a new map for each visit or make clear which registrations refer to which visit by using different letters for each (A for Visit 1, B for Visit 2, C for Visit 3).

The main objective of May visits is to record singing males or pairs, while that of the early and mid-June visits is to record birds carrying food to nests. Both male and female utter frequent flight-calls while carrying food, and give alarm-calls (sometimes performing a distraction display) when an observer or potential predator is near the nest. Pairs feeding young in the nest are thus often conspicuous and easy to locate, making survey work at this time very efficient.

The species can be semi-colonial, so spend some time observing areas where a male is singing or a pair is carrying food, as more than one pair may be present. The pair continue to feed the young once they have left the nest, so beware of misinterpreting a pair which is visiting fledged chicks in several different places as more than one pair visiting nests.

Summarise the registrations from the visit maps on to a summary map making clear (using A, B and C) which registration was from which visit. Estimate the number of territories over the three visits. Count the following as a single territory: a male singing on at least one visit; a male or pair present in suitable breeding habitat on at least two visits; a male, female or a pair seen carrying food on a single visit.

The density of breeding yellow wagtails can be high, with local densities of up to 400 pairs per km^2 (4 pairs per ha).

Contributed by Iain Gibson

Nightingale *Luscinia megarhynchos*

Status

Amber listed: BDM
SPEC 4 (S)

National monitoring

National survey: 1980 (BTO).

Population and distribution

Nightingales in Britain breed in the south and east of England. They occupy mainly lowland areas in sites with thick undergrowth (Hudson 1979). There was a contraction in range of breeding nightingales between the two atlases, most marked at the northern and western limits of the distribution (*88–91 Atlas*). There are estimated to be 5,000–6,000 pairs in Britain (*Population Estimates*), possibly fewer in recent years (A Henderson pers comm).

Ecology

Habitat varies from scrub and young conifer plantations to mature deciduous woodland. Although many good sites are in dampish places, most are in dry areas (*88–91 Atlas*). A clutch of 4–5 eggs is laid in May, and incubation last 13 days. The young are fed and cared for by both parents and fledge after 11 days (*BWP*).

Breeding season survey – population

This method is based on the Kent nightingale survey (Henderson 1996).

Information required
- number of singing males
- map of the survey boundary and summarised registrations.

Number and timing of visits

An initial visit to the site during daylight for familiarisation is advised. At least two (preferably four) subsequent visits in May (one during 7–15 May), about a week apart.

Time of day

Preferably midnight to dawn, but the first five hours of daylight are acceptable.

Weather constraints

Avoid counting in wet and windy weather.

Sites/areas to visit

All regularly occupied sites, formerly occupied sites and potential sites, including all habitats with a dense field or shrub layer (which could be thick, tall bramble, gorse thickets, thick hedgerows or managed coppice), eg forestry plantations and mixed woodland.

Equipment

- 1: 25,000 OS map
- clipboard and torch.

Safety reminders

Ensure someone knows where you are and when you are due back. If possible, all sites should be visited during the day so that landmarks can be noted on maps. Take a whistle and make sure you can find your way back to your car. Try to work in pairs, and if possible have a mobile phone or mobile radio.

Disturbance

This census method should not entail disturbance.

Methods

Map the boundary of the survey area. With the map, familiarise yourself with the site in daylight and work out the most efficient route to take you within about 100 m of each point.

During the night-time visits, walk at a steady pace along existing rides and paths, and stop every few minutes to listen for singing. Cupping your hands behind your ears and rotating your head can help to gauge sound direction. You should try to get within about 100 m of each point, except in densely wooded areas or in awkward topography, where up to 300 m should suffice, particularly in calm weather. Use a bike if you have to cover large distances. When birds are heard, follow them up to record locations accurately. Wherever possible, vary the direction of the survey route and use a new map on each visit.

Mark positions of singing male nightingales on the maps using standard BTO recording codes (see Appendix 1).

After the final visit to a site, transfer all records onto a single master map, making it clear which registrations belong to which visit (use a different colour or letter). Estimate the number of territories from clusters of song registrations. Circle those records thought to be the same bird. Outliers consisting of a single record should be treated as territories unless they are within 300 m of a cluster, and could represent the movement of a male. Report the result as the total number of singing males estimated from all visits.

References

Henderson, A (1996) Kent Nightingale Survey 1994. *Kent Bird Report* 43: 145–152.
Hudson, R (1979) Nightingales in Britain in 1976. *Bird Study* 26: 204–212.

Black redstart *Phoenicurus ochruros*

Status

Amber listed: BR
Non-SPEC
Schedule 1 of WCA 1981

National monitoring

National survey: 1977 (BTO).
Rare Breeding Birds Panel.

Population and distribution

Black redstarts have been slowly colonising Britain for about 70 years. They are concentrated in London, the Home Counties, Kent, west Midlands and Suffolk/Norfolk, and are thinly distributed elsewhere in England. The population has exceeded 100 occupied territories (Morgan and Glue 1981) but declined in the late 1980s and early 1990s before showing a slight recovery in recent years. There are currently estimated to be 27–74 pairs of black redstarts in the UK (*Population Estimates*). The population may fluctuate strongly from year to year.

Ecology

Breeding sites are typically in industrialised, built-up or derelict surroundings, although some coastal sites are on sea cliffs (Morgan and Glue 1981). A clutch of 4–6 eggs is laid between late April and the end of June. There are usually two broods and the young fledge by mid-July (*Red Data Birds*).

Breeding season survey – population

This method is based on the 1977 black redstart survey (Morgan and Glue 1981).

Information required

- number of proven, probable and possible breeding pairs
- map of the survey area boundary with a summary map of registrations.

Number and timing of visits

At least five fortnightly visits from mid-April to the end of June.

Time of day

Early morning (in the hours after sunrise) and/or evening (hours before sunset).

Weather constraints

Avoid cold, wet and windy conditions.

Sites/areas to visit

Urban and industrialised habitats, eg gasworks, docks and warehouses, power stations, railway sheds, derelict buildings, farm outbuildings, sites under construction and urban developments. A few birds nest on sea cliffs.

Equipment

- 1:10,000 OS map
- Schedule 1 licence.

Safety reminders

It is very important to gain access permission, even though this may prove quite difficult. Industrialised areas can be particularly dangerous, so please abide by any safety advice given by the landowner, eg wear a hard hat or stay away from unsafe buildings or machines.

Disturbance

The method involves minimal disturbance to nesting birds.

Methods

Map the boundary of the survey area. The survey route may depend on where access permission has been granted and where it is safe to survey. Taking these factors into consideration, walk a predetermined route through the area which allows you to approach to within 100 m of each accessible spot. Mark the route on to the map and use the same route every time, even between years. Take a new map on each visit, and clearly mark each map with the date. Alternate the direction of the route taken on each subsequent visit, so that you are not always starting and ending at the same place. Walk slowly, taking time to stop and listen for singing birds or to observe any suspected sightings through binoculars. Whenever a black redstart is seen or heard, follow this up immediately. This will inevitably mean deviating from the original survey route for a short time.

Males prefer to sing from a prominent position on a building – often a rooftop, gutter or chimney stack. The far-carrying (if scratchy) song is the best auditory cue. Map the location and behaviour of the bird(s) using conventional BTO codes (see Appendix 1) and then continue on the survey route.

From all the individual visit maps, create a summary map of registrations and use the following criteria to assess the number of proven, probable and possible breeding black redstarts that were present. Report these along with the summary map of registrations.

Breeding is *proved* if:

- a nest or used nest is found
- a nest with young is seen or heard
- recently fledged young are located
- adults are seen entering or leaving a nest-site, or an adult is seen incubating
- an adult is seen carrying a faecal sac or food for young.

Breeding is *probable* if:

- a pair of birds is seen in suitable nesting habitat during the breeding season
- a male is heard singing at the same place on two or more occasions
- courtship and/or display are seen
- a bird is seen visiting a probable nest-site
- birds exhibit agitated behaviour or give alarm-calls
- nest-building is observed.

Breeding is *possible* if:

- birds are seen in the breeding season
- birds are seen in possible nesting habitat during the breeding season
- a singing male is heard once during the breeding season.

Reference

Morgan, R A and Glue, D E (1981) Breeding survey of black redstarts in Britain, 1977. *Bird Study* 28: 163–168.

Whinchat
Saxicola rubetra

Status

SPEC 4 (S)

National monitoring

None.

Population and distribution

The whinchat is a summer visitor, widely distributed across the uplands of the UK. It is rare in the Midlands, the south and east of England and the flatter northern and eastern areas of Scotland, with few birds in Ireland (*88–91 Atlas*). Numbers have declined in the last 30 years, mainly due to loss of suitable habitat, and there are estimated to be 14,000–28,000 pairs breeding in the UK (*Population Estimates*).

Ecology

Adults generally arrive on territory in late April or early May. They are attracted to grassy areas, including some types of farmland. The majority of whinchats breed in the uplands on heaths and moors, including grass heaths with gorse, bracken, rush (*Juncus*) beds or open scrub. In the lowlands they are found on lowland wet grassland with open scrub. They forage and feed more on the ground than do stonechats, making less use of territorial perches. When nesting alongside whinchats, stonechats are always dominant. A clutch of 4–7 eggs is laid in May or June, and incubation takes 12–13 days. The young are fed and cared for by both parents; they leave the nest at 12–13 days if undisturbed and will fly at 17–19 days (*BWP*).

Breeding season survey – population

Information required

- number of territories
- map showing the boundary of the survey area, the survey route and a summary of registrations and territories.

Number and timing of visits

Three visits at least one week apart from mid-May to mid-June for upland sites; about one week earlier for lowland sites.

Time of day

Dawn to 1000 BST and in the hours before dusk.

Weather constraints

Fine, calm and warm sunny mornings and evenings are best.

Sites/areas to visit

Visit all suitable areas. Whinchats breed in farmland, upland and lowland grass and heath, and lowland wet grassland.

Equipment

- 1:10,000 or 1:25,000 OS map of the area to be visited.

Safety reminders

Nothing specific. See general health and safety guidelines in the *Introduction.*

Disturbance

Ensure as little disturbance as possible. Whinchats are more confiding than stonechats, returning readily to the nest after disturbance.

Methods

Mark the outline of the survey area on a map; this helps with year-to-year comparisons.

Work out a route that will take you to within 50 m of every point in the site. Map the route and always follow it (even between years), but alternate the direction of the route on each visit. Map the location and behaviour of all whinchats using standard BTO species and activity codes (see Appendix 1). Either use a separate map for each visit, or prefix each mapped registration with a visit-specific code (eg A, B, C).

After the third visit, estimate the number of territories. These are areas where one or more of the following was found:

- a singing or displaying male, where displaying includes calling
- a male and female together
- evidence of nest-building or of adults feeding young.

Neighbouring territories can be separated if singing or displaying males are heard/seen simultaneously or at points more than 200 m apart. Report the number of territories and provide a map showing the boundary of the site, the survey route and a summary of registrations and territories.

Stonechat
Saxicola torquata

Status

Amber listed: SPEC 3 (D)

National monitoring

None.

Population and distribution

In the UK, stonechats are in general loyal to well-defined territories throughout the year. They are distributed down the west coasts of Britain and Ireland and the south coast of England, but are most abundant in north and west Scotland and the west of Ireland (*88–91 Atlas*). Numbers of stonechats have declined in western Europe in the last 30 years; there are estimated to be 9,000–23,000 pairs breeding in the UK (*Population Estimates*).

Ecology

Stonechats are associated with coastal and lowland heath in the breeding season. A clutch of 4–6 eggs is laid between April and June, and incubation lasts 13–14 days. The young are cared for by both parents and the fledging period is about 14 days (*BWP*).

Breeding season survey – population

Information required

- number of territories
- map showing the boundary of the survey area, the survey route and a summary of registrations and territories.

Number and timing of visits

Three visits in April, at least one week apart.

Time of day

Dawn to 1000 BST.

Weather constraints

Fine, calm, sunny mornings are best.

Sites/areas to visit

Heather moor and coastal habitats. In particular: lowland heaths, coastal habitats with gorse *Ulex* and scrub, young forestry plantations, railway embankments and golf courses.

Equipment

- 1:10,000 or 1:25,000 OS map of the area to be visited.

Safety reminders

Nothing specific. See general health and safety guidelines in the *Introduction.*

Disturbance

Nothing specific, but ensure as little disturbance as possible.

Methods

Mark the boundary of the survey area on a map; this will help with year-to-year comparisons.

Work out a route that will take you to within 50 m of every point in the site. Map the route and always follow it (even between years), alternating the direction of the route on each visit. Map the location and behaviour of all stonechats on each visit using standard BTO species and activity codes (see Appendix 1). Either use a separate map for each visit, or prefix each mapped registration with a visit-specific code (eg A, B, C).

Stonechats are territorial, the boundaries of their territories usually following well-defined topographical features such as the sloping banks of a sand-pit, a wall, hedge or even a single strand of barbed wire (Johnson 1971). However, on some areas of heather moor, territory boundaries may be invisible to the human eye. A human or other intruder will cause the male to utter a warning 'chuck' to the female, and an intruding stonechat will be sung at by the resident male until an aerobatic chase ensues. A close approach to the nest by a potential predator (including human) will cause the male to start an elaborate distraction display.

After the third visit estimate the number of territories. These are areas where one or more of the following was found:

- a singing or displaying male, where displaying includes calling
- a male and female together
- evidence of nest-building or adults feeding young.

Neighbouring territories can be separated if males are heard/seen singing or displaying simultaneously, or at points more than 200 m apart. Report the number of territories and provide a map showing the boundary of the site, the survey route and a summary of registrations and territories.

Reference

Johnson, E D H (1971) Observations on a resident population of stonechats in Jersey. *British Birds* 64: 201–213.

Ring ouzel *Turdus torquatus*

Status

Amber listed: BDM
SPEC 4 (S)

National monitoring

National breeding survey in 1999.

Population and distribution

Ring ouzels occur above the 250 m contour. They are most abundant in the Pennines, the Lake District, the southern parts of the Grampian mountains, the western parts of the Highland region of Scotland and throughout the high ground of north and mid-Wales (*88–91 Atlas*). There has been a long-term decline in the UK population across its range. The causes of the decline are not clear, although increased disturbance, acidification and warmer temperatures have been suggested (*88–91 Atlas*). There are an estimated 5,500–11,000 pairs of ring ouzels in the UK (*Population Estimates*).

Ecology

Ring ouzels typically breed in steep-sided valleys in open country, but are also found in level moorland in Shropshire and young conifer plantations in south Scotland (I Sim pers comm). A clutch of 3–6 eggs is laid between mid-April and mid-June. Incubation lasts 12–14 days, and there may be one or two broods. The young are fed and cared for by both parents and fledge 14–16 days after hatching (*BWP*).

Breeding season survey – population

This method has been developed and used by the RSPB (I Sim pers comm).

Information required

- number of possible, probable and confirmed breeding pairs
- map of survey boundary and registrations.

Number and timing of visits

Known sites	Two visits, one mid-April to mid-May and one mid-May to June.
Unknown sites	Three visits, two between mid-April and mid-May, and a third between mid-May and the end of June.

Time of day

Mornings if possible. At known sites spend at least two hours locating birds on the first visit and an hour confirming breeding on the second visit.

Weather constraints

Avoid windy conditions during tape playback.

Sites/areas to visit

If surveying areas other than known breeding areas, search all habitats.

Equipment

- 1:25,000 OS map
- tape player and good-quality tape of male ring ouzel song (new areas only).

Safety reminders

A reliable person should know where you are and when you are due back. Carry a compass and know how to use it. If possible, do not work alone. In all upland areas, spare warm clothing, a plastic survival bag, whistle and food supplies should be carried.

Disturbance

There is no need to search for nests or to get particularly close to birds to prove breeding. The use of playback is purely to elicit a vocal response, and not to lure birds from one area to another. Follow the instructions for playback carefully.

Methods

Ring ouzels are difficult birds to survey accurately. They frequent remote upland regions and can be very elusive, especially in late summer when they skulk in bracken and other tall vegetation. Females are especially hard to locate.

At unknown sites (those from which ring ouzels have not been recorded before), play a tape of ring ouzel song at points within 500 m of all parts of the whole survey area. If the day is windy, reduce this distance by half. At each point, play the tape for one minute and listen for five minutes. If nothing is heard, move on; if a ring ouzel replies, follow this up immediately by locating the bird and observing its movements and behaviour. Map the location of any birds heard or seen and note their behaviour. Be very strict with the use of playback; the aim is simply to detect presence/absence, not to cause birds to change location.

Once presence has been established using playback, or when visiting a known site for the first time, spend two hours in mid-April to mid-May watching and listening from a few vantage points in the vicinity of

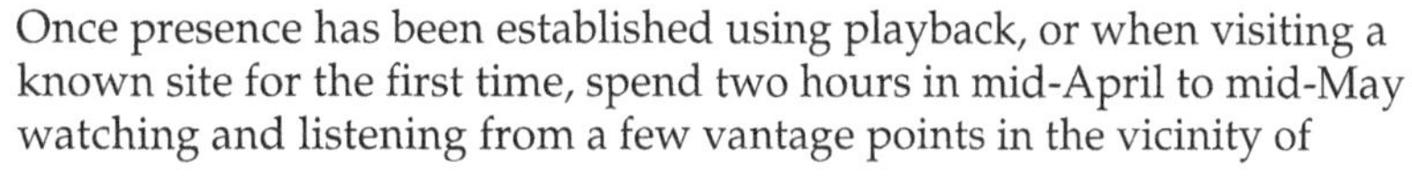

previous sightings or 100 m from the known breeding site. Check suitable song posts and feeding areas (eg grazed pasture). Record the location and behaviour of all ring ouzels using standard BTO notation (Appendix 1). Visit in mid-May to June to try to confirm breeding by locating fledged young. Do not continue visits into July, as single-brooded pairs with fledged young may have moved a considerable distance by then, and could be confused with birds from other sites.

Classify the breeding status of each record as follows:

Possible Observed in the breeding season in possible nesting habitat; singing male(s) present in breeding season.

Probable A pair observed in suitable nesting habitat in the breeding season; permanent territory presumed through registration of territorial behaviour (song, etc) on at least two different days, a week or more apart, at the same place; courtship and display; agitated behaviour or anxiety calls from adults; nest-building activity.

Confirmed Recently fledged young seen; adult seen carrying food.

Report the numbers of possible, probable and confirmed breeding pairs. Sum probable and confirmed when reporting the estimated number of breeding pairs.

Redwing
Turdus iliacus

Status

Amber listed: BR
SPEC 4^W (S^W)
Schedule 1 of WCA 1981
Annex II/2 of EC Wild Birds Directive

National monitoring

Rare Breeding Birds Panel.

Population and distribution

The redwing colonised Britain in the early part of the 20th century, first nesting in Sutherland in 1925 (*68–72 Atlas*). It is a rare breeding species, restricted for the most part to north and west Scotland, although breeding has been recorded as far north as Shetland, and as far south as Kent. The British breeding population, most of which breeds in the Highlands, is quoted as 40–80 pairs (*88–91 Atlas, Population Estimates*, Marchant et al. 1990), although this may well be a substantial underestimate (Ogilvie et al 1996). Redwings are common winter visitors from Iceland, Scandinavia and Finland (Goodacre 1960, Zink 1981). During this period, the British redwing population swells to about three-quarters of a million birds (*Population Estimates*), with the highest numbers in west and south-west Britain (*Winter Atlas*).

Ecology

Redwings breed in a variety of habitats, often close to water. These include hillside birchwoods, the edges of oakwoods, mature woodland and parkland surrounding estate houses, scrubby wooded areas along river valleys and plantations of spruce *Picea* (Spencer et al 1986). Characteristic of many of these sites is the presence of scrub with damp patches for feeding (Williamson 1973). Redwings generally breed solitarily, although loose colonial clusters have been recorded (*BWP*, Spencer et al 1990). Eggs are usually laid between May and mid-July, but occasionally earlier. Clutches are of 5–6 eggs, incubation lasts 12–13 days, young are in the nest for 10–11 days and fledglings become independent 14 days after leaving the nest. Pairs are frequently double-brooded (*BWP, Red Data Birds*).

Breeding season survey – population

This method is based on the territory-mapping approach of the CBC (see generic survey methods section). Because only a single, relatively rare species is recorded, the number of visits is reduced from that required by the CBC and the recommended plot size is increased.

Information required

- number of territories
- number of confirmed breeding pairs.

Number and timing of visits

At least three, between 1 May and 14 June with at least one week between each visit.

Time of day

Dawn to midday, preferably early morning.

Weather constraints

Avoid periods of heavy rain and winds of Beaufort force 5 and over.

Sites/areas to visit

Areas which have traditionally held breeding redwings. Also, areas of birch, oak and alder woodland, sitka spruce plantations, parkland and scrub, particularly close to water (eg river valleys) in north and west Scotland.

Equipment

- 1:25,000 OS map of the area
- larger-scale photocopied map (1:2,500 to 1:10,000)
- Schedule 1 licence.

Safety reminders

Always tell someone where you have gone and when you expect to return. Take a compass and always carry a survival bag, waterproofs, whistle, extra clothing, food and a first-aid kit whenever surveying in a remote area.

Disturbance

This method does not require nest visits and therefore causes little or no disturbance.

Methods

Define the boundary of your survey area, either by revisiting the birds' traditional haunts or by cold searching. To do this, drive, cycle or walk along minor roads and tracks through suitable habitat during early to mid-May. Stop for a few minutes every few hundred yards and listen for the characteristic, liquid, descending song of the male. If you hear or see a redwing, include the general area in which you located it within (one of) your study plot(s), bearing in mind that other pairs may be nearby.

Decide the overall boundary of your study plot and mark it clearly on a large-scale map (1:2,500–1:10,000 depending on plot size and availability of maps). This should incorporate as much suitable habitat as possible, but should not be so big that it cannot be surveyed in a morning. In relatively open terrain (eg scattered trees and scrub in a river valley) this should probably be less than 250 ha, and in a more closed habitat (eg woodland) less than 75 ha. These areas are bigger than are generally recommended for standard multi-species territory mapping, but are manageable because only a single species is being recorded.

Walk slowly around the study plot and mark on the map the location of all the redwings encountered, using a separate letter for each visit (A for 1st visit, B for 2nd visit, etc) and standard BTO codes (see Appendix 1) to indicate behaviour. For example, a male recorded singing on the second visit would be marked on the map as an encircled B. Some registration types are more useful than others in determining how many redwing territories there are in the survey plot. Records of two or more males singing simultaneously (denoted by two, or more, encircled visit letters joined by dotted lines) are particularly useful, as each represents a separate territory. Evidence of nesting, such as birds carrying nesting material or food, or alarm-calling adults, is also valuable.

Wherever possible, visit the study plot more than three times. In general, the more visits that are made, the more reliable are the results. However, if too many visits are made (eg more than 10) it is possible to become confused and to double-count the same pair/territory, particularly if the pair has moved between nesting attempts. Ideally, visits should be concentrated during the peak of song activity (RBBP records suggest this is the last three weeks of May and the first two weeks of June). This is the most efficient way of recording the species and gives a better estimate of the total potential breeding population than that given by surveys carried out later in the season, which only give an indication of the number of successful pairs.

At the end of the season, draw non-overlapping rings, representing territory boundaries, around the clusters of bird registrations. If only a few visits were made to each plot, each cluster will necessarily consist of only a few records. Simultaneous registrations indicate different individuals and should never be incorporated into the same cluster unless they are thought to be two adults of a pair. Records of nests can be counted as a cluster even in the absence of records of adults. When delineating territory boundaries, bear in mind that redwings do sometimes breed in loose colonies, so records from different pairs/territories may be clumped. Where individual clusters are difficult to differentiate, analyses of the map may inevitably become subjective.

Report the number of territories recorded on each study plot, and the number of these from which breeding evidence was obtained (eg alarm-calling adults, nesting material or food being carried, nest found, recent fledglings, etc).

References

Goodacre, M J (1960) The origin of winter visitors in the British Isles. 5. Redwing (*Turdus musicus*). *Bird Study* 7: 102–107.

Marchant, J H, Hudson, R, Carter, S P and Whittington, P (1990) *Population Trends in British Breeding Birds*. BTO, Tring.

Ogilvie, M A and Rare Breeding Birds Panel (1996) Rare breeding birds in the United Kingdom in 1993. *British Birds* 89: 61–91.

Spencer, R and Rare Breeding Birds Panel (1986) Rare breeding birds in the United Kingdom in 1984. *British Birds* 79: 470–495.

Spencer, R and Rare Breeding Birds Panel (1990) Rare breeding birds in the United Kingdom in 1988. *British Birds* 83: 353–390.

Williamson, K (1973) Habitat of redwings in Wester Ross. *Scottish Birds* 7: 268–269.

Zink, G (1981) *Der Zug Europäischer Singvögel: ein Atlas der Wiederfunde beringter Vögel*. Lieferung 3. Vogelwarte Radolfzell-Moggingen.

Cetti's warbler
Cettia cetti

Status

Amber listed: BR
Non-SPEC
Schedule 1 of WCA 1981

National monitoring

Rare Breeding Birds Panel.
National survey in 1996 (RSPB/EN).

Population and distribution

Cetti's warbler is a recent colonist of the UK. It was first recorded in Hampshire in 1961, although breeding was not confirmed until 1973 in the Stour valley, Kent (Wotton et al 1998). By 1984, the UK population had risen to 316 singing males, of which 114 were in Kent. Two cold winters in the mid-1980s reversed the species' fortunes, particularly in Kent, and by 1986 the UK population had declined to 211 singing males. By the late 1980s, the species had disappeared from Kent, and the centre of its population had shifted further towards south and south-west England. By the mid-1990s, the population had recovered and a national survey in 1996 located a total of 535–593 singing males at 168 sites in 26 counties across southern Britain (plus Jersey). Two-thirds of the population was recorded in only four counties: Hampshire, Dorset, Devon and Somerset (Wotton et al 1998).

Ecology

During the breeding season, Cetti's warblers are typically found around wet, swampy areas near the water's edge where there is low and fragmented scrubby cover. They prefer areas of scattered scrub in reedbed, in reed swamp and by open water. During the winter months, Cetti's warblers may move into reedbeds and open marshland (Bibby and Thomas 1984). Males patrol large territories, sometimes up to 450 m long, in which one or more females may breed (Bibby 1982). The breeding season lasts from late April to mid-July. The young fledge by mid- to late August. Clutches of 3–5 eggs are laid in well-concealed nests in dense vegetation.

Breeding season survey – population

Information required

- number of territories.

Number and timing of visits

Three visits: the first between the end of March and mid-April, the second between mid-April and mid-May, and the third between mid-May and early June.

Time of day

Dawn to 1100 BST.

Weather constraints

Fine, calm, sunny mornings are best.

Sites/areas to visit

Areas of scattered scrub in reedbeds, in reed swamp, in fen/marsh, by open water and along river systems.

Equipment

- 1:10,000 or 1:25,000 OS map of the area to be visited
- Schedule 1 licence.

Safety reminders

No specific advice. See the *Introduction* for general guidelines.

Disturbance

No disturbance to the birds is necessary.

Methods

Mark the boundary of the survey area on a large-scale field map.

Before starting each visit, listen for any calling or singing males. The male's explosive song is very distinctive, allowing it to be located readily in the field. In addition, males sing for prolonged periods which helps greatly to establish the number of territories present.

Walk around the site at a steady pace, using existing access routes or suitable vantage points from which to conduct the survey. Mark all records of Cetti's warblers on the field map using standard BTO (CBC) codes (Appendix 1). Males patrol their territory boundaries at about half-hourly intervals, and can move quite long distances, so beware of double-counting the same male. Linear territories (eg along river banks) can be up to 450 m long (Bibby 1982). Record birds moving from one location to another on the map using the standard codes. Try to obtain as many records of countersinging or simultaneously singing males as possible, as this will help when interpreting the maps later.

Following the third visit, estimate the number of territories on the site. To do this, count each separate singing male as a territory. Differentiating two neighbouring males from a single male that has moved can be difficult. If the two males were heard singing simultaneously, or if you were able to detect subtle differences in their songs (which is possible with practice), then they can be considered to be two separate males/territories. Note that some male territories may overlap, particularly in areas with a high density of Cetti's warblers. More than one female may nest within a single male's territory, therefore the total number of territories estimated cannot be considered to be the same as the number of pairs on the site.

Report the estimated number of territories on the site.

References

Bibby, C J (1982) Polygyny and the breeding ecology of the Cetti's warbler *Cettia cetti*. *Ibis* 124: 288–301.

Bibby, C J and Thomas, D K (1984) Sexual dimorphism in size, moult and movement of Cetti's warbler *Cettia cetti*. *Bird Study* 31: 28–34.

Wotton, S, Gibbons, D W, Dilger, M and Grice, P V (1998) *British Birds* 91: 77–88.

Marsh warbler *Acrocephalus palustris*

Status

Red listed: BD, BR
SPEC 4 (S)
Schedule 1 of WCA 1981

National monitoring

Rare Breeding Birds Panel.

Population and distribution

The marsh warbler has never been a common UK breeding bird, but has bred at various times in Gloucestershire, Worcestershire, Somerset, Sussex and Kent (*Red Data Birds*). The core European population in Germany, Russia and Romania remained stable from 1970 to 1990 whereas the UK population declined quite dramatically over the same period (*Birds in Europe*). The reason for the UK decline is unclear, but is most likely to be a combination of factors including habitat loss, weather and the isolation of the UK population (Kelsey et al 1989). An estimated 11–34 pairs of marsh warbler breed in the UK each year (*Population Estimates*) and there are signs that the population is increasing once again, albeit at new localities.

Ecology

Marsh warblers prefer habitat which combines tall, dense herbaceous vegetation, scattered shrubs or trees, fertile land with a history of disturbance, and the presence of plant species such as common nettle, meadowsweet and willowherb. Nesting has been recorded in small (a few square feet) patches of vegetation and arable crops. Nests are characteristically built around the stems of herbaceous plants or shrubs. There is a single brood, and the clutch of 3–5 eggs is laid in June or early July (*Red Data Birds*).

Breeding season survey – population

This survey method is adapted from Kelsey (1987).

Information required

- estimate of the number of males present for at least a week (ie territorial males)
- map showing registrations and the area covered.

Number and timing of visits

Weekly, to locate singing males, and at three-day intervals thereafter until territory occupancy has been confirmed. From last week in May to early July.

Time of day

Early morning, eg 0500–1000 BST.

Weather constraints

Avoid cold, wet and windy conditions.

Sites/areas to visit

Areas that have been occupied by breeding marsh warblers in the past and which still have suitable habitat.

Equipment

- 1:10,000 OS map (several copies, one for each visit)
- Schedule 1 licence.

Safety reminders

No specific advice. See the *Introduction* for general guidelines.

Disturbance

Do not disturb breeding birds and nest-sites. Never approach a nest and keep all marsh warbler sites strictly confidential.

Methods

This method involves mapping territories of singing males. However, marsh warblers only sing for a short period, their territories are relatively small (0.1 ha) and they do not always settle where they are first heard singing. Therefore some additions to the standard CBC method are required.

Mark the survey area on a map. If it is large, split it into more manageable sections (about 20 ha) and treat each as a separate survey area. Plan your route in advance and walk it systematically. Walk at a slow pace and get to within 50 m of each point within the survey area. Alternate the direction in which you walk the survey route through each area on each visit.

When marsh warblers are not singing, they are easily confused with reed warblers *Acrocephalus scirpaceus*. Ensure that you are confident with your identification. Reed warblers produce a fairly stereotypic monotonous song, interspersed with occasional mimicry; whereas marsh warbler song is delivered in a much more exuberant manner and is extremely varied. Marsh warblers frequently sing from bushes and trees or from the tall dead stems of herbaceous vegetation, while reed warblers sing from less exposed positions.

On the first visit, map the position of every marsh warbler seen and heard and make a note of the behaviour using standard BTO codes (Appendix 1). Wherever singing males are seen, revisit the location of each three days later and then again three days after that. Marsh warblers can be considered to be holding territory if they are present in the same area for at least a week. This rule of thumb avoids counting migrant birds – which might sing at a site for a few days and then move on – as territory-holders.

Beware of territory-holders moving to other areas in an attempt to attract new females, and mated males going quiet while the female builds the nest (which takes about four days). When returning to a site which has previously held a singing male, look for nesting activity. If there is evidence of this, the territory can be considered to be occupied, even if no male song is heard.

After your final visit, transfer all the information from the individual visit maps to a master map, annotating the records from each separate visit with a different letter (Visit 1 = A, Visit 2 = B, etc). Estimate the number of males which held territory for at least a week. At the end of the survey, report the number of occupied territories as the number of breeding pairs.

References

Kelsey, M G (1987) *The Ecology of Marsh Warblers*. D Phil Thesis, EGI, Oxford University.

Kelsey, M G, Green, G H, Garnett, M C and Hayman, P V (1989) Marsh warblers in Britain. *British Birds* 82: 239–255.

Dartford warbler *Sylvia undata*

Status

Red listed: HD, BL, SPEC 2 (V)
Schedule 1 of WCA 1981
Annex I of EC Wild Birds Directive

National monitoring

National surveys in 1974, 1984 and 1994; next will be 2004.
Rare Breeding Birds Panel.

Population and distribution

More than three-quarters of the Dartford warbler population is restricted to Dorset and Hampshire, with most of the rest in Devon and Surrey. A marked increase in the population was noted between 1984 and 1994. The population is prone to large fluctuations, depending on the harshness of the winter weather (birds are especially susceptible to prolonged snow cover). There are an estimated 1,600–1,890 pairs in the UK (*Population Estimates*, Gibbons and Wotton 1996).

Ecology

The main habitat is dry, mature lowland heath, with the nest either in heather *Calluna vulgaris* or gorse *Ulex europaea.* A clutch of 3–5 eggs is laid, generally between mid-April and early July, occasionally later; there can be one or two broods (*Red Data Birds*). Birds are most active feeding young from mid- to late May.

Breeding season survey – population

Information required
- number of occupied territories.

Number and timing of visits

Three visits: one from the beginning of April to mid-May, one between mid- and late May, and one in June.

Time of day

From about one hour after dawn onwards.

Weather constraints

Fine, calm days are best. Avoid cold, windy and rainy conditions, but males will sing in calm, warm conditions with light rain.

Sites/areas to visit

Anywhere with suitable nesting habitat (see *Ecology*, above).

Equipment

- 1:25,000 OS map of the area to be visited
- Schedule 1 licence.

Safety reminders

No specific advice. See the *Introduction* for general guidelines.

Disturbance

No disturbance is necessary.

Methods

Map the boundary of the survey area, identify suitable habitat within it and mark this on the site map. Recent aerial photographs may make it easier to map areas of gorse.

Dartford warblers are mainly sedentary and are site-faithful, so use the locations of previous years' territories to indicate where current territories might be. However, ensure that all suitable habitat in the survey area, including previous years' territories, is equally well covered.

During each visit, mark on the map any birds located, using standard BTO symbols (see Appendix 1). Use a separate map for each visit, or prefix each mapped registration with a visit-specific code (eg A, B, C).

The object of the survey is to estimate the number of occupied territories over three visits. A territory is considered occupied if a singing male, boundary dispute(s) or breeding activity are seen on at least one visit. Examples of breeding activity are a bird carrying food or behaving in an agitated manner, as if with young. The ability to record singing males is density- and weather-related. In mid- to late May, when adults are feeding young, they become agitated when their territory is entered, and are much more visible.

Territories (or home ranges) vary in size. Where densities of birds are high, home ranges may be 1–3 ha (Catchpole and Phillips 1992); in lower-density areas, home ranges may be up to 10 ha (Wotton pers comm). It is possible that males will not sing at low densities. If two (or more) males are heard singing simultaneously, treat them as separate territories. If two (or more) males are heard singing, but not simultaneously, treat them as being from the same territory if within 200 m of one another, or from separate territories if at a greater distance than this. Although this distance is likely to depend on the quality of the habitat, birds living at higher densities are more likely to be heard singing simultaneously anyway.

At the end of the third visit, combine the records of singing birds, those involved in boundary disputes and records of other breeding activity to obtain the number of territories. When reporting the total number of occupied territories, make an additional note of how many records were of singing males only.

References

Catchpole, C K and Phillips, J F (1992) Territory quality and reproductive success in the Dartford warbler *Sylvia undata* in Dorset, England. *Biological Conservation* 61: 209–215.

Gibbons, D W and Wotton, S (1996) The Dartford warbler in the United Kingdom in 1994. *British Birds* 89: 203–212.

Firecrest
Regulus ignicapillus

Status

Amber listed: BR
SPEC 4 (S)
Schedule 1 of WCA 1981

National monitoring

Rare Breeding Birds Panel.

Population and distribution

Firecrests were first recorded breeding in Britain in 1962. They are sparsely distributed in the south of England, in East Anglia and in Wales (*88–91 Atlas*). Breeding occurs in conifer plantations and in mixed or mainly broadleaved woodlands (*Red Data Birds*). The population in the UK appears to be slowly increasing and spreading, but is still vulnerable, and is estimated to number between 80 and 250 territory-holding males (*88–91 Atlas*), or perhaps fewer if all the main breeding sites have now been discovered. Around 200–400 birds, probably mainly of continental origin, overwinter.

Ecology

Firecrests are almost always summer migrants to their breeding areas, arriving mainly during April and early May and departing in July or August. Song may be heard on territory from late March to early October but peaks in mid-June. Some passage migrants also sing. Types of woodland occupied by firecrests are difficult to characterise. The UK sites with highest numbers have been mature plantations of Norway spruce. Other conifers, and perhaps particularly mixtures of ornamental trees, are also favoured. Some sing in mixed or predominantly deciduous woodland. In Norway spruce, males hold exclusive territories of about $\frac{1}{2}$–$1\frac{1}{2}$ ha, about 2–3 times larger than those of goldcrests in the same area; territories might be larger in other habitats. Because of this mobility, song locations within 200 m of one another should generally be expected to relate to the same male, unless there are other indications to the contrary.

Breeding season survey – population

This method is essentially the territory-mapping bird census method, as used by the BTO's Common Birds Census, with visits concentrated into the firecrest breeding season.

The species' rarity and its highly clumped distribution within suitable habitat makes surveys on a national or regional scale exceptionally difficult. Firecrests breed in dense woodland and are difficult to locate except by song; many singing males may fail to attract a mate. Observers should check that they are reliably able to hear high-pitched bird sounds, for example goldcrest song, before attempting to survey firecrests.

Information required

- number of territory-holding males
- any casual information on the number of territories in which breeding was proved, with date and a precise location, would be of value in addition to formal surveys.

Number and timing of visits

At least three thorough visits at intervals of at least ten days, spread between mid-May and early July. To prove breeding, extra visits would be needed in July.

Time of day

An early start is desirable but not essential, provided that the visit is completed by midday.

Weather constraints

Cold, wet or windy weather may depress song and should be avoided; song is most audible on calm days.

Sites/areas to visit

Any suitable habitat (see *Ecology*, above).

Equipment

- 1:10,000 OS map
- compass (optional)
- Schedule 1 licence.

Safety reminders

No specific advice. See the *Introduction* for general guidelines.

Disturbance

Disturbance at or near the nest is illegal unless licensed. Fieldwork for this method is unlikely to cause any disturbance.

Methods

Mark the boundary of the survey area on a map. On each survey visit, take a copy of the map and mark on it all firecrest registrations. Walk within 50 m of all points of suitable habitat on the plot. Firecrests with fewer neighbours might sing less frequently and be harder to locate. Use standard BTO species and activity codes (see Appendix 1) to record numbers, sexes and activities, with emphasis on song and contemporaneous encounters. Retrace your steps whenever it is necessary in order to confirm that new encounters are with different individuals. Spend several minutes with each bird to map its movements and any interactions with other firecrests. Once visits are completed, compile a species map and delineate territories in the standard way.

Report the number of territory-holding males, broken down into those which possibly, probably and definitely bred, defined as follows.

Possible Any observations of a bird apparently on territory, but seen only on one date, or a bird seen more than once in an area but not singing or showing other territorial behaviour.

Probable At least one bird recorded in the same territory on two or more visits over a period of at least ten days, where song or other territorial behaviour was observed.

Confirmed Observations of young, either fledged or still in the nest, or of adult behaviour such as a pair mating, nest-building or occupying a completed nest, or carrying food, eggshells or faecal sacs.

Contributed by John H Marchant

Bearded tit
Panurus biarmicus

Status

Amber listed: BL
Non-SPEC
Schedule 1 of WCA 1981

National monitoring

National surveys: 1992 (RSPB), 2002.

Population and distribution

Bearded tits breed in reedbeds, mainly in south-east England, with the highest concentration in East Anglia. In 1947 there were only 2–4 pairs of bearded tit breeding in England. By 1974, the British population had increased and was estimated to be at least 590 pairs in 11 counties (O'Sullivan 1976). Since 1974, the population has declined. The reasons for this decline are largely unknown, but may include: fewer birds immigrating from the continent; loss or degradation of habitat; and the severity of the winters. A survey in 1992 estimated that there were between 339 and 408 pairs of bearded tits breeding in Britain (Campbell et al 1996).

Ecology

Almost entirely confined to beds of reed *Phragmites australis* in the breeding season and only slightly less so in the winter. A clutch of 5–6 eggs is laid, and the breeding season can last from March to August, occasionally longer, with two, three or possibly four broods produced in a season (*Red Data Birds*).

Breeding season survey – population

Bearded tits are difficult birds to census. They do not sing, are non-territorial, and their reedbed habitat is often impenetrable. The method for surveying this species, developed for the 1992 national survey, is time-consuming and particular care is needed to follow the instructions closely so that the results are comparable between years and sites.

Information required
- estimated total number of breeding pairs
- total number of all confirmed and probable first-brood nests
- mapped locations of nests.

Number and timing of visits

A minimum of three visits, in the first three weeks of May. Preferably at least six, between mid-April and the end of May.

Time of day

Up to three hours after sunrise, but can be later for the detailed survey visits.

Weather constraints

All visits should be in calm weather.

Sites/areas to visit

All sites with a reasonable area of reedbed, particularly where bearded tits have been recorded previously.

Equipment

- 1:10,000 OS map of the area to be visited
- 1:25,000 OS map showing footpaths and rights of way
- telescope.

Safety reminders

Reedbeds are potentially hazardous habitats. Ensure someone knows where you are working and what time you expect to return. If entry into the reedbed is unavoidable, try to work in pairs and use a map showing safe and unsafe areas.

Disturbance

Reedbeds have restricted access, can be damaged by trampling, and new paths invite predators. In addition, the presence of rare breeding birds such as bitterns and marsh harriers may place restrictions on access to certain areas of reedbed at particular times of year.

Methods

Mark the extent of the reedbed on the map. Map the survey route and observation positions and shade any areas which could not be covered.

Cover the whole reedbed in the 2–3 hours after sunrise on three separate days. Split large sites into blocks which can be covered in the three-hour period. Each block may require more than one person to cover it, and, if this is the case, at the end of each visit produce a single visit map from the different observers' maps and cross-reference timed notes with mapped observations to reduce the chance of double-counting the same individual. On each visit, record sightings and birds heard calling on a map of the site using standard BTO symbols (Appendix 1). Use a separate map for each day.

Make observations from an elevated position wherever possible. The best observation position is for eye-level to be 1 m above reed height. If you can, use footpaths, elevated dykes, hides and any other access routes. It is difficult to observe behaviour accurately at a distance of more than about 250 m, even with a telescope, so observation points should be positioned to take account of this. If necessary, put out colour-coded marker posts (eg plastic drainpipes) in the reedbed before the breeding season. They should be visible above the reed growth at known distances and will allow more accurate mapping (Figure 1).

Areas where birds were regularly seen or heard in the first three visits should be targeted for a further detailed census, in order to produce a map of all 'confirmed' and 'probable' nest-sites. Observe areas of about 200 × 200 m for not less than one hour from a good vantage point.

A *confirmed* nest-site is one where either:

- adults are observed returning with food to the same specific location (about 10 × 10 m) on three or more separate observation periods, or
- an adult is seen returning with food and leaving the same small area with a faecal sac on the same day.

Figure 1
Two examples of ways in which reed areas can be split up to help observations on bearded tits.

A *probable* nest is one where either:

- adults have been observed returning to the same general location on three or more separate observation periods, or
- an accumulation of adult sightings, calls and observations of juveniles suggests that a pair are nesting or have nested in a particular part of the reedbed.

Care must be taken not to confuse adult feeding areas with nest locations. Adults are likely to return to the same site many times for food.

Transfer all observations from individual maps to one master map. Record the location of all confirmed and probable first-brood nests with the date on which each nesting area was determined. These detailed census visits should only be made in May. After May, some pairs produce second broods, and this confuses estimates of breeding numbers.

Report the combined total for confirmed and probable first-brood nests as the estimated total number of breeding pairs at a site.

If observations are carried out in June, then a separate map showing confirmed and probable second- and third-brood nests should be produced.

Breeding season survey - productivity

Information required

- maximum number of first-brood young observed in flocks in late June/July
- maximum number of birds observed in flocks in September
- maps of the area surveyed.

Number and timing of visits

Two or three visits in late June/July to record the maximum number of juveniles. Two or three visits in September to record the number of adults and post-moult juveniles.

Time of day

Preferably up to three hours after sunrise, otherwise later in the day.

Weather constraints, Sites/areas to visit, Equipment, Safety reminders, Disturbance

As for the population survey method (above).

Methods

On large sites, assessing productivity is difficult as flocks are very mobile. In this situation it may be more productive to use two teams of people to cover the area more quickly and accurately, and for the two teams to keep notes on times and flight directions of flocks.

Observe each area of reedbed for an hour, preferably from an elevated position (see the population survey method, above), marking the areas and sightings on the map and using a different map every day. Record the maximum number of juveniles on site in late June/July. At this time, they form small flocks which should be made up of first-brood young. Mark the numbers and location of first-brood young on the map.

Record the maximum number of birds (both adults and post-moult juveniles) in September in the same way. At this time, birds are conspicuous during 'high flying' displays prior to dispersal in late September.

References

Campbell, L H, Cayford, J T and Pearson, D J (1996) Bearded tits in Britain and Ireland. *British Birds* 89: 335–346.

O'Sullivan, J M (1976) Bearded tits in Britain and Ireland 1966–74. *British Birds* 69: 473–489.

Crested tit
Parus cristatus

Status

Amber listed: BL
SPEC 4 (S)
Schedule 1 of WCA 1981

National monitoring

National survey (Pinewoods Birds Project) 1992–95 (RSPB); next will be 2003–04.

Population and distribution

This species is found only in Europe and, in Britain, is resident in the Highlands of Scotland. The range has undoubtedly contracted with that of the preferred native pinewood habitat, although there has probably been a slight increase in the last 50 years, as birds now occur in some pine plantations (Cook 1982). There are estimated to be around 900 pairs of crested tit in Britain (*Population Estimates*).

Ecology

Crested tits are most abundant in semi-natural Scots pine forests and mature Scots pine plantations. Plantations are only suitable if they occur within existing areas of semi-mature Scots pine, on areas where older trees have been recently felled/removed and provide the necessary dead stump nest-sites, or where they are old enough to produce suitable deadwood habitat. A clutch of 3–9 eggs is laid in late April or early May and the young fledge from late May to early June (*Red Data Birds*).

Breeding season and winter survey - population

A sampling strategy and field method for covering large areas of woodland was devised by Buckland and Summers (1992) (see *Pinewood bird survey* in the generic survey methods section). This can be used to estimate breeding as well as winter densities, although it is important to distinguish between adults and fledged juveniles later in the season. This prevents the estimated density being inflated by young of the year.

References

Cook, M J H (1982) Breeding status of the Crested Tit. *Scottish Birds* 12 (4): 97–106.

Buckland, S T and Summers, R W (1992) *Pinewoods Birds Project: protocol for a sample survey of conifer woods.* RSPB unpubl.

Golden oriole *Oriolus oriolus*

Status

Amber listed: BR
Non-SPEC
Schedule 1 of WCA 1981

National monitoring

Rare Breeding Birds Panel.
National surveys in 1994 and 1995 (Golden Oriole Group).

Population and distribution

The golden oriole is a rare breeding species in the UK, and regular breeding is confined almost exclusively to the fenland basin in East Anglia. Numbers have remained relatively stable since 1985; there are estimated to be up to 35 pairs breeding in the UK (*Population Estimates*).

Ecology

Woods and belts of planted poplars *Populus* are the preferred nesting habitat in East Anglia, but oak and alder are also favoured elsewhere in Europe. It is not known why such a narrow range of tree species is chosen in England when other apparently suitable and widespread trees are available. A clutch of 3–5 eggs is laid between late May and early June, and one brood is usual (*Red Data Birds*).

Breeding season survey – population

Information required

- number of confirmed breeding pairs
- number of probable breeding pairs
- all other records.

Number and timing of visits

Two visits: one during the last ten days of May and the second between mid-June and the end of July.

Time of day

Visit in the two or three hours after dawn, for at least two hours.

Weather constraints

Do not visit during inclement weather, especially cold and windy conditions.

Sites/areas to visit

All areas that are known to have held orioles in the past.

Equipment

- 1:25,000 OS map
- Schedule 1 licence.

Safety reminders

No specific advice. See general guidelines in the *Introduction*.

Disturbance

Do not approach breeding sites without a Schedule 1 licence. A close approach is not necessary and may well be counter-productive, as orioles are liable to be quiet if there is a human nearby. Permission should be sought from landowners for access to all private land.

Methods

Mark the boundary of the survey area on a map. On each visit, get to within 100 m of all suitable habitat, make regular stops to watch for birds, and listen for calling males. By staying still and quiet, on a still day the song of the male can be heard from further away than 100 m. Beware that the male may give a warning note on seeing a human intruder and will then stay quiet. Isolated pairs are likely to be less vocal and more unobtrusive than those surrounded by neighbours. Record all golden orioles heard and seen on the map using standard BTO notation (Appendix 1) and use visit-specific codes (eg A and B) to clarify on which visit each record was made.

The male can range over several hundred metres through woodland as it sings, so take care not to overestimate the number of males present. Unpaired males also range over a wide area. To prove breeding, spend time observing any orioles recorded. A pair communicating with the 'cat-call' note is good evidence that they are holding territory and, if this is heard in late June, wait and watch carefully and you may see adults carrying food for their young. After the second visit, interpret the summary map as follows:

Confirmed breeding	Nest found (do not actively search for nests); adult seen carrying nest material; adult seen carrying food; recently fledged young, or family party seen.
Probable breeding pair	A pair present in suitable breeding habitat for at least one week; a cat-call heard from a pair; behaviour observed which is suggestive of breeding with no definite evidence (such as birds chasing predators).
Other records	A bird seen or heard in suitable habitat in the breeding season but no other evidence found.

Report the numbers of confirmed and probable breeding pairs and all other records. Provide a map of registrations.

Contributed by Peter Dolton/Golden Oriole Group

Chough *Pyrrhocorax pyrrhocorax*

Status

Amber listed: BL, SPEC 3 (V)
Schedule 1 of WCA 1981
Annex I of EC Wild Birds Directive

National monitoring

National surveys: 1992 (RSPB/JNCC/IWC/Manx Chough Study Group), 2002.

Population and distribution

Breeding chough are found off the west coasts of Britain and Ireland with concentrations on Islay, the Isle of Man, Pembrokeshire and Caernarvonshire (*88–91 Atlas*). The breeding population in Britain is estimated at about 315 pairs, and there are about 830 pairs in Ireland (*88–91 Atlas*).

Ecology

Choughs require a mix of suitable nest-sites on cliffs, caves or old buildings, and feeding areas on pasture and low-intensity cattle-grazed land (Monaghan et al 1989). They begin nest-building in March, and incubate mainly from the beginning of April to the beginning of May; incubation takes about 21 days and most eggs hatch from the last week in April to the first week of May; young are fed in the nest mainly from the beginning of May to mid-June. Young leave the nest from the first week of June onwards but are dependent on their parents for 6–8 weeks after this.

Breeding season survey – population

Information required

- numbers of confirmed, probable and possible breeding pairs
- map of the survey boundary, sites, registrations and route walked.

Number and timing of visits

At least two visits (more if two visits are insufficient to classify the breeding attempt as probable or confirmed). First visit: end of first week in April to the end of first week of May. Second visit: end of first week of May to mid-June.

Time of day

Any time of the day.

Weather constraints

Avoid adverse weather: strong winds, persistent rain or poor visibility.

Sites/areas to visit

Anywhere with a suitable mix of nesting and feeding habitat (see Ecology, above).

Equipment

- 1:25,000 OS and field maps
- Schedule 1 licence.

Safety reminders

On extended coastal or mountain walks tell someone where you are going and when you intend to return. Do not attempt to climb cliffs or enter caves or crevices. Take extra care around quarries and mine shafts, and always heed the owner's advice on which areas are safe to visit.

Disturbance

No visits to nest-sites are necessary; observe from a distance. Do not disturb any easily accessible nest-sites. Breeding should be confirmed without visiting nest-sites.

Methods

Map the boundary of the survey area. Identify all potential breeding habitat from an OS map. If potential breeding habitat is scattered (eg quarries, buildings, discontinuous cliffs, rocky outcrops, etc) then treat each as a separate site. On each survey visit, spend at least an hour in the vicinity of each site. Increase this to 1.5–2 hours for those that may hold several pairs. In coastal or mountainous habitat which is more continuous, follow a route which goes past all potential and/or known sites. Walk the route at a slow pace, stopping regularly for periods of 10–15 minutes to scan cliffs and listen for choughs. Follow the same route on all visits.

If any feeding choughs are encountered (away from nest-sites) watch these for about an hour or until they visit a likely nest-site. If feeding bird(s) head to a cliff/quarry, etc, and the nest-site is not located, then sit and watch the site for a further hour until the bird(s) return.

Figure 1 shows a recording form used in previous chough surveys. The form is designed to be used in conjunction with a field map of the survey area. Complete a new recording form for each visit to a site, even if no choughs are found. For each visit, record the start and finishing points of the survey by entering names of sites and their six-figure grid reference. Enter on the form the start and finish time of each 1-km square surveyed and the total duration of time spent in each.

Mark all chough registrations on the map. To help you determine the category of proof of breeding, record one of the two-letter codes given below alongside each registration, both on the form and on the map. Use only one code per registration.

Confirmed	NE	Nest with eggs.
	NY	Nest with young.
	FE	Adults seen leaving nest with faecal sac or eggshell.
	NH	Nestlings heard begging.
	FY	Brood of dependent young seen with parent(s).
Probable	CM	Adults carrying nest material.
	PV	Pair visiting likely nest-site.
Possible	SV	Single bird visiting likely nest-site.
	SF	Single bird feeding in suitable habitat.
	PF	Pair feeding in suitable habitat.

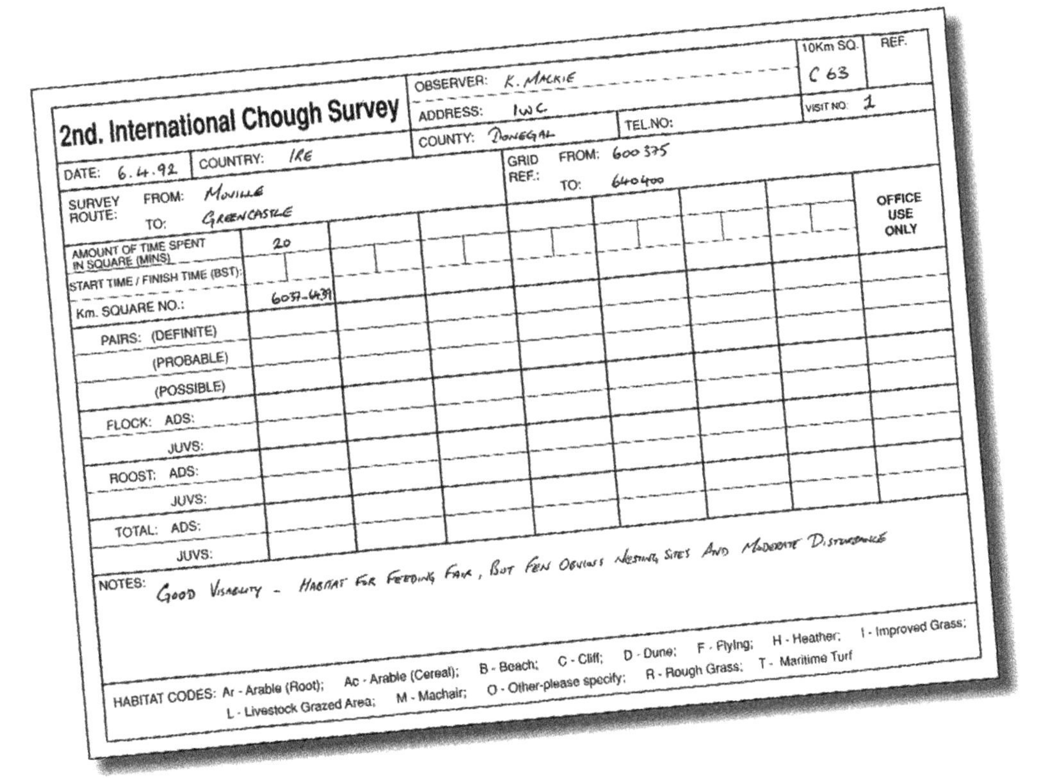

2nd. International Chough Survey

OBSERVER: K. MACKIE
ADDRESS: IWC
COUNTY: DONEGAL
TEL.NO:
10Km SQ.: C 63
REF.:
VISIT NO: 1
DATE: 6.4.92
COUNTRY: IRE
GRID REF.: FROM: 600 375 TO: 640400
SURVEY ROUTE: FROM: MOVILLE TO: GREENCASTLE
OFFICE USE ONLY

AMOUNT OF TIME SPENT IN SQUARE (MINS)	20							
START TIME / FINISH TIME (BST):								
Km. SQUARE NO.:	6037-6439							
PAIRS: (DEFINITE)								
(PROBABLE)								
(POSSIBLE)								
FLOCK: ADS:								
JUVS:								
ROOST: ADS:								
JUVS:								
TOTAL: ADS:								
JUVS:								

NOTES: GOOD VISIBILITY – HABITAT FOR FEEDING FAIR, BUT FEW OBVIOUS NESTING SITES AND MODERATE DISTURBANCE

HABITAT CODES: Ar - Arable (Root); Ac - Arable (Cereal); B - Beach; C - Cliff; D - Dune; F - Flying; H - Heather; I - Improved Grass; L - Livestock Grazed Area; M - Machair; O - Other-please specify; R - Rough Grass; T - Maritime Turf

Figure 1 Recording form used for the Second International Chough Survey (RSPB 1992).

If you are unable to categorise records into probable or confirmed breeding, make further visits to upgrade all possible breeding records. Confirmation of breeding is relatively easy after incubation and when young leave the nest. Young choughs spend a lot of time begging for food from their parents, calling and 'shimmering' their wings vigorously. By the end of July, the situation can become somewhat confused as parents, young and non-breeding birds combine into flocks.

Note any flocks of choughs seen as you pass through each 1-km square. A single bird can constitute a flock if you believe that it is not a member of a breeding pair.

On the first visit, most records are likely to be of adults, but this will not be the case on the second visit. Wherever possible, try to distinguish between adults and juveniles – young choughs have yellow-orange legs and bill and shrill cries. If this is not possible, indicate this in the 'notes' section of the recording form and bracket the two age-categories together. It is unlikely that you will come across roost-sites, but, if you do, record the numbers of adults and juveniles seen. At the end of each visit, record the total numbers of birds seen.

Report the numbers of confirmed, probable and possible breeding pairs of choughs estimated over all visits.

References

Monaghan, P, Bignal, E, Bignal, S, Easterbee, N and McKay, C R(1989) The distribution and status of the chough in Scotland in 1986. *Scottish Birds* 15: 114–118.

RSPB (1992) *Second International Chough Survey: instructions and recording forms.* RSPB unpubl.

Tree sparrow *Passer montanus*

Status

Red listed: BD
Non-SPEC

National monitoring

BBS.

Population and distribution

In the UK, tree sparrow distribution is very patchy at all scales. Their stronghold is in the central counties of England and adjoining east Wales. They are scarce in the southern counties of England along the channel coast from Sussex to the extreme south-west. In the north they are largely confined to the east coast (*88–91 Atlas*). The size of the tree sparrow population fluctuates in an irregular cyclic manner, but there has been an overall decline in numbers which is thought to be linked to changes in agricultural practices (*88–91 Atlas*). There are an estimated 110,000 territories in the UK (*Population Estimates*).

Ecology

Tree sparrows are hole-nesters which are found at a range of densities, from tight-knit colonies in nest-boxes, woodland edges or outbuildings, to single pairs nesting solitarily in hedgerow trees, but with large areas of apparently suitable habitat unoccupied. Solitary-breeding tree sparrows may be quite unobtrusive, since the species has no distinctive song or other territorial behaviour and is shy of humans (Summers-Smith 1995). With the remainder of the population breeding colonially, this makes it difficult to monitor absolute numbers accurately using conventional census methods (CBC/BBS), although an index of year-to-year changes may still be obtained. Eggs are laid from mid-April to mid-August.

Breeding season survey – population

If an accurate assessment of numbers is needed at a particular site, the following intensive method is suggested. Note that the method has not been field-tested.

Information required

- number of occupied nest-sites.

Number and timing of visits

An initial pre-breeding visit (before April) to locate potential nest-holes, then one visit to each potential nest-site every two weeks from early April to late June. Total of seven visits (assuming a Midland/southern England study area).

Time of day

Morning.

Weather constraints

Avoid very windy or wet weather.

Sites/areas to visit

All buildings, mature hedgerow trees, and mature trees along woodland edges, including nest-boxes with a hole size of 30–35 mm.

Equipment

- 1:10,000 OS map.

Safety reminders

No specific advice. See the *Introduction* for general guidelines.

Disturbance

Tree sparrows are prone to desert nest-sites if disturbed during nest-building or incubation. All observation of nest-sites should take place at a distance which does not disturb the birds, and the contents of nest-boxes should not be inspected unless there are specific reasons for collecting data on other aspects of breeding biology.

Method

Mark the boundary of the survey area on a map. On the first, pre-breeding visit locate every potential nest-hole and mark them on the same map. On all subsequent visits, watch each potential nest-hole for ten minutes. Unfortunately, there are no data on the proportion of time tree sparrows are present and visible at the nest-site at different stages of the nesting cycle. To ensure a consistent level of accuracy in detecting occupied sites, the amount of time spent watching each potential nest-site should be fixed. A minimum is probably 10 minutes, recognising that at colonies it may be possible to watch several nest-sites simultaneously.

Record a potential nest-site as occupied if a tree sparrow is seen entering or leaving.

Given that tree sparrows are usually monogamous, and that one pair usually occupies the same nest-site for all breeding attempts in one season (Summers-Smith 1988), the number of occupied nest-sites will give a good approximation of the number of breeding pairs.

Contributed by Jeremy Wilson

References

Summers-Smith, J D (1988) *The Sparrows*. Poyser, Calton.

Summers-Smith, J D (1995) *The Tree Sparrow*. Privately published.

Linnet
Carduelis cannabina

Status

Red listed: BD
SPEC 4 (S)

National monitoring

CBC, BBS.

Population and distribution

Widespread throughout the UK except in north-west Scotland and Shetland, linnets nest in scrub and hedgerows. Numbers have declined in the past 30 years, mainly due to the loss of food plants and nest-sites (*88–91 Atlas*). There are estimated to be 540,000 linnet territories in the UK (*Population Estimates*).

Ecology

Male linnets have a distinctive song, and sing persistently and conspicuously, although the song is used only to defend the female and the area immediately around the nest. Pairs may nest solitarily or, more often, in loose, semi-colonial clusters, with nests of neighbouring pairs often only a few metres apart. Despite this, most linnets are believed to be monogamous. At a local scale, therefore, linnets are often very patchily distributed. For example, one hedgerow may harbour several pairs of nesting linnets, whereas a similar hedge on the opposite side of the field may have none. Linnets feed almost exclusively on seeds, and adults routinely fly distances of 1 km or more to feed on patches of weeds (especially Compositae) and in fields of ripening rape-seed. Feeding areas are often communal, attracting birds from a wide surrounding area, and birds may often fly to and from feeding sites communally, resulting in considerable periods of absence when a colony may appear to hold few or no breeding linnets. Eggs are laid from mid-April to early August.

Breeding season survey – population

Patchy distribution, lack of territoriality and wide-ranging foraging behaviour with relatively long absences from the nesting area make absolute abundance of linnets difficult to census accurately using conventional CBC or BBS methods, although year-to-year changes may still be monitored well using these means.

If an accurate assessment of numbers is needed at a particular site, the following, more intensive method is suggested. Note, however, that this method has not been field-tested.

Information required
- number of breeding pairs.

Number and timing of visits

Make a minimum of one visit to the study area every week from mid-

April to the end of May (assuming a Midland/southern England study area). Total six visits.

Time of day

Morning.

Weather constraints

Avoid very windy or wet weather.

Sites/areas to visit

All areas of scrub and hedgerow in the study area. Clumps of ivy and other evergreen shrubs or small trees (eg gorse) are often favoured early in the season when deciduous vegetation provides poor nesting cover.

Equipment

- 1:10,000 OS base map.

Safety reminders

No specific advice. See general guidelines in the *Introduction.*

Disturbance

Linnets are very tolerant of human presence near nests – but there may be some risk of causing desertion during nest-building and incubation, so do not get close to known or suspected nests at these times.

Methods

Walk alongside all hedgerows and isolated bushes or scrub patches until linnet activity is noted. Stop and watch all suitable habitat in that area for a fixed period (probably a minimum of 10 minutes but longer if the size of the study area allows, up to a maximum of 30 minutes) to establish the number of pairs present. Move on, and repeat this process whenever linnet activity is noted. Location and activity of single birds or pairs should be recorded accurately on a 1:10,000 base map using standard BTO notation (Appendix 1). Use a separate base map for each visit, or use a visit-specific code (eg A, B, C, etc) to clarify on which visit each registration was made. Nest-building linnets can frequently be seen carrying nesting material into bushes. Food-carrying is not usually seen because parents regurgitate seeds to nestlings from the gullet. Birds seen repeatedly carrying straw or entering the same site in scrub or a hedge are almost certainly visiting an active nest. Males usually sing from a vantage point close to the female (before nest-building begins) and close to the nest itself once nesting has begun.

In a 'colony' area, accuracy in recording is essential in order to establish how many pairs are present. Ideally, observe from a relatively high vantage point, or from gaps in hedgerows so that activity on both sides of hedges can be seen from one point. Collation of records over all visits should allow reasonably accurate determination of the number of breeding pairs present. Note, however, that linnet pairs often make two or more nesting attempts per season. Little is known about the distance moved between successive nesting attempts (although over 100 m is not uncommon). This method thus restricts visits to the early part of the breeding season (breeding may continue until early August) to reduce the risk of double-counting of pairs moving between successive nesting attempts, and thereby assumes that all linnets initiate their first nesting attempt of the season in April or May.

Contributed by Jeremy Wilson

Twite
Carduelis flavirostris

Status

Red listed: HD
Non-SPEC

National monitoring

National survey: 1999.

Population and distribution

Most UK breeding twite occur in Scotland, with much smaller numbers in Wales and the Pennines. In Scotland, the bulk of the population is distributed along the north-west coasts and on the Western Isles, Orkney and Shetland. The breeding twite population in Britain and Ireland has undergone a long-term decline (*88–91 Atlas*) due, mainly, to the degradation and loss of habitat. The breeding population of twite in the UK is estimated at about 66,000 pairs (*Population Estimates*). The Irish and Scottish populations are thought to be fairly sedentary (Davies 1988), whereas the Pennine population is mostly migratory (McGhie et al 1994).

Ecology

Twite breed in upland and coastal areas, in Shetland typically on well-vegetated sea cliffs and uninhabited islands, while in the Pennines they nest on moorland edges. Herb-rich meadows are heavily used for feeding from late May to the end of July. Twite commonly nest in heather or bracken, but other sheltering vegetation can suffice and they have been known to nest in bushes such as gorse. When nesting on heather moor, they favour tall heather, while on grass moor they nest in bracken or surrounding long grass (McGhie et al 1994). Twite can feed up to 3 km from their nest-site, on past or present agricultural land, pastures, hay meadows, near water, on the moorland itself, along seashores and in areas where their favoured food plants (eg sorrel, dandelion, thistle) are present. Breeding twite are difficult to survey. Small parties of non-breeding twite move over the hills as late as the middle of June and some breeding birds may leave their breeding area before the end of July (Orford 1973). Because of this, non-breeding birds may be encountered during mid-May to mid-July, and breeding birds can easily be missed. A clutch of 5–6 eggs is laid from late May to mid-June with some second clutches in July. The young fledge during June–August and form post-breeding flocks with adults (*Red Data Birds*).

Breeding season survey – population

This method relies on obtaining behavioural indications of breeding behaviour and may be time-consuming. A more rapid assessment of twite populations could be obtained using the method designed for surveying upland breeding waders (Brown and Shepherd 1993, Stillman and Brown 1994; see generic survey methods section) although this is likely to yield an index of abundance rather than a comprehensive population estimate. An alternative method, advocated by H A McGhie (pers comm), involving counts of pre-breeding flocks in early May (divided by two) could be used; this method is not given here.

Information required
- estimated number of confirmed breeding pairs and other birds
- summary map showing the survey boundary and twite registrations.

Number and timing of visits

Three visits: one in mid- or late May, one in early June and another in early or mid-July.

Time of day

Between 0830 and 1800 BST.

Weather constraints

Avoid poor weather, particularly persistent precipitation, poor visibility or winds above Beaufort force 4 .

Sites/areas to visit

Any areas with a mixture of apparently suitable nesting and feeding habitat, eg moorland and in-bye land (see *Ecology*, above).

Equipment

- 1:25,000 OS map of the survey area
- compass.

Safety reminders

Ensure a reliable person knows where you are and when you are due back. Carry a compass and know how to use it and, if possible, work in a team. In all upland areas, carry spare warm clothing, a plastic survival bag, whistle and food supplies.

Disturbance

Do not search for nests during the survey.

Methods

Map the boundary of the survey area. On each visit, cover the survey area systematically, getting to within 100 m of all suitable habitat. To save time, cover feeding habitat (eg hay meadows) first as this may help to confirm whether twite are present. Note where any birds fly off to (this may give clues to the locations of nesting areas) then cover the nesting habitat.

Stop to listen and watch for birds as often as possible. Watch or follow individual twite, even if this means doubling back along the survey route. Mark any twite registrations on the field map, using standard BTO codes (see Appendix 1).

The following gives an indication of situations in which twite will be found:
- Birds(s) feeding in nesting areas or, more commonly, fields. Generally, these birds will be in ones, twos or small flocks. They will fly off when approached.
- Bird(s) song-flighting above these feeding groups.
- Males singing ('chortling') in nesting areas only. When twite are nesting, the male may 'chortle' a quiet song from a boulder, patch of tall heather or bracken frond. Such a bird can be approached closely and does not flee from the observer.

- An agitated male in potential nesting habitat, eg perched or in circling flight, calling persistently or bill-wiping, and reluctant to leave the immediate area.
- A bird sitting in one spot until flushed and then circling the surveyor, reluctant to leave the immediate area.
- Bird(s) flying from fields to nests. Note flight directions and subsequently follow these up to try to locate 'chortling' nesting birds.

Mark all registrations on your field map, distinguishing clearly between these behaviours, so that confirmed breeding pairs can be separated from other sightings. In study areas with many twite, use a new field map for each visit but carry a copy of the earlier maps with you into the field as this will help you interpret your observations. Numbering or naming confirmed nest-sites may help to avoid confusion between visits. In areas where you are likely to record few twite, it may be easier to use a single map for all visits. In such cases, write a visit-specific code (eg A, B, C) in front of each registration on the map to clarify on which visit each record was made.

At the end of the field season, produce a single summary map showing all registrations and the survey area boundary, and estimate the numbers of confirmed breeding pairs and other records. Use the rules below to distinguish between these categories.

A confirmed breeding pair is present if one or more of the following are seen or found:
- A nest (do not actively search for these).
- A pair with young.
- A male 'chortling' in suitable breeding habitat from a prominent feature in a manner suggesting that a nest or young are nearby (eg birds reluctant to leave the immediate area). Male twite often sing from a favourite post, fence, wall, rock or bush close to an incubating female.
- An agitated male in potential nesting habitat (eg the male may be perched or in circling flight, calling persistently or bill wiping, and reluctant to leave the immediate area).
- A bird sitting in one spot until flushed and then circling the surveyor, reluctant to leave the immediate area.
- A bird seen carrying nesting material into potentially suitable nesting habitat.

For many species, several of the following lines of evidence would be sufficient to suggest that breeding was probably occurring. However, among twite these behaviours may occur at feeding areas at a distance from nesting areas. Because of this, the following observations cannot be included in counts of breeding pairs unless followed up and breeding is subsequently confirmed (eg by following feeding birds back to their nesting areas):
- A ground display between male and female. In this, the male lowers his wings and moves in a semi-circle around a female, with his tail twisted so as to expose his pink rump to her (Marler and Mundinger 1975, McGhie et al 1994).
- An adult seen carrying nesting material but not in potentially suitable nesting habitat.
- A male in song-flight. During this a male will fly to a height of about 10 m and sing a scratchy, highly modulated song. While singing, he will stop flapping his wings, hold them stiffly in a downward and forward position and spread his tail; he may twist and turn as he descends, displaying his white wing flashes and pink rump (McGhie

et al 1994). The song flight only lasts for a few seconds, but has been recorded in situations that suggested breeding was in progress or intended (McGhie et al 1994).
- Birds flying to and from different areas (eg, in the Pennines, feeding and nesting areas).

Report the number of confirmed breeding pairs and records of other birds. Give details of flocks and flock sizes where relevant.

Twite may produce two broods and can move nest-sites more than 400 m between broods (Marler and Mundinger 1975). Because of this it is likely that the estimate of confirmed breeding pairs will be an overestimate and so should be considered a maximum figure.

References

Brown, A F and Shepherd, K B (1993) A method for censusing upland breeding waders. *Bird Study* 40: 189–195.
Davies, M (1988) The importance of Britain's twites. *RSPB Conservation Review* 2: 91–94.
Marler, P and Mundinger, P C (1975) Vocalisations, social organisation and breeding biology of the twite *Acanthis flavirostris*. *Ibis* 117: 1–17.
McGhie, H A, Brown, A F, Reed, S and Bates, S M (1994) *Aspects of the Breeding Ecology of the Twite in the South Pennines*. English Nature Research Rep. 118. Peterborough.
Orford, N (1973) Breeding distribution of the twite in central Britain. *Bird Study* 20: 50–62.
Stillman, R A and Brown, A F (1994) Population sizes and habitat associations of upland breeding birds in the south Pennines, England. *Biological Conservation* 69: 307–314.

Scottish crossbill *Loxia scotica*

Status

Red listed: SPEC 1 (Ins), BI, BL
Schedule 1 of WCA 1981
Annex I of EC Wild Birds Directive

National monitoring

National survey (Pinewoods Birds Project) 1992–95 (RSPB); next will be 2003–04.

Population and distribution

Possibly Britain's only endemic bird species. Scottish crossbills occur in Scotland: the Great Glen, Strathspey, Deeside and Perthshire, mostly in the remaining fragments of Caledonian pinewood. Close links with native pinewoods suggest that the present population must be only a small fraction of what it once was. There is very little information on trends in the population other than that there are marked local fluctuations due to food availability (*88–91 Atlas*). Present numbers are estimated at 300–1,250 pairs (*Population Estimates*).

Ecology

The Scottish crossbill is presently regarded as specifically distinct from other crossbills. It is difficult to separate this species from congeneric species in the field, which makes surveying difficult. A clutch of 2–5 eggs is laid between mid-February and May, and the young mostly fledge before the end of June (*Red Data Birds*).

Breeding season and winter survey – population

A sampling strategy and field method for covering large areas of woodland was devised by Buckland and Summers (1992) (see *Pinewood bird survey* in the generic survey methods section). This can be used to estimate breeding as well as winter densities, although it is important to distinguish between adults and fledged juveniles later in the season. This prevents the estimated density being inflated by young of the year.

Reference

Buckland, S T and Summers, R W (1992) *Pinewoods Birds Project: protocol for a sample survey of conifer woods.* RSPB unpubl.

Hawfinch *Coccothraustes coccothraustes*

Status

Amber listed: BDM
Non-SPEC

National monitoring

None.

Population and distribution

Hawfinches have a sparse distribution in the UK, concentrated mainly in the south and east of England. There are small populations throughout the rest of England, some in Wales (mainly in Gwent), a few in south-east Scotland and none in Ireland (*88–91 Atlas*). Between the two atlas periods the breeding population apparently declined, partly due to the loss of mature trees in the 1987 gale (*68–72 Atlas, 88–91 Atlas*). The current breeding population is estimated at 3,000–6,500 pairs (*Population Estimates*).

Ecology

Hawfinches favour a variety of habitats, from broadleaf and mixed woods to large gardens, coppices, thickets, orchards and hedges. Examples of typical habitats are: natural open mixed oak and hornbeam forest, other tall deciduous trees which carry fruits (such as beech, ash, wych elm, sycamore and maple) or areas near other fruit trees such as cherry and apple. In Gwent and the Forest of Dean (Gloucestershire) there is high dependence on wild cherry and ivy. Many nests are built in ivy, so areas containing ivy-covered trees, particularly slender cherry, ash, birch, etc, should be covered. Pairs are also regularly found in conifer and mature ivy-covered larch (Bealey and Sutherland 1982, Roberts and Lewis 1988). Hawfinches are elusive and secretive birds, preferring the thick cover of the high canopy. Eggs are laid from the first half of April to late June.

Breeding season survey – population

This species is sparsely distributed and difficult to locate. As a result, this is an intensive method (requiring knowledge of vocalisations) which may be difficult to carry out in larger woodland areas.

Information required

- maximum number of individuals recorded on any one visit, divided by two
- map of the survey area with the boundary clearly marked.

Number and timing of visits

At least three visits from mid-March to mid-May.

Time of day

In March and April, when the canopy is relatively leafless, any time of day will do. In May, when leaf cover makes observation difficult, mornings are best.

Weather constraints

Poor weather should be avoided, especially moderate or high winds (greater than Beaufort force 4) and heavy rain.

Sites/areas to visit

Areas containing any of the following: broadleaved and mixed woodland, large gardens, thickets, orchards and hedges.

Equipment

- 1:10,000 OS map of the survey area
- compass (optional).

Safety reminders

No specific advice. See the *Introduction* for general guidelines.

Disturbance

Hawfinches are extremely wary and if disturbed will immediately fly to cover in the canopy. Although this method only involves observations from the ground, be as unobtrusive as possible as this helps in obtaining accurate observations.

Methods

Mark the boundary of the survey area on the map. Decide on a route which enables you to walk within 50 m of every point in the survey area. Because of this distance rule, it is likely that you will only be able to cover about half of a 1-km square on a single visit, although it would be possible to cover a larger area in a shorter time with a coordinated team.

Walk through the area as quietly as possible, paying particular attention to bird movement in the tree canopy, checking out any movements with binoculars. Stop every 50 paces (or less) and scan the trees around you.

Hawfinches often nest in loose colonies, with several pairs close together, and these groups are easier to locate than single pairs. In March and April, there is much territorial and courtship activity in the leafless canopy, and pairs will chase each other at high speed. Birds can be particularly noisy when flocks break up into their constituent breeding pairs, whereas single pairs quickly fall silent after an initial burst of courtship and are less easily found. In May birds may be busy nest-building and egg-laying and this can again be a noisy period in a colony. At this time, courtship feeding is accompanied by feverish calling from females and much calling and singing from males. Leaf cover in May makes observation more difficult.

It is difficult to discriminate between male and female hawfinches unless they are seen at very close quarters, so count all individual birds. It is *essential* to be able to recognise hawfinch vocalisations. Unfortunately, the song and calls are neither loud nor of striking quality, and the most characteristic vocalisation is the flight-call, a single explosive 'tzick' (Mountfort 1957) repeated at intervals of 2–4 seconds in leisurely flight (*BWP*).

On each visit, use a different field map with the route marked on it, but alternate the direction of the route on each visit. Mark any hawfinch registrations on the map using standard BTO notation (Appendix 1). It is particularly important to note when two or more individuals are seen or heard simultaneously. Cross-reference the notations to a notebook allowing more room for comments.

At the end of each visit estimate the total number of hawfinches present from the mapped registrations. Report the maximum number of individuals counted on any one visit, divided by two. This will give a rough approximation of the number of pairs of hawfinches present.

References

Bealey, C E and Sutherland, M P (1982) Woodland birds of the West Sussex Weald. *Sussex Bird Report* 35: 69–73.
Mountfort, G (1957) *The Hawfinch*. Collins, London.
Roberts, S and Lewis, J (1988) Observations on the sensitivity of nesting hawfinch. *Gwent Bird Report* 23: 7–10.

Snow bunting
Plectrophenax nivalis

Status

Amber listed: BR
Non-SPEC
Schedule 1 of WCA 1981

National monitoring

National survey: 1999.
Rare Breeding Birds Panel.

Population and distribution

In the UK, snow buntings breed above the treeline in montane plateaux and corries of the Scottish Highlands (*88–91 Atlas*). The population has apparently increased in the last twenty years, and this is thought to be due both to increased observer effort and to a genuine increase in numbers (*88–91 Atlas*, Watson and Smith 1991). The current breeding population is estimated at 70–100 pairs (*Population Estimates*).

Ecology

In the breeding season snow buntings feed mainly on insects and their larvae. Nests are well hidden under rocks or scree boulders and contain 4–6 eggs, laid between late May and mid-July. There are often two broods, the young fledging in June–August (*Red Data Birds*).

Breeding season survey – population

This method is based on Amphlett and Smith (1995) and was designed to monitor large-scale changes between years for large sites with about five or more snow bunting territories. For one site, at least, the method has been shown to provide a reliable estimate of the population.

Information required
- number of singing males
- summary map of all vantage points and registrations.

Number and timing of visits

At least five visits during mid-May to mid-July, with at least five days between each. Complete each visit on the same day. If the area is relatively unknown, make an additional early-season visit to choose suitable vantage points; for such sites you may need to make more than five bird survey visits as birds may be at low density and hard to detect.

Time of day

Avoid the period two hours after dawn, and two hours before sunset. Each visit to each vantage point should last 30 minutes.

Weather constraints

All counts should be made in settled conditions. Wind speed must be low enough to allow singing birds to be heard.

Sites/areas to visit

Wherever summering snow buntings have previously been reported, or where suitable habitat occurs: such areas will be montane, mostly on hills with some ground above 3,000 feet (lower in the north), and consist of rocky terrain (scree and boulder fields) on plateaux and in corries, often near snow and ice.

Equipment

- 1:10,000 OS map of the area
- prepared field maps and a notebook
- compass and safety equipment
- Schedule 1 licence.

Safety reminders

Ensure that a reliable person knows where you are and when you are due back. Carry a compass and know how to use it and, if possible, work in a team. In all upland areas carry spare warm clothing, a plastic survival bag, whistle and food supplies.

Disturbance

Egg-collectors may be a threat to this species so keep all records confidential. During the survey there should be few problems with disturbance. Even though snow buntings are quite confiding, there is no need to get close to nests or to individual birds.

Methods

Mark the boundary of the entire survey area and the positions of the fixed vantage points on the map. The same points (or at least within 50 m of these points) will be visited in subsequent years to allow between-year monitoring, so maintain a record of their precise locations.

If the survey area is relatively unknown, make a visit early in the year (before bird surveying starts) to choose suitable vantage points. These should give a good view of the surrounding area and should be no more than 1 km apart.

On each of the five survey visits, spend 30 minutes at each vantage point. Sit down and look/listen for any snow buntings. Beware that often there may be no activity. Take a new field map with you on each visit and mark on it the locations and sex of any snow buntings recorded. Other particulars, such as behaviour, and presence of nest or young, should all be mapped if discovered. Use standard BTO codes (Appendix 1). Cross-reference all records to a notebook where you can give further details (eg observer names, times, weather).

The same male can often sing from rocks several hundred metres apart, so try to obtain records of birds singing simultaneously. This will help to determine whether two male song registrations are from one male or from two neighbouring males. It can also be helpful to sketch the black-and-white plumage patterns (especially the head, mantle and wing-coverts) of each male to facilitate individual recognition.

On smaller sites, it might be better to reduce the total number of visits to three but to spend longer (1 hour) at each vantage point (R Smith pers comm).

At the end of all visits report the number of singing males recorded at each vantage point on each visit and the mean number of singing males recorded across all visits to the site (the latter is the most useful figure for between-year comparisons).

Also provide a map showing the survey of the boundary area, the location of the vantage points and a summary of bird registrations. Clarify which records were from which visit by writing a visit-specific code (eg A, B, C, etc) against each.

References

Amphlett, A and Smith, R (1995) *A Method for Monitoring the Snow Bunting Breeding Population in the Central Cairngorms*. Confidential report to RSPB.

Watson, A and Smith, R (1991) Scottish snow bunting numbers in summer 1970–87. *Scottish Birds* 16: 53–56.

Cirl bunting *Emberiza cirlus*

Status

Red listed: BD, BR
SPEC 4 (S)
Schedule 1 of WCA 1981

National monitoring

National breeding surveys in 1982 (BTO) and 1989 (RSPB and Devon Bird Watching and Preservation Society).
Complete annual censuses 1990–93 and sample censuses 1994–97 (RSPB/EN).
Full national census 1998 (RSPB/EN); next will be 2003.
Rare Breeding Birds Panel.

Population and distribution

One of the UK's rarest resident passerines, currently distributed along the south Devon coast between Plymouth and Exeter (*88–91 Atlas*). The population in central and south-east Europe appears to have remained stable whereas there has been a long-term range contraction in north-west Europe (*Birds in Europe*). The UK range and population declined considerably up to 1989, when there were an estimated 120 pairs (Evans 1992). Habitat change and changing farming practices were the main causes of decline. The population has recently increased to about 350 breeding pairs, which may be attributable to the introduction of set-aside (A D Evans pers comm). This species is thought to be mostly sedentary in the UK, with the wintering range very similar to that in the breeding season (Cole 1993).

Ecology

Breeding habitat preferences are for coastal scrub adjacent to arable land, or lowland agricultural land with hedgerows. A diversity of land-use including arable, pasture, horticulture and hedgerows appears to be important within a territory. A clutch of 3–4 eggs is laid from the end of April to August. Second broods are common and third broods are likely (*Red Data Birds*). The preferred winter habitat is sheltered stubble, such as barley stubble, particularly that with hedge or tree cover. Sites with insubstantial hedges and trees are less likely to hold wintering cirl buntings. Other suitable sites are grass slopes and meadows which have not been treated with fertilisers or sprays. Stock feeding areas often attract cirl buntings, as may garden bird-feeders.

Breeding season survey – population

Information required

- estimated number of possible, probable and confirmed breeding pairs
- map of the locations.

Number and timing of visits

A minimum of two and a maximum of four visits. One visit should be in late April or early May, one in late May or early June, and the remainder in June–August.

Time of day

Between dawn and 1100 BST, or from mid- to late afternoon.

Weather constraints

Avoid poor weather.

Sites/areas to visit

Any suitable habitat (see *Ecology*, above).

Equipment

- 1:25,000 OS map
- Schedule 1 licence.

Safety reminders

Tell a reliable person where you are and when you are due back.

Disturbance

Do not disturb breeding birds and nest-sites; under no circumstances approach a nest. It is not necessary to find a nest to prove breeding. The locations of all sites should be treated in strictest confidence.

Methods

Map the survey boundary. Decide on a route which gets to within 150 m of every point. Walk systematically along the route at a slow pace. Wherever access allows, try to vary the route between visits to increase the overall amount of ground covered.

Cirl buntings are most easily located in late April/early May when males are singing. From early May to late May/early June birds are elusive. Subsequently, failed pairs and birds feeding their first broods become more obvious.

The song of the male is audible from up to 500 m on a good day and can resemble that of greenfinch, wren, yellowhammer or lesser whitethroat, with individual males varying their song within a song bout.

On each visit, map the position of every cirl bunting, indicating its behaviour with a code (see below). Clarify which observations refer to which visit, eg with a small '1' for the first visit, '2' for the second, etc, preferably in red. After your final visit, circle those sightings from a single territory and give the dates for each visit.

Behavioural codes and their interpretation are as follows:

Possible breeding	✓	Bird recorded in breeding season in suitable habitat
	S	Male singing
Probable breeding	P	Pair in suitable nesting habitat
	T	Territorial behaviour
	D	Display
	N	Visiting probable nest-site
	A	Agitated behaviour
	BB	Carrying nest material

Confirmed breeding FL Recently fledged young
FY Carrying food for young

If you see any birds with colour rings, please note the colour combinations, and the legs the rings were on, and report the sightings to the RSPB.

Report the number of possible, probable and confirmed breeding pairs within the survey area.

Winter survey

Information required

- maximum number of individuals found on four visits
- map showing the boundary of the survey area and all bird registrations.

Number and timing of visits

Four visits, one in each month from December to March.

Time of day

Start early in the morning.

Weather constraints

Avoid poor weather, in particular high winds and heavy rain.

Sites/areas to visit

Any suitable winter habitat. Winter sites are often close to the breeding areas.

Equipment

- 1:25,000 OS map.

Safety reminders

As for the population survey (above).

Disturbance

Obtain permission to enter agricultural fields, even if permission was obtained in the summer.

Methods

Locating wintering flocks is difficult and initially time-consuming. Once a flock has been located, however, it should be relatively easy to find again. Mark the area to be surveyed on a map and decide beforehand on a route to walk. Ensure the route takes you through each field and to within 30 m of each field boundary. Randomise the starting point and direction of each visit. Take careful notes on where any flushed birds land, and check birds for colour rings. This will help to ensure that the same flocks are not counted more than once.

There are likely to be many species of bird feeding in the same field, although cirl buntings will stick together and will not feed out in the centre of the field. In stubble fields it may be necessary to flush birds before they are seen. The distinctive flight-call will be heard and the flock can then be counted.

The maximum number of birds seen in an area on any one visit should be reported along with a map of their location. Note the field type in which any flocks are seen feeding.

Reference

Cole, A (1993) *In Search of the Cirl Bunting*. Alan Sutton, Stroud.
Evans, A D (1992) The number and distribution of cirl buntings *Emberiza cirlus* breeding in Britain in 1989. *Bird Study* 39: 17–22.

Corn bunting *Miliaria calandra*

Status

Red listed: BD, HD
SPEC 4 (S)

National monitoring

BBS.
National breeding survey 1993 (BTO/JNCC).

Population and distribution

Corn buntings are found predominantly in the cereal-producing areas of the UK, ie mainly in the south and east of England and the east of Scotland, but with outlying populations in Ireland, south-west England and the Outer Hebrides. The decline in numbers and range of the corn bunting since the mid-1970s has been well-documented both in Britain and in Europe. In Britain, the number of occupied 10-km squares fell from 1,358 in 1968–72 to 921 in 1988–91 (*88–91 Atlas*), representing a contraction in breeding range of 32%. Several possible contributory factors have been suggested. These include changes in the times of sowing and harvesting of both cereal and pasture crops, declines in the cultivated area of barley *Hordeum* (particularly spring-sown varieties), increased pesticide usage and increasing regional specialisation leading to the loss of traditional rotations. There are an estimated 19,800 corn bunting territories in the UK (*Population Estimates*).

Ecology

Corn buntings are found in open habitats, usually associated with arable land. They need perches overlooking their territory, but there are only minimal requirements for cover. Corn buntings are polygamous, although the extent of polygamy varies greatly between regions. Even in polygamous populations, however, the sex ratio approximates equality. In north-west Scotland eggs are laid in early June to mid-July, and in England in May and June, with a mean date of 25 May. Incubation lasts 12–14 days and is by the female only. The young are fed and cared for by the female, and fledge after 9–13 days. The song is produced by the male only; it begins with an ascending trill, is basically segmented in structure and has been described as sounding like a jangling bunch of keys (*BWP*).

Breeding season survey - population

Information required

- maximum number of singing males
- map of survey boundary and all registrations.

Number and timing of visits

One visit per month in May, June and July, at least two weeks apart.

Time of day

Morning.

Weather constraints

Very windy or wet weather should be avoided.

Sites/areas to visit

Any area of open farmland, grassland or machair. A reasonable-sized survey area is about 50 ha. This is large enough to detect meaningful changes in population size , but not so large that it cannot be covered in a single morning.

Equipment

- 1:10,000 OS map.

Safety reminders

No specific advice; see general guidelines in the *Introduction*.

Disturbance

No disturbance is necessary.

Methods

Map the survey boundary and work out a route which will allow you to cover the entire study area. Within a field system, for example, this would normally involve walking round the edge of each field. Walk at a steady speed, marking all birds seen or heard on the map using standard BTO notation (Appendix 1). The same map can be used for each visit, either by using different coloured pens or by using the letters A, B and C for the May, June and July visits respectively.

Report the maximum number of territorial (singing) males recorded on any one visit.

Generic survey methods

Generic breeding bird monitoring methods

Common Birds Census (CBC)

The Common Birds Census (CBC), run by the British Trust for Ornithology for more than 35 years, is a partnership between the BTO and JNCC. It has been the main scheme by which populations of common breeding birds are monitored in the UK. The CBC began in 1962 and is based on a survey method known as 'territory mapping'. It has proved highly valuable in revealing population fluctuations and trends among British birds and in helping to understand their causes. CBC data have played a key role in drawing up the listings of Birds of Conservation Concern (Gibbons et al 1996). Territory mapping methods provide full site coverage, they require many visits and provide quality information. However, because of the time-consuming nature of the fieldwork and the complex analysis required by the mapping method, an alternative scheme for national bird monitoring, the Breeding Bird Survey (BBS), has been developed. CBC will be continued in parallel with BBS for some time so that the results from the two can be properly calibrated for national monitoring purposes.

Where a complete census of birds at a site during the breeding season is required, the CBC method is the most accurate and practical way. Some CBC guidelines are provided below (after Marchant 1983, Marchant et al 1990, *Census Techniques*).

Information required

- number of territories of each species
- map for each species showing the registrations from all ten visits.

Number and timing of visits

Ten visits, March–July, ideally with at least ten days between each visit.

Time of day

Early morning. Avoid the first hour before sunrise. Up to two evening visits can be helpful.

Weather constraints

Avoid days of high winds (greater than Beaufort force 5) and poor visibility.

Sites/areas to visit

CBC covers farmland and woodland but the methods can be adapted to most habitats.

Equipment

- 1:2,500 OS map (see *General field guidelines* in the *Introduction* for information on these large-scale maps)
- clipboard
- two pencils.

Safety reminders

No specific advice. See the *Introduction* for general guidelines.

Disturbance

Disturbance to birds is minimal.

Methods

The size of the study plot will depend on the objective of the study. The area should be large enough to include sufficient numbers of any scarcer species of particular interest. A single visit should take 3–4 hours. In woodland with high bird densities, choose a plot of about 10–20 ha. On farmland, the plot should be about 50–100 ha depending on the number of hedges and woody areas. Be careful when choosing the boundaries of a plot; it is better to have plots that are roughly square or round, rather than ones which are long and thin or which have complicated edges. Using hedgerows as boundaries may be unavoidable but it exaggerates the density of birds on farmland, since the bulk of the birds are in the hedges. These 'edge' effects become relatively less important the larger the area.

Map the boundary of the survey area. Take a new field map with you on each visit. Walk the area at a slow pace so that all birds detected can be identified and located. Choose a route which gets you to within 50 m of every point. Where vegetation is thick, a closer approach is desirable. Walk all hedgerows on farmland. Vary your route and direction between visits so that there is no systematic tendency for any part of the plot to be visited later or earlier in the day. Complete a single visit in a single morning.

Map the identity and activity of all birds with small and tidy writing in pencil or ball-point pen. Use standard BTO codes for all species and behaviour (Appendix 1). Record as much detail as possible, such as the age and sex of each bird. The most important point to concentrate on is the location of individuals of the same species which can be heard or seen simultaneously. Be sure to note on each visit map the visit, date, times, observer and weather (eg Visit F; 6/6/99; 0730–1025 BST; observer: JHM; Weather: cool, overcast and wind speed force 3).

When the visits have been completed, transfer all the information obtained from each species to a separate map, the 'species map'. Registrations on the species maps will fall more or less neatly into clusters indicating the activity of particular birds or pairs throughout the season. The maps can then be 'analysed' to determine the number of territories present. Example maps and analysis guidelines are given in Marchant (1983), Marchant et al (1990) and *Census Techniques*.

A cluster is in general a spatially distinct group of registrations in which not more than one male and one female (or two adults) are represented. Depending on the biology of the species, it usually relates directly to a breeding territory. Ideal clusters show a series of registrations of territorial behaviour spanning most of the visits, and dotted lines radiating out to neighbouring clusters. In practice, a small amount of confusion is to be expected, particularly for species such as dunnock or long-tailed tit which do not always hold territory in pairs, and for smaller birds such as the tits and *Phylloscopus* warblers which are mobile and inconspicuous.

Some territories will overlap the plot boundary, so it is usually necessary to map records just outside the survey area. In the field, this translates as recording all birds up to 50 m outside the plot boundary in order to ensure that all territories straddling the boundary are mapped. Clusters are often difficult to differentiate and may sometimes overlap. As a consequence, map analysis can involve a certain amount of subjectivity in interpretation.

Report the number of territories of each species.

References

Gibbons, D W, Avery, M I, Baillie, S, Gregory, R D, Kirby, J, Porter, R, Tucker, G and Williams, G (1996) Bird Species of Conservation Concern in the United Kingdom, Channel Islands and Isle of Man: revising the Red Data List. *RSPB Conservation Review* 10: 7–18.

Marchant, J H (1983) *BTO Common Birds Census Instructions*. BTO, Tring.

Marchant, J H, Hudson, R, Carter, S P and Whittington, P (1990) *Population Trends in British Breeding Birds*. BTO, Tring.

Breeding Bird Survey (BBS)

The UK Breeding Bird Survey is a partnership between the BTO, JNCC and RSPB and was introduced in the breeding season of 1994 as an annual survey of widespread and abundant terrestrial birds in the UK. The BBS runs in parallel with the Common Birds Census (CBC; see above), but will eventually take over as the UK's main scheme for national bird monitoring, having been set up to improve geographical and habitat representation and to increase species coverage. The BBS is a sample survey in which observers walk two 1-km transects within randomly allocated 1-km squares. The methods provide much more reliable information on UK-wide population trends than does the CBC, although the BBS does not provide as detailed full census information at the plot level as does the CBC.

In terms of national population monitoring, the BBS will provide:

- trends for many species for the UK as a whole
- trends for individual countries within the UK
- trends for European Union regions within the UK
- trends by habitat type.

In addition, the conservation of particular species and habitats will be greatly improved by a more complete understanding of the relationships between birds and broad habitat types, both of which are recorded by the BBS (Gregory et al 1996, 1997).

The BBS is not only a scheme, it is also a method. There are many circumstances in which using the line-transect method of the BBS may be the best way to monitor widespread and common breeding species in a particular area, even though these areas may not be within the formal sampling design of the BBS. In general, this would only work well for reasonably large sites, for example Environmentally Sensitive Areas (ESAs) or National Parks. In practice, such monitoring would probably be done by randomly selecting a sample of 1-km squares within the overall area, using the BBS method within each of these squares and repeating between years. Users of this approach should be warned that the analysis of the data collected using the BBS method can become quite complex and it is recommended that specialist advice is sought. This is particularly the case if the method is to be used to estimate absolute breeding densities and population sizes of individual species, rather than merely performing a between-year monitoring function. To estimate densities and population sizes it is necessary to use the distance sampling methods of Buckland et al (1993) and, ideally, the DISTANCE software specially written for such analyses.

The following is taken from 1998 Breeding Bird Survey Instructions.

Information required

- number of individual birds (excluding juveniles) of all species that were recorded in each 200-m section of a 2-km-long transect, in each of several distance bands, on each of two visits.

Number and timing of visits

Three: first visit March–April (set up route and record habitat); one between early April and mid-May (early transect count); one mid-May to late June (late transect count). The two count visits must be *at least four weeks apart*. NB Fieldwork should begin and end later in more northerly parts of the UK.

Time of day

Morning, beginning ideally 0600–0700 BST, and no later than 0900 BST.

Weather constraints

Do not attempt to census in heavy rain, poor visibility or strong winds.

Sites/areas to visit

All accessible habitat types (except large expanses of water). UK-wide scheme: 1-km squares allocated to you by your BTO Regional Representative. BBS as a method outside the UK-wide scheme: randomly selected 1-km squares within the study area.

Equipment

- recording forms (see Figures 1 and 2)
- 1:50,000 and 1:25,000 OS maps of the area.

NB When contributing to the national BBS always obtain instructions and recording forms from: Census Unit, British Trust for Ornithology, The Nunnery, Thetford, Norfolk IP24 2PU.

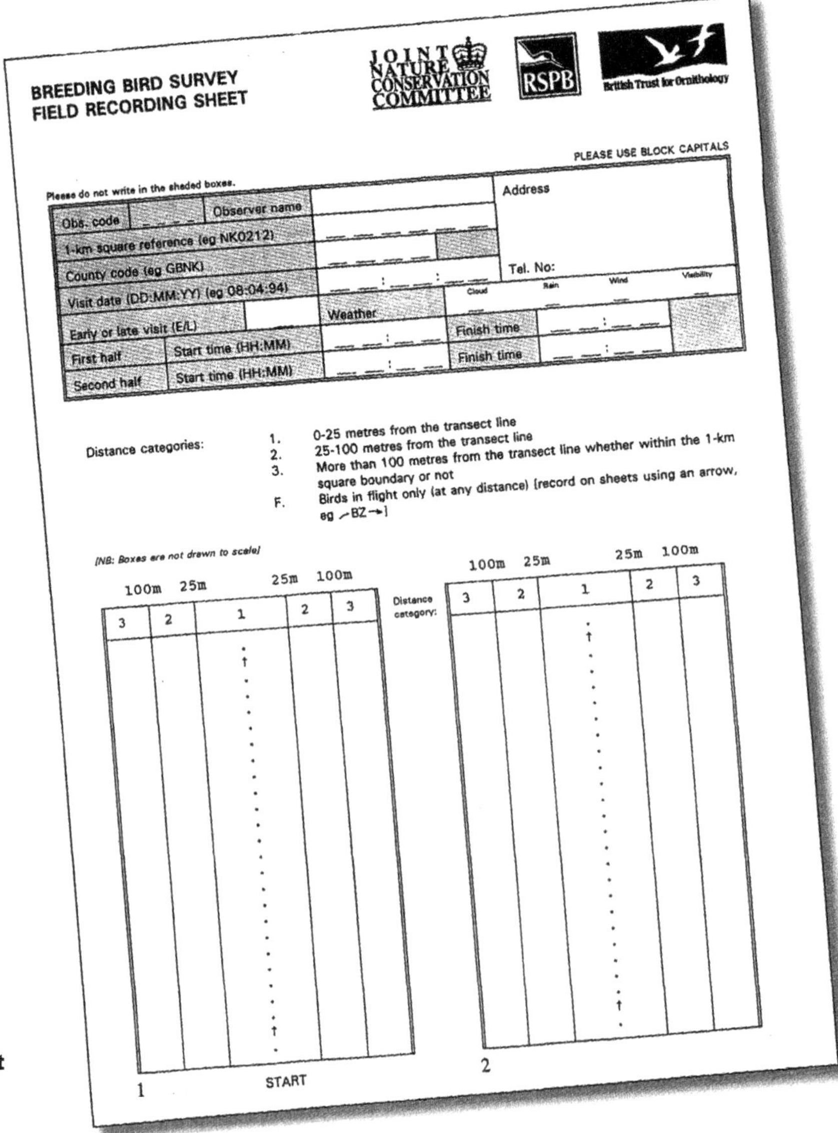

BREEDING BIRD SURVEY
FIELD RECORDING SHEET

JOINT NATURE CONSERVATION COMMITTEE | RSPB | British Trust for Ornithology

PLEASE USE BLOCK CAPITALS

Please do not write in the shaded boxes.

Obs. code — Observer name — Address
1-km square reference (eg NK0212)
County code (eg GBNK) — Tel. No:
Visit date (DD:MM:YY) (eg 08:04:94)
Early or late visit (E/L) — Weather: Cloud, Rain, Wind, Visibility
First half — Start time (HH:MM) — Finish time
Second half — Start time (HH:MM) — Finish time

Distance categories:
1. 0-25 metres from the transect line
2. 25-100 metres from the transect line
3. More than 100 metres from the transect line whether within the 1-km square boundary or not

F. Birds in flight only (at any distance) [record on sheets using an arrow, eg ↗BZ→]

[NB: Boxes are not drawn to scale]

100m 25m 25m 100m

Distance category: 3 | 2 | 1 | 2 | 3

1 START 2

Figure 1
Example of a field recording sheet as used in the UK Breeding Bird Survey.

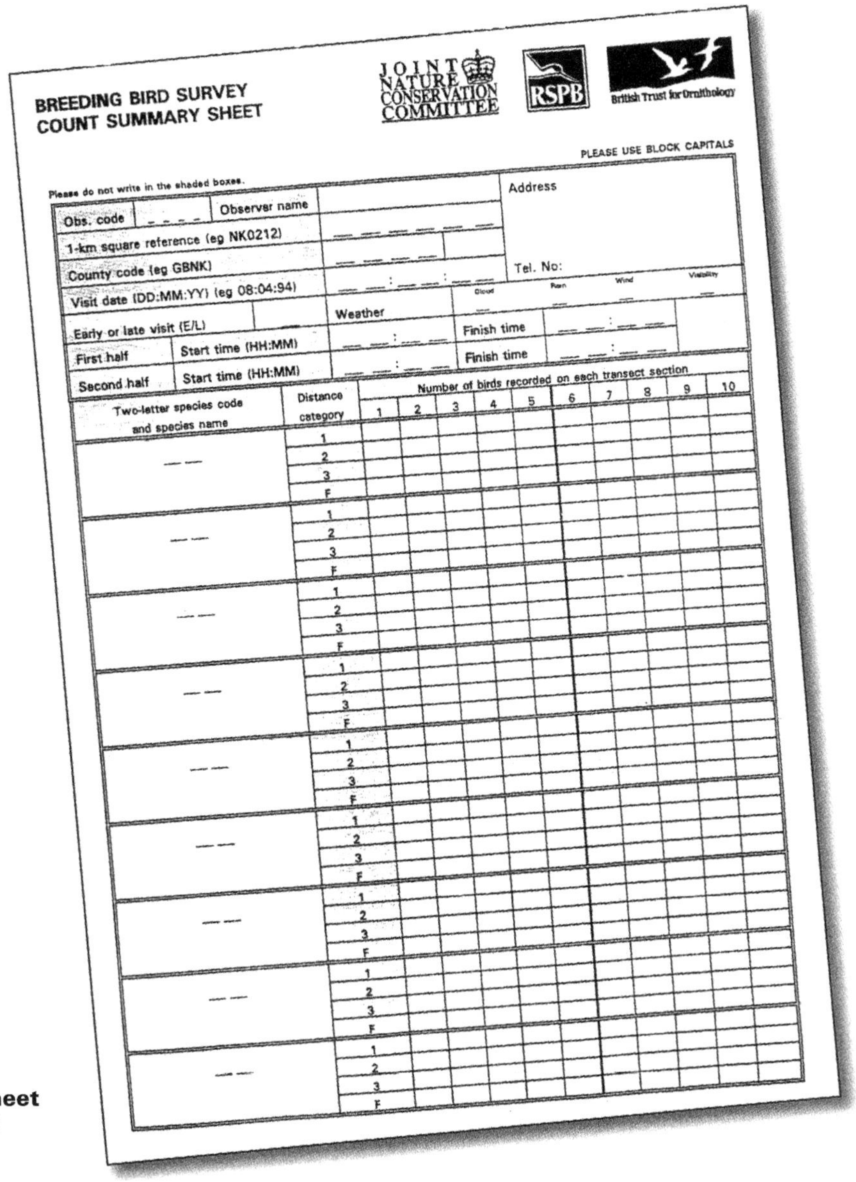

BREEDING BIRD SURVEY
COUNT SUMMARY SHEET

JOINT NATURE CONSERVATION COMMITTEE
RSPB
British Trust for Ornithology

PLEASE USE BLOCK CAPITALS

Please do not write in the shaded boxes.

Obs. code
Observer name
1-km square reference (eg NK0212)
County code (eg GBNK)
Visit date (DD:MM:YY) (eg 08:04:94)
Early or late visit (E/L)
First half — Start time (HH:MM)
Second half — Start time (HH:MM)
Address
Tel. No:
Weather — Cloud, Rain, Wind, Visibility
Finish time
Finish time

Two-letter species code and species name
Distance category: 1, 2, 3, F
Number of birds recorded on each transect section: 1, 2, 3, 4, 5, 6, 7, 8, 9, 10

Figure 2
Example of a count summary sheet as used in the UK Breeding Bird Survey.

Safety reminders

No specific advice. See general guidelines in the *Introduction*.

Disturbance

This method involves little disturbance to breeding birds.

Methods

If you are contributing to the national scheme, your 1-km square(s) will be allocated for you. If you are using BBS as a survey method outside the national scheme, randomly select several tens of squares within your study area.

Once you know which 1-km square(s) to survey, map your transect route through each square on a habitat recording form. The transect route should consist of two parallel lines, north–south (preferably) or east–west, each 1 km long. It is important to use the same route each year. Transect lines should be 500 m apart and 250 m from the edge of the square. Each transect line should be divided into five equal 200-m-long sections, making a total of ten 200-m transect sections, numbered 1 to 10 (see Figure 3). It is important to note the starting points of each transect section either by using permanent landmarks or temporary markers. In practice, transect lines are likely to deviate from the 'ideal' because of problems with access, or barriers such as roads, etc. However, at no point should the two transect lines be less than 200 m apart. Minor intrusions into adjacent squares may be unavoidable. It is imperative that the same route be followed year after year.

From your chosen starting point, walk the first half of your transect route at a slow pace, pausing briefly at intervals to listen for song and to scan for birds flying overhead. Using the standard recording form (Figure 1), record all birds seen and heard in the appropriate transect sections and distance categories (see below). At the end of section 5, stop recording, go to the start of section 6 and begin recording sections 6–10. Try to avoid double-counting the same individual. Use standard BTO codes (Appendix 1) and distinguish adults from juveniles.

Record birds in one of the following four distance categories when first noted:

1. Within 25 m either side of the transect line.
2. Between 25 and 100 m either side of the transect line.
3. More than 100 m either side of the transect line (including birds outside the 1-km square boundary).
F. Birds in flight only (at any distance).

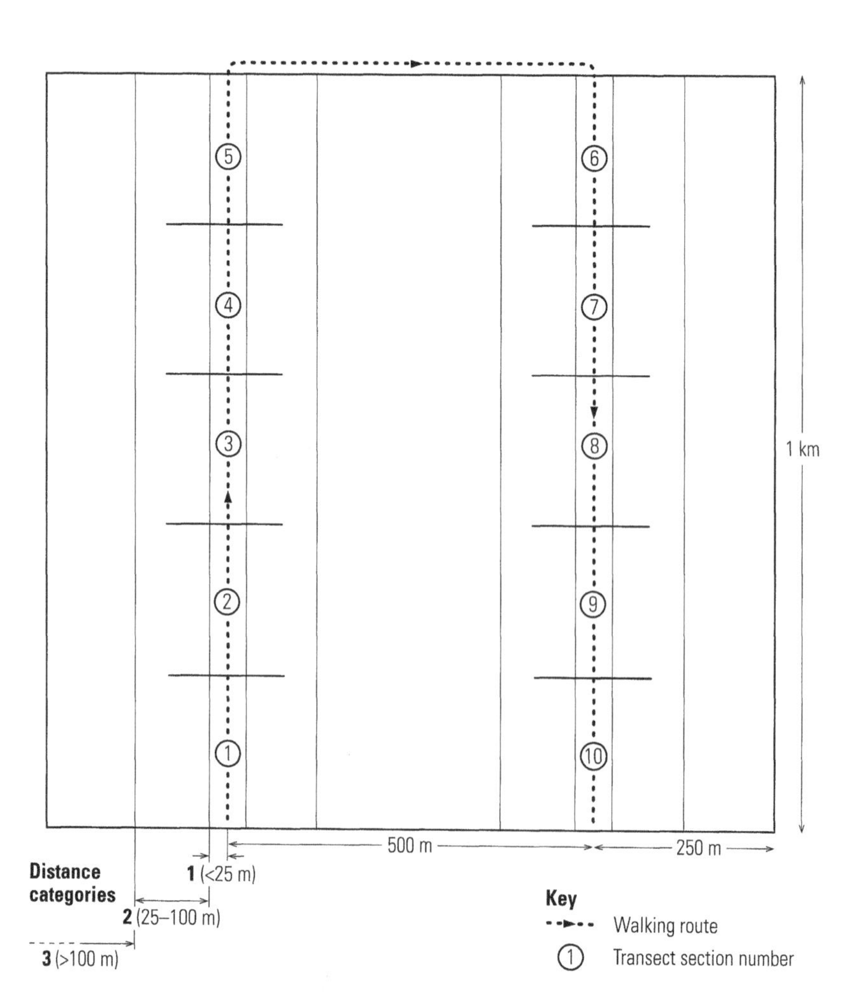

Figure 3
Transect line route, distances and numbered transect sections for a 1-km square in the UK Breeding Bird Survey.

Distances are measured perpendicular to the transect line. A bird seen 200 m ahead of the observer but within 25 m of the transect line should be recorded in category 1. To familiarise yourself with judging 25 m and 100 m distances, pace these out before starting the survey. For category F (birds in flight), draw an arrow through the species' two-letter code to indicate that it is in flight. If a bird is seen to take off or land it should be recorded in the appropriate distance category (1–3) at that position. NB skylarks in display flight should be recorded in the relevant distance category.

Complete the summary sheets (Figure 2) as soon as possible after each field visit. Transfer the number of individuals (excluding juveniles) that were recorded in each section of the transect, 1–10, on each visit, in each distance band. Habitat (and possibly other species) recording are essential parts of the national scheme; details of these will be included with the official forms and instructions received from the BTO.

If the survey is being undertaken as part of the UK-wide scheme, return all recording forms to the BTO via your Regional Representative, who will analyse and report the results. If you are using the BBS as a method rather than as part of the UK-wide scheme, you will probably have to undertake all analyses yourself. It is recommended that you seek specialist help for these analyses: try the BTO Census Unit in the first instance.

References

Buckland, S T, Anderson, D R, Burnham, K P and Laake, J L (1993) *Distance Sampling: Estimating Abundance of Biological Populations.* Chapman and Hall, London.

Gregory, R D, Bashford, R I, Balmer, D E, Marchant, J H, Wilson, A M and Baillie, S R (1996) *The Breeding Bird Survey 1994–1995.* BTO, Thetford.

Gregory, R D, Bashford, R I, Balmer, D E, Marchant, J H, Wilson, A M and Baillie, S R(1997) *The Breeding Bird Survey 1995–1996.* BTO, Thetford.

Waders

These are standard methods of surveying a variety of breeding waders in the UK. While not necessarily the best methods for each individual species, they combine ease of use and compatibility with detailed nest-finding studies, and provide data comparable with other surveys in the UK.

Although a number of methods for surveying breeding waders have been developed, two are currently the most widely used: those of O'Brien and Smith (1992) and Brown and Shepherd (1993). Both methods were designed to survey a number of species simultaneously; thus they are generic, rather than species-specific, schemes. The method of O'Brien and Smith was originally developed for surveying breeding waders on lowland farmland/enclosed areas, and as a consequence is based on field-by-field observations. The O'Brien and Smith method is, however, still recommended for grasslands in Scotland and northern England which are sometimes considered as upland (Grant et al in prep). Provided that these areas are enclosed and below the moorland wall (eg as for previous surveys of Baldersdale, Allendale, etc, in the northern Pennines: M C Grant pers comm), then O'Brien and Smith should be used. By contrast, the method of Brown and Shepherd, which involves timed observations in 500 × 500 m grid cells, was originally developed for breeding waders in upland areas, particularly if open and unenclosed. In principle the Brown and Shepherd method could be extended into enclosed lowland areas, but, given that the majority of lowland wader surveys have followed O'Brien and Smith, and that the lowlands are mostly divided into fields, we recommend that O'Brien and Smith's method is adhered to in the lowlands for ease of data collection and comparability. Details of these two methods are outlined below.

A third method, that of Reed and Fuller (1983) has been developed for censusing waders breeding at high densities, for example on the machair of the Uists in the Outer Hebrides. Once again, though the methods of O'Brien and Smith and Brown and Shepherd could be used under these circumstances, we recommend that those of Reed and Fuller are used in order to maintain comparability with earlier surveys.

The Brown and Shepherd (1993) method for censusing upland breeding wader populations

This method is used to census upland breeding waders, principally golden plover, dunlin, oystercatcher, lapwing, curlew and redshank.

Information required

- estimated number of breeding pairs of each species
- final visit maps showing registrations and the boundaries of the areas covered.

Number and timing of visits

A minimum of two, early April to late June. First visit, early April to mid-May; second visit, mid-May to late June.

Time of day

0830–1800 BST.

Weather constraints

Avoid high winds (greater than Beaufort force 5) and other poor weather conditions.

Sites/areas to visit

This method is recommended for use in open upland moor, although it can also be used in enclosed field systems. Vegetation may range from montane to semi-improved grassland, encompassing boulder-strewn blanket bog, bare stony ground, heath, etc.

Equipment

- 1:25,000 OS maps
- 1:25,000 field maps of the area
- compass.

Safety reminders

Ensure someone knows where you are and when you are due back. Always carry a compass. If possible, surveyors should not work alone. In more remote upland areas, spare warm clothing, a first-aid kit, a plastic survival bag, whistle and food should be carried.

Disturbance

Some disturbance may be unavoidable but can be kept to a minimum; there is no need to search for nests or to get too close to adults.

Methods

Surveys of upland waders should be carried out by observers with experience of their behaviour, calls and songs. This survey method is timed, therefore it is important for the observer to have had a trial run before attempting the real thing. The method is based on constant search effort, so keeping to within the specified times for given areas is important. These times are: 20–25 minutes in each 500 × 500 m quadrat of open land and 0.8–1.0 minute per ha for enclosed fields. Practice at covering quadrats in these times is essential to ensure even coverage.

Clearly mark the boundary of the survey area and the quadrat areas to be covered on a map. Large sites will require a team of people working on adjacent quadrats; this is safer, more enjoyable and reduces the chances of double-counting individual birds moving between quadrats.

Follow a predetermined route through each square so that (most important of all) all parts of each quadrat are approached to within at least 100 m. Although some areas may look more or less attractive to breeding waders (eg lake shores and pools are good for feeding birds), it is important that all areas are covered equally. If possible, walk in the opposite direction through the quadrats on the second visit.

At regular intervals (at least every 100 m) scan round in every direction as far as the terrain or weather allows and also listen for calls and songs. If necessary, scan for a short time from a rock or hillock to get a better view. For some species, eg golden plover, males and females can be distinguished with practice, which helps the interpretation of results. As each individual or pair is encountered, decide whether these are new birds, using individual characteristics such as the amount of black in the plumage of golden plover. However, it may still be necessary to retrace your steps to check on the continued presence of any birds previously located.

Record the location and activities of all wader species seen on 1:25,000 field maps using the standard BTO symbols (see Appendix 1). Make additional notes and cross-reference them with mapped symbols to avoid confusion. These should include information such as the time of the observation (this is particularly important if you are surveying an area adjacent to another person's quadrat), and the behaviour and flight line of the birds.

At the end of each visit, all the observers involved with the survey should get together and put all their registrations on a final visit map. At this stage, duplicate registrations of the same bird made in adjacent survey areas by different people can be removed.

Birds can be said to be breeding if:
- they are observed displaying or singing
- nests, eggs or young are located
- adults repeatedly alarm-call
- distraction displays are seen
- territorial disputes are seen.

Other records are considered to be of non-breeding birds, failed breeders or birds loafing, feeding or on passage to other areas.

Sometimes two birds may be seen close to one another but it may not be clear if they are two members of the same pair, or birds from different pairs. Where this is the case, birds separated by less than 500 m (or 200 m for dunlin) on a given visit are arbitrarily considered to be from the same pair; those separated by more than 500 m are treated as being from two different pairs.

At the end of the season, observers will end up with two final maps of the site, one from each visit. On these maps those registrations which represent pairs and those which do not should be clearly indicated. Estimates of the numbers of pairs at each site or survey area are derived from both the final visit maps. Pairs are considered to be separate from one another only if they are at least 1,000 m apart (500 m for dunlin) on the different visit maps.

The O'Brien and Smith (1992) method for censusing lowland breeding wader populations

This is a generic method covering several species of lowland wader, but there are differences between species in what to record (pairs, displaying males, etc) and in the interpretation of what is recorded. The timing of visits differs, depending on the geographical location of the area to be covered.

Information required

- *Oystercatcher* peak number of pairs.
- *Lapwing* peak number of birds.
- *Snipe* peak number of drumming plus chipping birds.
- *Curlew* mean number of birds.
- *Redshank* mean number of birds.

Number and timing of visits

Three visits, at least one week apart; 15 April to 19 June overall, but with some geographical variation. Note that the third visit is technically redundant for redshank, and can be for lapwing, but these species should nevertheless be recorded on each visit.

England and Wales	First visit 15 April - 30 April Second visit 1–21 May Third visit 22 May - 18 June
N England and Scotland (lowland)	First visit 18 April - 8 May Second visit 9–29 May Third visit 30 May - 19 June

Time of day

Oystercatcher, lapwing, snipe	During the three hours after dawn or the three hours before dusk. It is very important that counts of snipe are not continued more than three hours after dawn.
Redshank	Between dawn and 1200 BST.
Curlew	At least one of the three counts within three hours of either dawn or dusk. The remaining counts can be performed between 0900 and 1700 BST.

Weather constraints

Avoid cold, wet and windy conditions (do not survey when the wind exceeds Beaufort force 3).

Sites/areas to visit

Oystercatcher

Coastal breeding sites include saltmarsh, coastal grazing marshes, shingle beaches, dunes and rocky shores. Inland they breed in a variety of habitats, especially on arable along river valleys and around lakes.

Lapwing

Damp lowland grassland subject to freshwater flooding or waterlogging such as floodplain grasslands, coastal grazing marshes, washlands and isolated pockets of poorly drained grassland. Lapwing also breed on moorland and the surrounding in-bye land. Large numbers also occur on tilled land.

Snipe

Coastal grazing marshes, damp lowland grassland subject to freshwater flooding or waterlogging such as floodplain grasslands, washlands and isolated pockets of poorly drained grassland. Snipe also breed on moorland bogs and marshy rough pasture, and on in-bye/marginal grassland.

Curlew

Found generally on damp upland and northern moorlands and areas of rough grazing, although curlews also breed on some lowland and agricultural sites, particularly in northern Britain.

Redshank

Saltmarsh, coastal grazing marshes, damp lowland grassland subject to freshwater flooding or waterlogging such as floodplain grasslands, washlands and isolated pockets of poorly drained grassland.

Equipment

- 1:10,000 map of the area to be visited.

Safety reminders

No specific advice. See the *Introduction* for general guidelines.

Disturbance

Keep to a minimum. There is no need to search for nests or to get too close to adults.

Methods

All fields on the site should be numbered, as information is collected on a field-by-field basis. Note, however, that the results are analysed and reported on a *site* basis, defined as the amount of ground that can be covered in one visit. If the whole survey area is split into sites, make the boundaries very clear for comparable analysis in future years. Ensure that a site map (1:10,000) showing field numbers is held with the results so that sites/fields can be located easily in the future. Saltmarsh can be divided into 'fields' by using the channels or creeks as boundaries.

Each field should be walked so that the observer gets to within 100 m of every point. Record on the site map the location, movement and behaviour of all waders using standard BTO codes (Appendix 1). This is important as it will help to avoid double-counting the same individual.

Allocate each bird to a single field – the first field in which it was recorded. If the bird was first observed in display-flight it should be allocated to the field in the centre of its flight. Conspicuous waders such as lapwing are best assigned to a specific field by scanning with binoculars and looking 200–400 m ahead, in order to record them before they are disturbed.

Oystercatcher

Record the total number of *pairs* where a pair is taken as:

- number of paired individuals/2
- a displaying individual
- a single bird (not birds in flocks)
- a nest
- a brood.

Record the number of pairs seen in each field. Birds in flocks should *not* be included in this figure and should be recorded separately. The estimate used for year-to-year comparisons is the peak number of pairs per site seen on any one visit.

Lapwing

Record the total number of *birds* seen in each field. Report the maximum number of individuals recorded on the site *between mid-April and late May* divided by two as the total number of pairs.

Snipe

Record the total number of *birds* heard drumming or chipping in each field and report the maximum number for the site over the three visits as the number of pairs. However, if no snipe were recorded during May the figure should *not* be reported.

Curlew

Record the total number of *individuals* (excluding birds in flocks) in each field on each visit. Transform the mean number of individuals (excluding birds apparently in flocks) counted over the three visits at each site using the formula (0.71 × mean count) + 0.10 to give the estimated number of pairs per site. NB if no curlew were recorded at all, the figure should be reported as 0 pairs and not 0.10 (Grant et al in prep).

The previously recommended count unit was the number of apparent pairs (definition of a pair as for oystercatcher, above). This should also be recorded to enable comparisons to be made with past survey data.

Redshank

Record the total number of *birds* seen in each field on each visit. Report the mean number of birds found on the site over both visits (or all visits). This is taken as the number of breeding pairs. Give details if more than two visits are made.

The Reed and Fuller (1983) method for surveying machair breeding wader (high-density) populations

This method is used for censusing waders breeding *at relatively high densities on machair* in the Hebrides. It can be used for dunlin, oystercatcher, ringed plover, lapwing, snipe and redshank.

Information required

- maximum number of breeding pairs of each species from one visit
- maps showing registrations and the boundaries of the areas covered.

Number and timing of visits

At least one visit (maximum four) in the first three weeks of June.

Time of day

0830–1800 BST.

Weather constraints

Avoid persistent rain (any more than a drizzle) or high wind (greater than Beaufort force 5).

Sites/areas to visit

Machair is mainly found on the west coast of the Outer Hebrides. It is a flat sandy plain which can include strip rotation farming, hay and silage fields, machair marshes and small machair lochs. Other areas of damp grassland and marsh on adjacent 'blackland' should be included.

Equipment

- 1:10,000 OS maps of the area
- A4 photocopied field maps.

Safety reminders

Ensure someone knows where you are and when you are due back.

Disturbance

May be unavoidable but should be kept to a minimum. Do not search for nests or get too close to adults.

Methods

Mark the boundary of the survey area clearly on a map and split it into manageable areas. Two people should be able to cover about 4 km^2 in one visit. Working as a pair, walk parallel transect lines 150 m apart (lower bird densities) or 100 m apart (high bird densities, usually wetter habitat). Record locations and behaviour of all waders seen on 1:10,000 field maps, using BTO symbols (see Appendix 1). On subsequent visits, walk the same transect lines, but from the opposite direction to the previous visit (see Figure 4). If more than one visit is made, any possible differences between observers should be minimised by walking different lines on two visits. However, in broken or difficult terrain it is often easier to walk the same line as on the previous visit. If necessary, put out marker posts, sticks, etc, to keep transect lines accurate in high-density areas where there are few geographical features.

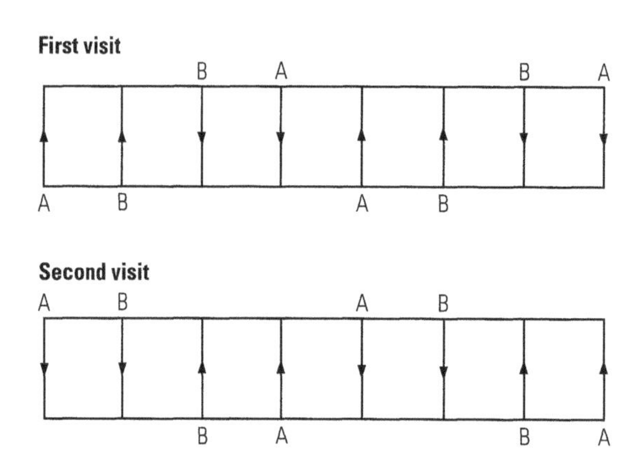

Figure 4
Diagram to illustrate the directions walked by two observers, A and B, on consecutive visits to a hypothetical study site. Any possible differences between observers could be minimised by walking different lines on two visits. However, in broken or difficult terrain it is often easier to walk the same line as on the previous visit (Reed and Fuller 1983).

On the field maps, record the start and finish time, and the weather conditions for each transect. At the end of each pair of transect lines, cross-check with your partner's records to reduce the likelihood of double-recording. A single summary map of the visit should be made as soon as possible; it should include the weather details, the times and the direction of the lines walked.

If several individuals are seen in an area, it may be difficult to determine the number of breeding pairs. Males and females can be distinguished with practice, which helps with interpretation.

The following represent breeding pairs.

Ringed plover and dunlin

(a) One bird recorded alone 50 m or more from other birds = one pair.
(b) Two individual birds within 50 m of each other = one pair.
(c) Two birds together, or two birds recorded as a pair = one pair.
(d) 3–4 birds together = two pairs.
(e) 1–4 birds flying into, out of, or through the area, or into the site = 1–2 pairs.
(f) Five or more birds remaining in the area, either on the ground or circling around (vocal birds only) = three+ pairs.

The following are *excluded* from estimates of breeding dunlin and ringed plover populations.

(g) Five or more birds in a flock on the ground without vocal registrations (assumed non-breeding).
(h) Five or more birds in a flock flying into, out of, or through the area or site.
(i) Any bird(s) which fly out of or through the site in one direction for more than 150 m without landing.
(j) Nests are not included in the estimate of breeding numbers. Some nests discovered lack adults in the immediate area, and inclusion of the nest may lead to inflated population estimates if birds already recorded elsewhere on the site are a nesting pair.

Oystercatcher

(a)–(d) As for dunlin and ringed plover, with the exception that the distance for single birds or two individual birds (rules a and b) is 125 m.
(e) 1–2 birds flying into, out of, or through the area, or into the site = one pair.
(f) Three or more birds remaining in the area either on the ground or circling around (vocal birds only) = 2+ pairs.

The following are *excluded* from estimates of breeding oystercatcher. (g)–(j) are excluded, with the exception that three or more birds (vocal or non-vocal) flying into, out of or through the site are excluded, compared with five or more birds for dunlin and ringed plover.

Redshank and lapwing

Lapwing pairs tend to form mobbing groups, and records are best made by scanning ahead so that pairs are recognised before groups form. Estimates for lapwing and redshank populations are made the same way as dunlin and ringed plover, except that the distance for single birds (rules a and b) is 75 m, rather than 50 m.

Snipe

Snipe counts derived from this count method are undoubtedly an underestimate.

For all species, if more than one visit is made to a site, then the maximum number of breeding pairs for that site should be taken from any one visit.

References

Brown, A F and Shepherd, K B (1993) A method for censusing upland breeding waders. *Bird Study* 40: 189–195.

Grant, M C, Lodge, C, Moore, N, Easton, J, Orsman, C and Smith, M (in prep) Estimating the abundance and hatching success of breeding curlew using survey data.

O'Brien, M and Smith, K W (1992) Changes in the status of waders breeding on wet lowland grassland in England and Wales between 1982 and 1989. *Bird Study* 39: 165–176.

Reed, T M and Fuller, R J (1983) Methods used to assess populations of breeding waders on machair in the Outer Hebrides. *Wader Study Group Bull.* 39: 14–16.

Dabbling and diving ducks

The similarities in behaviour, ecology and habitat preferences of dabbling and diving ducks mean that several species can be surveyed simultaneously using similar methods, even though there are differences between species in the timing of breeding, choice of habitat, etc.

In Europe, the most detailed consideration given to the survey of breeding duck populations, including validation of survey methods, has occurred in Finland (eg Koskimies and Poysa 1985, 1987, 1989, Koskimies and Vaisanen 1991, Poysa and Nummi 1992, Poysa et al 1993, Poysa 1996). Although there have been relevant studies elsewhere – eg Iceland (Gardarsson 1979), Scotland (Boyd and Campbell 1967, Newton and Campbell 1975), Czech Republic (Musil 1995, 1996) – the methods developed in Finland allow surveys which are rapid and easy, with fieldwork which is systematic and standardised.

Standing waters (eg reservoirs, lakes) require slightly different survey techniques to those needed for drier wetland habitats (eg wet grassland, marsh, fen). The method suggested here for standing waters is based on the work conducted so far in Finland; that for wet grassland, etc, is based on the RSPB's reserve monitoring programme. These methods are recommended for wigeon, gadwall, teal, pintail, garganey, shoveler, pochard and goldeneye.

Breeding season survey – population

Information required

- maximum number of males, females or pairs, alone or in groups
- map showing the boundary of the survey area.

Number and timing of visits

Three visits, about one month apart: early to mid-April, early to mid-May and early to mid-June.

Time of day

Early morning, visits to be completed by 1000 BST.

Weather constraints

Do not survey when visibility is poor, or in high winds when large expanses of water are very choppy.

Sites/areas to visit

Standing waters, usually fringed with emergent vegetation. Also wet grassland, marshes, fens, etc, occupied by breeding ducks.

Equipment

- 1:25,000 OS map of the survey area
- enlarged map showing the most important landmarks and the shape of the shoreline (optional)
- boat (optional)
- telescope (optional).

Safety reminders

Take extra care when working close to water and, if any boat trips are necessary, make sure at least two people are present and that life-jackets are worn.

Disturbance

For smaller sites with good vantage points and hides, systematic scanning of all habitat using a telescope causes less disturbance and produces less confusing results.

Methods

When surveying standing water, map the boundary of the survey area and mark on the map the survey route. Walk as close to all suitable habitat as is (safely) possible, paying particular attention to ditches, small bays and reedbed edges. Use suitable vantage points to count the diving species on open water. It may not be necessary to visit some parts of the site if they are easily observed from a distance with binoculars or a telescope. To see all of the shoreline, however, you may need to walk or row around most of the waterbody keeping close to the shore. Two observers are essential when censusing large stretches of water by boat.

When surveying grassland/marsh habitats, set and mark a transect route on a map of the site which will take you to within 100 m of every point within suitable habitat. Walk along any ditches present as these birds could go unnoticed.

In both cases, reverse the direction of the route on the second visit to avoid visiting the same part of the site at the same time of day (particularly important at large sites), and use the same route each year.

Identify the species and sex of individuals and groups. Either record observations directly on a map or cross-reference notebook records to a map. Record the birds according to the following example:

Teal ♂♀ + ♂ + ♂♀ + ♂ + 3♂♂ + 2♂♂ + ♀
(♂ = single male, 3♂♂ = group of three males, ♂♀ = pair, etc)

Include all large groups, eg 5♂♂3♀♀ (or simply write the total number of individuals), but be careful to distinguish between breeding individuals and flocks of non-breeding or late wintering individuals, eg teal, which may be present late into the breeding season.

Try to avoid overlooking or double-counting birds which have flown or swum from one place to another by writing down the direction of flight and landing place of any birds seen in flight, especially if adjacent waterbodies are counted in succession on the same day.

Interpretation of census results

Count the following as breeding pairs:

For wigeon, gadwall, teal, pintail, garganey, pochard and shoveler:

- single pair (♀♂)
- lone male (♂)
- males in groups of 2–4 (2–4 ♂♂ = 2–4 pairs)
- small male groups chasing a female (2–4 ♂♂1♀ = 2–4 pairs)
- *lone* females (♀), *if* their total number is larger than that of males (♂).

For goldeneye:

- adult male (♂)
- pair (♀♂).

Exclude groups of five or more males in the estimates of breeding pairs. Larger groups and flocks are probably non-breeding or wintering flocks.

For each visit, calculate the number of pairs for each species and use the maximum number of pairs recorded during any of the visits for year-to-year comparisons.

Breeding season survey - productivity

A quarter to a third of the UK goldeneye population nests in artificial nestboxes. Many of these sites are well-monitored by the Scottish Goldeneye Study Group. The following method will give an *index* of productivity for a range of dabbling and diving ducks.

Information required

- number of males, females or pairs, alone or in groups
- number of ducklings, broods and brood sizes (also size of young).

Number and timing of visits

Two visits, one in mid-June and one in mid-July. The first visit should, if possible, correspond with the third visit of the population survey (above).

Class I A Down-covered; 1–7 days old.

Class I B Down-covered but colour fading; 8–13 days old.

Class I C Down-covered but colour faded, body elongated; 14–18 days old.

Class II A First feathers appear, replacing down on sides and tail; 19–27 days old.

Class II B Over half of body covered with feathers; 28–36 days old.

Class II C Small amount of down remains, among feathers of back; 37–42 days old.

Class III Fully feathered but incapable of flight; 43–55 days old, flying at 56–60 days.

Adult dabbling duck.

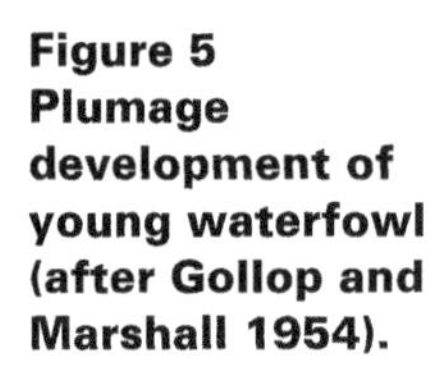

Figure 5 Plumage development of young waterfowl (after Gollop and Marshall 1954).

Time of day , Weather constraints, Sites/areas to visit, Equipment, Safety reminders, Disturbance

As for population survey (above).

Methods

Use the same survey area and survey route adopted for the population survey (above). Record all ducks seen, noting their sex and whether they were individuals or groups, as for the population survey. Record the number of lone adults, the number of young and the number of broods attended by adults. Record the age class of the young as a fraction (eg ¼, ½, ¾, etc) of the adult size. Figure 5 shows the plumage development of young waterfowl and will help with ageing. This can help distinguish between different broods on subsequent visits. Pay particular attention to places where duck broods may seek concealment. Record the additional information according to the following example:

	Teal ♀ + 8 (¼)	Female with eight quarter-grown young
or	Teal ♀ + 8 (¾)	Female with eight three-quarters-grown young

For each species, report productivity as the maximum number of young of at least ¾ adult size (about three weeks old) seen on any one visit, divided by the number of breeding pairs (from the population survey, above).

References

Boyd, H and Campbell, C R G (1967) A survey of the ducks breeding at Loch Leven. *Ann. Rep. Wildfowl Trust* 18: 36–42.

Gardarsson, A (1979) Waterfowl populations of Lake Myvatn and recent changes in numbers and food habits. *Oikos* 32: 250–270.

Gollop, J B and Marshall, W H (1954) *A Guide for Ageing Duck Broods in the Field*. Miss. Flyway Coun. Tech. Sect. mimeo.

Koskimies, P and Poysa, H (1985) Monitoring waterfowl populations in Finland: methodological remarks. *Lintumies* 20: 270–279.

Koskimies, P and Poysa, H (1987) Vesilinnuston seuranta ja laskentamenetelmat. *Suomen Riista* 34: 31–41.

Koskimies, P and Poysa, H (1989) Waterfowl censusing in environmental monitoring: a comparison between point and round counts. *Ann. Zool. Fennici* 26: 201–206.

Koskimies, P and Vaisanen, R A, eds (1991) *Monitoring Bird Populations*. Zool. Mus., Finnish Mus. Nat. Hist., Univ. Helsinki, Helsinki.

Musil, P (1995) Five years of monitoring water birds breeding populations by the two-check method in Czechoslovakia (1988–1992). *Proc. 12th Int. Conf. IBCC & EOAC, 'Bird Numbers', Noordwijkerhout, 14–19 September 1992*, pp. 219–226.

Musil, P (1996) Monitoring water birds breeding population in 1995. *Zpravy CSO* 42: 12–18.

Newton, I and Campbell, C R G (1975) Breeding of ducks at Loch Leven, Kinross. *Wildfowl* 26: 83–103.

Poysa, H (1996) Population estimates and the timing of waterfowl censuses. *Ornis Fennica* 73: 60–68.

Poysa, H, Lammi, E, Vaisanen, R A and Wikman, M (1993) Monitoring of waterbirds in the breeding season: the programme used in Finland since 1986–92. Pp 7–12 in Moser, M, Prentice, R C and van Vessem, J, eds. *Waterfowl and Wetland Conservation in the 1990s: a Global Perspective*. IWRB Spec. Publ. 26. Slimbridge.

Poysa, H and Nummi, P (1992) Comparing two methods of data collection in waterfowl habitat use studies. *Bird Study* 39: 124–131.

Gull populations

The following five methods, all taken from the *Seabird Monitoring Handbook*, are suitable for common, lesser black-backed, herring and great black-backed gulls. All of these species present survey problems because of the wide variation within species in nesting density and colony size, and because of the species' extended breeding seasons. In addition, it is not possible to distinguish accurately between the nests of herring and lesser black-backed gulls. For methods based on direct observation (especially Method 1), separating the two species is generally not difficult. For Methods 2 and 3, the number of active nests belonging to each species is estimated by determining the ratio of adult herring to lesser black-backed gulls and partitioning the nest-count using the same ratio, although the error involved in this method may be considerable. To do this, walk through all parts of the colony (as breeding pairs of the two species may be clumped) and note the numbers of birds of each species seen. Alternatively, view from a distance, provided most of the colony is visible.

This problem may also occur with black-headed gulls *Larus ridibundus* and common gulls, and a similar approach can be taken. Eggs of these species are, with experience, somewhat easier to distinguish from each other than are those of herring and lesser black-backed gulls, although there is some overlap in colour, weight and measurements. Assessing the species proportions from adult birds is probably more practical than checking a sample of clutches in detail (which poses sampling problems if species are unevenly mixed within the colony).

The recommended census unit is the Apparently Occupied Nest (AON) which is defined as a well-constructed nest attended by an adult and capable of holding eggs or, if it is not possible to see the nest (for example if it is obscured by vegetation) an apparently incubating adult. Some count methods use slight variations on this, for example active nests containing eggs or with other signs of use, counted during transect or quadrat surveys, when it is not possible to record attendance by adults. Others use 'apparently occupied territories' (AOTs), based on the spacing of birds or pairs viewed from a vantage point, where actual nests or incubation cannot be discerned. Counts of individual adults may also prove necessary on occasion.

1. Counts from vantage points

Information required

- maximum number of Apparently Occupied Nests (AONs) or Apparently Occupied Territories (AOTs)
- map of the survey area with the boundaries and quadrats/transects marked.

Number and timing of visits

At least one but preferably more, each separated by a few days. Late May to early June is ideal, but counts may also be made later in June.

Time of day

Between 0900 and 1600 BST.

Weather constraints

Avoid counting in heavy rain, fog or high winds.

Sites/areas to visit

Any colony not already covered.

Equipment

- 1:10,000 OS map of the area
- telescope (optional).

Safety reminders

When working alone in remote areas ensure someone knows where you are going and when you will return. When working from a boat, ensure a second person is always present; experience of the local sea conditions and boat handling skills are essential. Rocky shores are always hazardous because of their uneven, slippery surfaces, weed-covered rocks and fissures. On cliffs and crags beware of slippery vegetation at cliff edges and of undercutting or loose strata.

Disturbance

When counting from vantage points ensure that the count position is not so close as to cause disturbance.

Methods

This method involves observation of the colony from one or more vantage points. It is most suitable for colonies on cliffs and rocky islets visible from the cliff-top and for small ground-nesting colonies which can be viewed well from a distance (ie without many nests likely to be hidden by tall vegetation or undulations in terrain). In general, this method also applies to counts of roof-nesting gulls in urban or industrial situations, where access to a large number of vantage points (eg tall buildings) may be required for good coverage.

Count AONs, including nests which are obscured by vegetation but where sitting birds are visible. Keep a separate note of the number of these assumed incubating birds. Beware of counting both members of a pair sitting in close proximity as two AONs.

Where observations are made from more than one vantage point, avoid counting the same nests twice by noting their positions in relation to natural features within the colony. Sketches or photographs can also be useful for this, especially where boundaries are poorly defined.

Counts made later in June are also useful but more difficult, as many chicks will have hatched and careful scanning will be needed to locate nests and chicks. Most nests will have an adult reasonably close by, but some adults may have left the colony after failing to nest. If you have difficulty locating nests, or lack time for detailed scans, keep a note of numbers of adults on the ground, on cliffs, or flying agitatedly over the colony. If nests without incubating adults are hidden by vegetation but standing adults are visible, count them and estimate the number of AOTs additional to AONs. Quote this as a minimum–maximum range (to allow for members of the same pair standing a distance apart); the errors involved in such figures are likely to be high.

If several counts are made, use the maximum number of AONs counted on a single visit as the population estimate, but report all counts. Report AOTs separately.

If parts of the colony are hidden, estimate (minimum–maximum) the likely number of nests hidden, based on nest density in the rest of the colony and the area involved; append the estimate to your actual count.

2. Ground-nesting gulls with few vantage points

Information required
- number of active nests per random quadrat
- number and size of quadrats
- total area of the colony
- map of the survey area boundaries with the locations of quadrats marked.

Number and timing of visits
A minimum of one visit before egg-laying and another during late incubation. Late April or early May to July.

Time of day, Weather constraints, Sites/areas to visit, Safety reminders
As for Method 1.

Equipment
- 1:10,000 OS map of the area
- stakes
- 9.77-m rope fixed to a pole
- random number generator or tables.

Disturbance
This method entails considerable disturbance of birds in survey plots. It may therefore not be appropriate where nest predation risk is high (almost always the case in gull colonies) and if gulls are of particularly high conservation value at the colony. If you use this method, arrange the order in which quadrats or transects are surveyed so that individual birds are not disturbed for longer than 30 minutes at a time. Counts of large areas will need a team of several people.

Methods
This method is suitable for colonies that can safely be covered on foot. It is accurate but entails considerable disturbance to birds in survey plots. For small numbers of great black-backed gulls nesting among other gull species, it may be better to attempt to map individual nest-sites or territories throughout the colony, preferably from vantage points.

Count active nests (equivalent, in this case, to an AON), defined as a fully constructed nest containing eggs or chicks, or with signs of recent use such as fresh soil or nest material.

Early in the breeding season, mark the boundaries of the colony on a map. For small colonies, it may be possible to survey all nests. Otherwise, superimpose a grid on this map, and randomly select 30 or more of the grid intersections; these will form the centres of quadrats. Further explanation of sampling is given in the *Introduction*.

Locate quadrat centres by reference to natural features, using triangulation and/or angles and distances where necessary. Mark these points with stakes (permanently if repeat counts are planned). This should preferably be done before egg-laying starts.

Quadrats can be of any shape (provided they can be easily defined and are all of equal size), but should be fairly large because of the relatively low nest density typical of gulls. Circular quadrats of 300 m^2 are generally suitable for the smaller species (provided nest density is fairly high), which can conveniently be measured using a rope 9.77 m long fixed to a pole at the centre of the quadrat. Larger quadrats (up to 5,000 m^2) may be necessary for the larger species, especially where nesting density is low, as at some inland colonies on moors. Square or rectangular quadrats marked by stakes are more convenient for these larger sizes.

Visit the quadrats during the late incubation period (usually late May to early June). Count the number of active nests in each quadrat and record the clutch size (to assess the stage of breeding; many small clutches will suggest the count date is too early). Be sure to cover the ground carefully, as nests are easily overlooked; even experienced teams can miss 5–20% of nests in dense vegetation, where special care is needed (Ferns and Mudge 1981).

If more time is available, count all active nests in the quadrats every few days in the laying period. In estimating colony size use the maximum number of active nests recorded in each quadrat on any one visit.

Estimate the total colony size (total number of active nests) as:

(Sum of maximum number of nests from each quadrat/No. of quadrats) × (Total area of colony/Area of quadrat)

Recounts of the same quadrats in future years allow comparison with this figure, but it will be insensitive to changes in the area of the colony. Preferably, re-survey the colony extent each count-year and re-establish a different random sample of quadrats.

The error associated with this estimate can be calculated in much the same way as given for puffin Method 1 (p 286).

3. Transect counts

Information required

- number of active nests marked
- proportion of all active nests that were marked
- an estimate of the total number of active nests in each area and in the colony as a whole
- map of the survey area with the boundaries and any sample areas marked.

Number and timing of visits

Either every few days during the laying period in late April, and then a single whole-colony count when laying is completed (this may take more than one visit) in early to mid-May, *or* a single whole-colony count in the last week of May.

Time of day, Weather constraints, Sites/areas to visit, Safety reminders

As for Method 1.

Equipment

- 1:10,000 OS map of the area
- spray paint (not red) or bamboo canes
- stakes or rope to mark transects.

Disturbance

This method entails considerable disturbance to birds. Thus it may not be appropriate where nest predation risk is high (almost always the case in gull colonies) and if gulls are of particularly high conservation value at the colony. If you use this method, arrange the order in which transects are surveyed so that individual birds are not disturbed for longer than 30 minutes at a time. Counts of large areas will need a team of several people.

Methods

This method is based on Wanless and Harris (1984) and is generally used to provide counts of entire colonies, ie it is not based on sampling. Modifications of this method are given in the *Seabird Monitoring Handbook*. Like Method 2, it is suitable for colonies which can be safely covered on foot. If the colony is large, and the time or the number of fieldworkers available is limited, it may be more sensible to use sample quadrat counts (Method 2).

The count unit is the active nest (equivalent to an AON), defined as a fully constructed nest containing eggs and/or chicks (in or near the nest), or empty but judged capable of holding a clutch (ie well constructed). NB This is slightly different to the definition given in Method 2.

Determine the pattern of laying by counting complete nests and clutches in sample areas (preferably randomly selected) every few days and make the complete colony census when laying is completed – *or* simply undertake the complete census in the last week of May.

Small colonies can be dealt with as a whole; large colonies should be divided into a number of areas using unambiguous landscape features (or, if necessary, rope boundary markers). The entire colony (where small) or each sample area should ideally be covered by a number of observers, each standing (initially) 10 m apart. The counters should then move in parallel through the colony, zigzagging in order to cover the whole area; the whole width of the 'strip' along which they walk should be about 10 m. They should mark, count and note the contents of every active nest that falls within the walked transect. Mark the nests either by spraying a little paint on the side of the nest (avoid red paint or spraying the eggs), or with bamboo canes. If canes are used, count them before you start and subtract the number of canes left over at the end to arrive at your transect totals. Calculate the total number of marked active nests in the area/colony by summing the results of the individual transects.

Estimate the proportion of nests that were marked by getting one or more observers (or even better, someone who had not taken part in the count) to count all the active nests in each area. This is best done by walking back and forth across the area at 90° to the route taken during the original count. Repeat this procedure for each area.

Calculate the number of active nests in each area as:
no. of active nests marked × (total no. of active nests on recount/no. of marked nests on recount)

The total population is then the sum of the active nests in each area.

Modifications of this method are given in the *Seabird Monitoring Handbook*.

4. Flush counts

Information required

- number of individual adults flushed into the air during late incubation.

Number and timing of visits

At least one visit between early May and July, but ideally late May to late June.

Time of day, Weather constraints, Sites/areas to visit, Safety reminders

As for Method 1.

Equipment

- 1:10,000 OS map of the area.

Disturbance

This method involves flushing breeding birds and should only be used where the terrain is very difficult or time is limited. It is only suitable for small colonies where all the birds can be flushed at once. Considerable disturbance is caused, and, wherever possible, it is recommended that Methods 1–3 are used instead.

Methods

This method is rapid, but the count obtained only provides an approximate guide to colony size. It is recommended that Methods 1–3 are used wherever possible.

Count individual adult birds during the period from early incubation until early fledging, although the best time is during late incubation (usually late May/early June).

Before disturbing the gulls, count or estimate the number of adults visible on the ground and in the air. Then flush adults from the colony and count them several times as they fly above the colony, separating the species as necessary. It is best if birds can be flushed from a distance using a horn or other loud noise but, in some cases, it may be necessary to approach the colony more closely. Standing in a prominent position overlooking the colony, or suddenly appearing above the skyline, will often flush most of the adults, while a sudden movement or sound may help flush those remaining.

Record the number of individuals counted on each visit and, if more than one visit is made, report the mean (with standard errror) of these values.

5. Counts from photographs

Information required

- estimated number of apparently incubating birds
- the number of likely additional territories
- the method of counting
- map of the survey area with the boundaries marked.

Number and timing of visits

At least one visit, late May to late June.

Time of day, Weather constraints, Sites/areas to visit, Safety reminders

As for Method 1.

Equipment

- 1:10,000 OS map of the area
- camera with black-and-white film
- projector and slide film (optional)
- aircraft (optional).

Disturbance

When counting from vantage points, ensure the count position is not so close as to cause disturbance.

Methods

Good-quality photographs taken from suitable vantage points (or from an aircraft) are potentially useful for counts of large colonies of ground-nesting gulls which are viewable from a distance but are not accessible on foot or sub-dividable for direct field counts. See Method 1 for appropriate timing of counts.

Using projected slides, or enlarged prints, it may be possible to distinguish and mark the positions of incubating gulls, although other occupied nests are unlikely to be discernible. At best, all that can be achieved is a count of apparently incubating birds plus likely additional territories. Photographic counts are most suited to single-species colonies because of problems of species identification.

Direct counts from the air, combined with photography, may be appropriate in some cases, such as inland areas with large gull populations scattered in many colonies (eg Bourne and Harris 1979). In such cases, species separation is more straightforward than from photos alone, although distinguishing AONs or apparent territories will still be difficult.

Aerial or photographic counts will, in general, be of little use for detailed population monitoring. However, they have potential for use in large-scale, basic surveys of inland gulls, and in identifying colony locations for further survey, particularly in remote areas.

References

Bourne, W R P and Harris, M P (1979) Birds of the Hebrides: seabirds. *Proc. Royal Soc. Edinburgh* 77B: 445–475.

Ferns, P N and Mudge, G P (1981) Accuracy of nest counts at a mixed colony of herring and lesser black-backed gulls. *Bird Study* 28: 224–246.

Wanless, S and Harris, M P (1984) Effect of date on counts of nests of herring and lesser black-backed gulls. *Ornis Scandinavica* 15: 89–94.

Gull productivity

These five methods are taken from the *Seabird Monitoring Handbook* which highlights the fact that the major practical difficulty in productivity monitoring is chick mobility. The standard method of controlling for this is to select naturally isolated small patches of breeding habitat (eg islets) or to restrict the movements of chicks from a number of nests by enclosing small areas of the colony with low fences. Alternatively, ring/recapture methods can be used to estimate the total number of young produced.

Methods 1–2, variants of a relatively simple capture/recapture approach, or *Method 3*, based on fencing plots within the colony, are recommended. Raw counts of ringed chicks (*Method 4*) provide some information but are crude. For groups of birds nesting in inaccessible locations such as cliffs or stacks, a direct observational method (*Method 5*) may be necessary. All these methods involve some degree of inaccuracy. In particular, the assumptions involved in making capture/recapture estimates mean that results should be treated with particular caution. It is often difficult to separate chicks and fledglings of lesser black-backed and herring gulls in mixed colonies of these species. The *Seabird Monitoring Handbook* gives details of how to do this.

1. Capture/recapture of large chicks

Information required

- total estimated number of young fledged.

Number and timing of visits

Two visits two weeks apart, the first one week before the first chicks fledge.

The dates may change slightly from year to year, but can be worked out from population count observations (see *Gull populations*, above) and from incubation and fledging times as given in Table 1.

Table 1
Incubation and fledging periods for gulls; overall ranges in parentheses (taken from the *Seabird Monitoring Handbook*, after *BWP*).

	Incubation period (days)	Fledging period (days)
Common gull	(22–)24–27(–28)	c. 35
Lesser black-backed gull	24–27	30–40
Herring gull	(26–)28–30(–32)	35–40
Great black-backed gull	27–28	7–8 weeks

Time of day

Between 0800 and 1600 BST.

Weather constraints

Do not visit any colony if the weather is foggy, wet or windy.

Sites/areas to visit

Any colony not already covered.

Equipment

- 1:10,000 OS map of the site
- ringing licence
- ringing equipment.

Safety reminders

See *Gull populations* (above). Also be aware of the signs of Lyme's disease which can be carried by seabird ticks.

Disturbance

See *Gull populations*. Observers should not remain in a colony for more than half an hour at a time to prevent eggs or small chicks chilling.

Methods

This method estimates the number of chicks that fledge from the colony. To express productivity as fledglings/breeding pair, an earlier count of apparently occupied or active nests will need to have been undertaken (see *Gull populations*).

If possible, cover the whole colony as a unit. If this is not practical, divide larger colonies into areas separated, where possible, by stretches of unused ground or natural barriers so that movement of chicks between areas is limited.

To obtain an accurate estimate, visit on two dates, a week or more apart, during the chick-rearing period, with two ringing 'runs' on each date. On the first visit, move through the colony ringing chicks of at least two weeks old (common gulls) or at least three weeks old (larger gulls), about a week before the first birds are due to fledge.

Later the same day, repeat the ringing exercise, aiming to capture all chicks of the requisite age; keep a note of how many recaptured birds are already ringed, and ring any 'new' birds. At this stage ignore any chicks below ringing age (these will be dealt with on the next date).

After about two weeks, repeat the above ring/recapture procedure for chicks which would have been less than two weeks old (common gulls) or three weeks old (larger species) at the time of the first count. This will allow estimation of the numbers of large chicks which have appeared since the first date. Do not handle any chicks which look as if they might have been of ringable age during the first count as, in theory, they will have been ringed on the first visit. Keep a note of any ringed chicks which have died since the first date.

On each ringing run (both visits) it is important that chicks are encountered at random. Do not select ringed chicks (rather than unringed chicks) for recapture. It is also important that all parts of the colony are sampled evenly on both runs on each visit. Thus it is better to move smoothly through a colony knowingly missing some chicks in all parts of it, rather than cover some areas very effectively and others not at all. A useful method for achieving this is to zigzag through the colony in a haphazard fashion. Even coverage is easy to accomplish in small colonies but requires a conscious effort in large ones.

In some seasons there may be a substantial number of chicks too young to ring at the second visit. If so, a third visit may be necessary.

The number of 'ringable' chicks present on the first visit is estimated as follows:

Total no. ringed on first run × (No. of birds handled on recapture run/ No. of these that were already ringed)

For example, if 90 chicks were ringed on the first run, and 100 large chicks, including 50 already ringed, were recaptured on the second run, the total number of large chicks present is estimated as 90 × (100/50) = 180 chicks (of which 140 were actually ringed over the two runs).

The number of ringable chicks present on the second date is calculated the same way, but is based only on chicks that were not of ringable age on the first date.

If some of the chicks ringed on the first visit were found dead on the second, correct for this mortality according to the following example. If 10 ringed chicks were found dead (of 140 ringed), the estimated number of large chicks which died = 180 × (10/140) = 13. The number of large chicks surviving from the first date is thus estimated as 180 minus 13 = c. 167 chicks.

The total estimated number of chicks fledged is the sum of the calculated number of chicks from the first and second visits.

If resources are lacking it is possible to obtain a crude estimate of the number of chicks fledged by following the same procedure but only on one visit about a week before the first chicks fledge. To do this, ring all chicks of ringable age encountered and make a note of the ratio of small (not ringable) to large (ringable) chicks. Make a capture–recapture estimate as outlined above and correct as per the following example:

If the capture–recapture estimate is of 150 'ringable' chicks and a ratio of 20 small to 80 large chicks is recorded on the first run, the total estimate for chicks present is 150 + [150 + (20/80)] = 187 chicks. Because many of the small, unringed chicks are likely to die before fledging (as are many of the chicks large enough to be ringed), this estimate of chick production may be considerably higher than the actual number of chicks which fledge.

2. Assessing the ratio of ringed to unringed fledglings

Information required

- estimated number of fledged young.

Number and timing of visits

Every few days from the first date of ringable chicks (two weeks old for common gulls and three weeks old for larger gulls) to about a week after seeing the first fledged young followed by one additional visit.

The dates may change slightly from year to year, but can be worked out from population count observations (see *Gull populations*, above) and from the incubation and fledging times given in Table 1 (above).

Time of day

Between 0800 and 1600 BST.

Weather constraints, Sites/areas to visit, Safety reminders, Disturbance

As for Method 1.

Equipment

- 1:10,000 OS map of the site
- ringing licence
- ringing equipment
- telescope.

Methods

This method estimates the number of chicks fledged from the colony. To express productivity as fledglings/breeding pair, an earlier count of apparently occupied or active nests will need to have been undertaken (see *Gull populations*, above). This method is not suitable for mixed colonies of herring and lesser black-backed gulls unless observers are experienced at distinguishing between their very similar plumages. See the *Seabird Monitoring Handbook* for notes on separating these species.

As the method involves checking fledglings which wander widely, the whole colony must be treated as a single unit.

Ring as many large chicks as possible from the date ringable chicks first appear until about one week after the first flying young are seen. Although some small chicks may be ringable, they are less likely to survive to fledging, so only ring chicks that are at least two weeks old (common gulls) or at least three weeks old (larger gulls). Keep a note of any ringed birds subsequently found dead in the colony.

Re-check the entire colony after the majority of chicks have fledged, and check fledglings and near-fledglings for rings, using telescopes and binoculars. Note the relative numbers of ringed and unringed birds. A total count of fledglings is not needed, but you should attempt to check as much of the colony and as many birds as possible.

Where movements of fledged birds to or from other colonies is not a problem (eg where a colony is well separated from others and it is not too late in the season), fledglings present on tidal rocks, on the sea close inshore, or in other habitats close to actual breeding areas can be included to improve sample sizes. If there is any doubt, however, confine checks of fledged birds to those present within known breeding areas.

The number of fledged young is estimated as:

(No. of chicks ringed – No. of ringed chicks known dead) × (Total of fledglings or near-fledglings checked/No. of these bearing rings)

For example, if 500 chicks have been ringed, of which 20 are known to have died before fledging, 480 ringed chicks are estimated to have survived; if only half of the fledglings checked bear rings, then the estimate of surviving chicks can be doubled, to 960.

3. Use of enclosures or other confined plots

Information required

- mean number of chicks fledged per breeding pair.

Number and timing of visits

Every few days, throughout incubation and from before the first chicks fledge until most have fledged.

The dates may change slightly from year to year, but can be worked out from population count observations (see *Gull populations*, above) and from incubation and fledging times in Table 1 (above).

Time of day, Weather constraints, Sites/areas to visit, Safety reminders

As for Method 1.

Equipment

- 1:10,000 OS map of the site
- ringing licence
- ringing equipment
- stakes
- chicken wire or (preferably) plastic-coated weld-mesh about 35 cm high.

Disturbance

See *Gull populations* (above). To prevent eggs or small chicks chilling, observers should not remain in a colony for more than half an hour at a time. Construct and erect the enclosures before the breeding season starts in order to minimise disturbance.

Methods

The method involves monitoring the survival of young in accessible, small, discrete patches of breeding habitat from which chicks cannot disperse while flightless (eg artificial or natural islets), or in enclosures where chick dispersal is prevented. Direct counts of unfenced plots can considerably underestimate the true chick population.

Ideally, enclosure sites should be selected at random using a grid superimposed on a map of the colony. Often, however, logistical restrictions or the convenience of using naturally isolated units result in haphazard selection of sites. If this is the case, bear in mind that nesting success often varies between different parts of the colony due to factors such as density, exposure, and the age of the subcolony.

Most studies have for logistical reasons used a small number (1–5) of relatively large enclosures, each enclosing up to 40 nests. A larger number of smaller enclosures is preferable on statistical grounds, but may result in less representative data if predation occurs within the colony. Enclosed plots should not be more or less susceptible to predation than unenclosed sections of the colony. This can pose methodological problems at colonies where mammalian predators (eg rats *Rattus*) are active. Enclosures should be constructed using stakes and chicken wire or (preferably) plastic-coated weld-mesh about 35 cm high. Care should be taken to ensure that chicks will be unable to force their way under or through seams. If possible, construct the enclosures before the breeding season, to minimise disturbance.

Assess the average productivity of the pairs in each enclosure by counting the number of Apparently Occupied Nests (AONs) during incubation, noting any additions during the season (if nest positions have been numbered or mapped), and counting chicks just before the first ones fledge. If chicks are ringed or individually marked, and fairly frequent visits are made, it should be possible to follow each individual to fledging. For each enclosure, productivity is expressed as the number of chicks fledged divided by the total number of breeding pairs.

Overall productivity can be expressed as the mean (with standard error) of the figures from individual enclosures.

4. Chick ringing totals

Information required

- number of chicks ringed
- number of AONs in the ringed area.

Number and timing of visits

Preferably several, during the chick-rearing period. Otherwise, a single visit before the first chicks fledge.

The dates may change slightly from year to year, but can be worked out from population count observations (see *Gull populations*, above) and from the incubation and fledging times given in Table 1 (above).

Time of day, Weather constraints, Sites/areas to visit, Safety reminders, Disturbance

As for Method 1.

Equipment

- 1:10,000 OS map of the site
- ringing licence
- ringing equipment.

Methods

This is a simple comparison between the number of chicks ringed and the number of apparently occupied or active nests counted. Errors may be large, as many chicks will be missed and some ringed chicks will die, so productivity may be substantially overestimated in a poor season or underestimated in a successful season. Accuracy improves with increasing effort, and Method 2, which requires minimal additional effort, should be used instead when possible.

Ring over the entire colony. If this is not practical, average productivity for the colony can be estimated based on defined areas (preferably selected randomly).

Attempt to ring every chick. Several visits over the course of the season are greatly preferable, but a single visit just before the first chicks fledge is better than nothing. Keep a note of the number of chicks too small to ring, and of any eggs still present.

Productivity is expressed as the number of chicks ringed divided by the number of pairs breeding in the ringed area. Any information to suggest that the productivity figure is an overestimate (eg casual reports of many dead chicks later in the season) should be noted; equally, in a good season, the number of chicks ringed may underestimate numbers fledged.

5. Observation of mapped nests

Information required
- number of AONs visible from a suitable vantage point
- number of large chicks, number of small chicks and number of incubating adults from those nests seen a few days prior to fledging of the first chicks.

Number and timing of visits

Preferably weekly or fortnightly, from the start of incubation until the first chicks are about to fledge.

The dates may change slightly from year to year, but can be worked out from population count observations (see *Gull populations*, above) and from the incubation and fledging times given in Table 1 (above).

Time of day

Between 0800 and 1600 BST.

Weather constraints, Sites/areas to visit, Safety reminders, Disturbance

As for Method 1.

Equipment

- 1:10,000 OS map of the site
- telescope.

Methods

This method may be used for groups of gulls (usually herring gulls) nesting on cliffs, or in other inaccessible locations (eg offshore stacks) where nests can be observed directly but not visited. It is based on the methods used for kittiwake. Accuracy is limited if visits are brief or few, but can be improved if there is time to sit and watch the colony at weekly or two-weekly intervals.

Locate nests from vantage points during the incubation period. Mark their positions on sketch maps or photographs.

A few days before you estimate the first chicks are due to fledge, count all the visible large chicks (three weeks old or more for the three large gulls). Note the number of small chicks separately and any nests with apparently incubating adults.

Estimate productivity as the number of large chicks divided by the number of nests. Also report the number of small chicks and incubating adults (preferably re-check these on later dates).

Tern populations

These three methods – taken from the *Seabird Monitoring Handbook* – are appropriate for sandwich, roseate, arctic and little terns (also for common tern *Sterna hirundo*).

Breeding terns can be very mobile in comparison to other seabirds. In some cases, whole colonies may shift location from year to year, or a large proportion of one colony may move to a different colony (not always nearby) in a different year. Pairs which fail at one colony early in the breeding season may even move to a different colony in the same season. This makes it particularly important to try to fill any gaps between currently monitored colonies.

Method 1 should always be used where possible. It can be used at colonies where 80% or more of the occupied area can be viewed from vantage points; it involves no disturbance and produces reasonable estimates. If it is not possible to use Method 1, use *Method 2*, which is a useful method for colonies where it is only possible to see up to 80% of the occupied area from suitable vantage points. If this method cannot be used either, *Method 3* provides a quick estimate of the number of adults present; it is designed for use particularly in colonies where counts of apparently incubating adults or active clutches are difficult.

1. Counts of apparently incubating adults

Information required

- maximum number of Apparently Incubating Adults (AIAs).

Number and timing of visits

One to three visits (one week apart), between mid-May and late June. At least one should be made during the late incubation period, approximately 3½ weeks after the first incubating bird is seen or the first egg is laid (usually early June).

Time of day

Any time.

Weather constraints

Avoid cold, wet and windy days.

Sites/areas to visit

Any colonies not currently being monitored.

Equipment

- nest markers, eg pasta shells.

Safety reminders

If working alone, always ensure someone knows where you have gone and when you intend to return. If cliffs or steep slopes are to be climbed to gain access to beaches, use the correct safety equipment and do not work alone. If working with boats, never work alone. The boat should be operated by an experienced and trained boat handler and life-jackets should be worn at all times. Take the necessary equipment to deal with any emergency. Be aware of the signs of Lyme's disease which can be caught from seabird ticks.

Disturbance

All tern colonies are very sensitive to disturbance so try to keep this to a minimum. Terns should never be flushed from the nest in rain or strong winds. When counting nests, visits should not be made to colonies in poor weather or in very hot conditions. To minimise disturbance it may be best to make several short visits rather than one long one. As a rule, not more than 20 minutes should be spent in a colony. Care should be taken that predators such as gulls and skuas do not take advantage of your presence to rob unguarded nests. If in doubt, leave the colony and return at a later date.

Methods

Apparently Incubating Adults (AIAs) can be counted in colonies where most (not less than 80%) of occupied areas can be viewed from suitable vantage points. When choosing vantage points, it is important to allow for vegetation growth as the season progresses. Large and complex colonies will require more than one vantage point. Ensure that parts of the colony are not double-counted or missed. Use physical features, if possible, to divide the colony into subsections to be counted from each vantage point. Mark these subdivisions and vantage points on a sketch map of the colony. Large colonies which cannot be counted in one day should be covered on consecutive days (equivalent to one visit), in order to avoid double-counting failed breeders making a second attempt in a different part of the colony.

Count the number of birds that appear to be incubating a clutch of eggs, whether or not nest material is visible. Incubating birds can be distinguished from off-duty resting birds by their posture: a bird incubating will be sitting in a hollow or scrape and thus be partly hidden with its tail pointing up at a sharp angle; resting birds which are not incubating are more visible and their tail is held at a shallower angle, although this might not be so where birds are sitting in slightly undulating terrain. Where nests are spaced widely enough it is usually possible to distinguish members of the same pair (one sitting, one standing nearby). Using a sketch map, record all nests or clutches seen.

Attempt to estimate the number of nests in any parts of the colony that cannot be seen from the vantage points (minimum and maximum) from the density of nests in the visible part of the colony. If more than 20% of the colony cannot be observed from vantage points, use Methods 2 or 3 (see below).

Keep a note of the counts made on each date and report the maximum count of AIAs for the whole colony on any one visit. If more than one visit has been made, do not sum the maximum counts for each subsection from different visits, as birds which move after a failed breeding attempt might be counted twice.

2. Counts of apparently occupied nests, with eggs or nest material

Information required

- maximum count of 'active' nests.

Number and timing of visits

One to three visits (one week apart), between mid-May and late June. At least one should be made during the late incubation period, approximately 3½ weeks after the first incubating bird is seen or the first egg is laid (usually early June).

Time of day, Weather constraints, Sites/areas to visit, Equipment, Safety reminders, Disturbance

As for Method 1.

Methods

Active nest counts should be used for colonies where more than 20% of the occupied area is not visible from suitable vantage points. Colonies should be small enough, or divisible into small enough sections, that a complete 'ground' survey can be made without prolonged disturbance.

Determine the timing of the breeding season by casual observations to record the first incubating birds or eggs laid. Counts should be made late in the incubation period about 3½ weeks after the first egg is seen in the colony. Counts in early June are usually suitable if the timing of the season is not known in detail. Keep a note of obvious empty nest-scrapes. If there are many empty scrapes or single eggs it might be more appropriate to count a week later. Record any signs of predation as this will also produce a high number of empty scrapes.

For large colonies which cannot be subdivided and visited on consecutive days, it will be necessary to use several observers walking in a line through the colony. Count accuracy will be improved by the use of markers (these do not have to be numbered). These should be inconspicuous to avoid attracting predators, but should be robust enough to survive a few weeks (eg small stakes). Observers should be close enough to each other to ensure that few, if any, nests are missed. If possible, a second count should be made a week later and a third count two weeks later. The cumulative number of clutches recorded over three dates will provide an alternative measure of the population.

Correct for count efficiency on each visit, especially if the terrain is complex or if more than one observer is used. To do this, walk through the colony on the same day but in a different direction to that used in the initial count and record the ratio of marked to unmarked nests.

Correct for count efficiency as in the following example:

250 nests initially marked
230 marked nests and 15 unmarked nests found on re-check carried out on the same date
Corrected count = (245/230) × 250 = 266 clutches

A rapid, accurate count of active nests should be the priority, but, if possible, record the clutch size on each visit. Keep a note of any obviously deserted eggs, ie those which are displaced, broken, excessively dirty and/or coated in droppings. If in doubt, check whether the eggs are cold.

Report results as:

- maximum count of active clutches on any one visit (corrected for count efficiency where possible)
- cumulative total of active clutches recorded over all count dates
- counts of active clutches (both corrected and uncorrected, ie raw data) on each date
- additional numbers of empty nests (with material) recorded on each date.

Comparisons between years or colonies are best based on peak counts. Although cumulative totals allow for the spread of breeding, they may also include repeat clutches by failed pairs in new scrapes. This means they are less suitable for comparisons because of double-counting.

3. Flush counts of individual adults

Information required

- maximum mean flush count.

Number and timing of visits

One to three visits (one week apart) during the last two weeks of incubation and first week after hatching (usually early June).

Time of day

Preferably 1000–1200 BST, and not outside the period 0800–1600 BST.

Weather constraints, Sites/areas to visit, Equipment, Safety reminders, Disturbance

As for Method 1.

Methods

Flush counts are the quickest but least accurate way of counting incubating adults or nests. Flush counts should, if possible, be made on three dates, one week apart, during the last two weeks of incubation and first week after hatching. If the timing of the breeding season is not known and/or only one visit is possible the count should be made in early June. On each date a minimum estimate of the number of birds present should be obtained by counting the number of birds visible on the ground and in the air before flushing. Also count birds resting on the edge of the colony before flushing birds from the nesting area.

Flush the birds from some distance away using a loud noise such as a horn. It might be necessary to get closer for large colonies.

Flush counts provide two different population estimates depending on the time of day, so always record the time. Counts between 1000 and 1200 BST provide a good estimate of the number of breeding adults as most non-breeders are absent at this time. If it is impossible to count at this time then ensure that counts are at least confined to the period 0800–1600 BST, as most non-breeders are present just before dusk.

Count the flushed adults several times as they wheel around in a tight flock over the colony (little terns flock less tightly than other species). At colonies with more than one tern species, flush counts are more difficult, particularly when common and arctic terns are present. Where possible, accurate counts should be made of each species. In mixed colonies flushed flocks may not be single-species, and at large colonies it may be impractical to count each species accurately. Where this is the case, flush count the whole colony and obtain population figures by estimating the proportions of each species from sample counts.

The flush count method is quick, quantitative and repeatable. Without calibration with nest counts it may give inaccurate estimates of breeding numbers, but these estimates will still be useful in assessing changes in numbers from year to year. In Orkney and Shetland it has been shown that the number of flushed birds was equivalent to about 1.5 times the

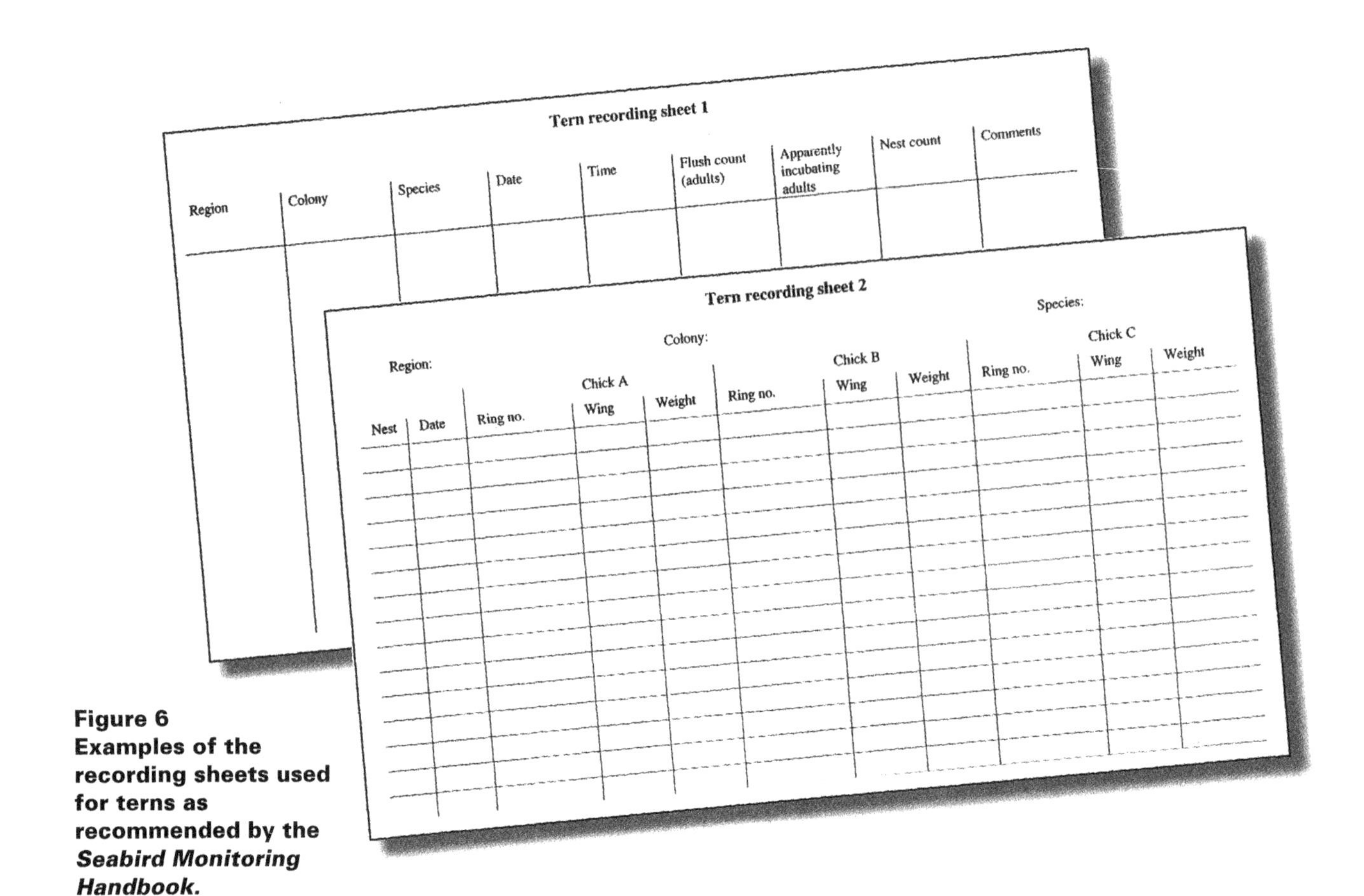

Tern recording sheet 1

Region	Colony	Species	Date	Time	Flush count (adults)	Apparently incubating adults	Nest count	Comments

Tern recording sheet 2

Region: Colony: Species:

Nest	Date	Chick A			Chick B			Chick C		
		Ring no.	Wing	Weight	Ring no.	Wing	Weight	Ring no.	Wing	Weight

Figure 6
Examples of the recording sheets used for terns as recommended by the *Seabird Monitoring Handbook.*

number of nests in a colony around midday in the main incubation period (Bullock and Gomersall 1980). However, this correction factor may vary according to species, time of day, state of incubation, weather, food availability or other factors.

Wherever possible make both nest counts and flush counts to allow comparison of calibration factors between different colonies and species.

Calculate the average count on each visit. Report the maximum of these averages. See Figure 6 for an example of the recommended recording sheet for tern population counts.

Reference

Bullock, I D and Gomersall, C H (1980) *The Breeding Population of Terns in Orkney and Shetland in 1980.* RSPB.

Tern productivity

These two methods are taken from the *Seabird Monitoring Handbook*, and are appropriate for sandwich, roseate, common, arctic and little terns.

Several methods exist to monitor tern productivity. Three of these, which involve ringing and chick mark/recapture, are not given here but are documented in the *Seabird Monitoring Handbook*. Two which do not involve ringing are presented here: *Method 1* involves counting nesting/incubating adults, with a single count of large chicks; *Method 2* involves flush counts of adults with single or multiple counts of large chicks.

These two methods, although less accurate than those involving mark and recapture, can be undertaken without having to catch and ring any birds. They are thus more applicable to most fieldworkers and cause much less disturbance to the terns. The estimate based on flush counts is the least accurate of the two methods presented.

1. Nest/incubating adult count, with single count of large chicks

Information required

- maximum count of apparently incubating adults or active nests
- estimated number of large chicks and fledged young.

Number and timing of visits

One visit at first fledging, usually early July (NB a population survey must be carried out earlier in the season; see *Tern populations*, above).

Time of day

Midday.

Weather constraints, Sites/areas to visit, Equipment, Safety reminders, Disturbance

As for the population survey (see *Tern populations*).

Methods

This may be the only possible method for colonies that are visited infrequently. It is most suitable for small colonies or those where pairs breed synchronously.

Count either apparently incubating adults (population survey Method 1) or active nests (population survey Method 2). Around the date of first fledging (±1 week), usually early July, count large chicks (10–14 days old, depending on species; see Tables 2–3), including any nearby fledglings which are associated with the colony. Chicks may be counted from a suitable vantage point at some small colonies, although some may be missed. Keep a separate note of numbers of smaller chicks and unhatched eggs.

Estimate productivity as the number of large chicks plus fledged young divided by the maximum count of apparently incubating adults or active nests. This may substantially underestimate productivity, as some large chicks may be missed and smaller/unhatched chicks not included in the count may survive to fledging. On the other hand, some large chicks may die before fledging.

Table 2
The average incubation and fledging periods for terns (overall ranges in parentheses) and ages at which tern chicks should be recorded as potentially fledged (*BWP*, cited in the *Seabird Monitoring Handbook*).

	Incubation period (days)	Fledging period (days)	'Potentially fledged'
Sandwich tern	(21–)25(–29)	28–30	2 weeks
Roseate tern	(21–)23(–26)	(22–)27–30(–31)	2 weeks
Common tern	21–22	(22–)25–26(–33)	2 weeks
Arctic tern	(20–)22(–24)	21–24	2 weeks
Little tern	(19–)21–22	(15–)19–20	10 days

Table 3
Plumage classes and ageing characters defined by Nisbet and Drury (1972, cited in the *Seabird Monitoring Handbook*). Data for a few retarded chicks (third chicks in common tern broods and second chicks in roseate tern) are given in parentheses. In little tern, pin feathers (primaries) erupt when the chick is between 7–8 and 12 days old (Davies 1981, Norman 1992).

Plumage class	Ageing characters	Age (days)	
		Common tern	Roseate tern
1	Newly hatched. Legs short, fat. Chin black.	0–1 (2)	0–2 (2)
2A	Legs elongated, narrow shank between foot and joint. No pin feathers on outer wing.	2–5 (6)	few data
2B	Pin feathers present on outer wing but not erupted. Black chin almost gone.	6–9 (11–12)	few data (12)
3A	Pin feathers erupted on outer wing.	8–12 (13–19)	11–13 (14)
3B	Tail feathers erupted (shaft visible), but <6 mm long (white not visible). Black feathers not visible on nape.	12–15 (13–20)	13–16
4A	Tail feathers >6 mm (white visible),but down still on tips. A few speckles of black show through down on nape when brushed.	15–18 (22–23)	15–20 (21)
4B	No down on tips of tail, but down on tail coverts. Black appearing on nape. Mantle feathered with some down tips.	17–23 (21–28)	18–22 (23)
5A	Nape black with speckles. No down on back, but a little down on tail coverts. Older birds fly when frightened.	21–25 (21–31)	20–24 (28)
5B	Fully feathered, free flying. No down except on forehead.	24 onwards	23 onwards

Common and arctic tern chicks will need to be counted separately in mixed-species colonies; see Table 4 for identification features.

Table 4
Some plumage features for distinguishing chicks of common and arctic terns (Craik 1985, cited in the *Seabird Monitoring Handbook* which gives further details).

	Common tern	Arctic tern
Underwing pattern in chicks near fledging (with well-developed juvenile plumage)	Black carpal bar.	Grey carpal bar.
Colour of dorsal down (head and body) in younger chicks	Some variation between individuals, but always some shade of brown (cinnamon-brown to 'house mouse' brown); apparently never grey.	Much more variable than common tern, from walnut-brown to silver-grey.
Pattern of spotting on upper body in small downy chicks (0–6 days)	Black spots on back are large and few.	Black spots on back are small, many, and elongated along body axis.
Colour of down on belly in younger chicks	Invariably pure white (unless wet or dirty)	Varies individually, from dark (white with a dark cast which may be intense or slight) to pure white.

2. Flush counts of adults, with single or multiple counts of large chicks

Information required

- number of Apparently Occupied Nests (AONs)
- estimated number of large chicks and fledged young.

Number and timing of visits

One visit at first fledging, usually early July (NB a population survey must be carried out earlier in the season; see *Tern populations,* above).

Time of day

Midday.

Weather constraints, Sites/areas to visit, Equipment, Safety reminders, Disturbance

As for the population survey (see *Monitoring breeding tern populations*).

Method

Estimate breeding numbers of terns from flush counts of adults (population survey Method 3). Apply the correction factor of 1.5.

Count the number of large plus fledged chicks as in productivity Method 1, above. Express results as follows: 50 large chicks and fledged young were recorded in a colony from which 300 adults had earlier been flush counted. Estimate of AONs using correction factor: 300/1.5 = 200; estimate of chicks fledged per AON = 50/200 = 0.25.

When reporting results, make it clear that the estimates are based on flush counts, as the errors are potentially high.

Only use this method to estimate productivity when time or logistical constraints prevent more accurate censusing or when the population has already been censused using a flush count.

References

Craik, J C A (1985) Chicks of common and arctic terns. *Ringers Bull.* 6: 92.
Davies, S (1981) Development and behaviour of little tern chicks. *British Birds* 74: 291–298.
Nisbet, I C T and Drury, W H (1972) Measuring breeding success in common and roseate terns. *Bird Banding* 43: 97–106.
Norman, D (1992) The growth of little tern *Sterna albifrons* chicks. *Ringing and Migration* 13: 98–102.

Generic wintering bird monitoring methods

Non-breeding waterfowl: general

The distribution of most species of waterfowl (principally swans, geese, ducks and waders) during the non-breeding period is restricted largely or wholly to wetland habitats. Many wetland sites represent relatively discrete areas and, with most species readily visible within these areas, regular monitoring of total numbers is/can be relatively easy. Simple 'look-see' methods, whereby all birds present within a pre-defined area are counted, are thus usually employed for the majority of surveys.

The objectives of surveys of non-breeding waterfowl are normally threefold:

- to determine population size
- to determine trends in numbers and distribution
- to identify important sites.

The Wetland Bird Survey (WeBS) Core Counts represent a generic approach that, given sufficient coverage of sites, provides adequate data to fulfil these objectives satisfactorily for a large proportion of the species concerned. Details of methods and requirements of WeBS Core Counts are given below.

The different habits of some species, and their use of habitats not regularly covered by WeBS Core Counts, necessitate additional surveys to properly fulfil the objectives. These are complementary to WeBS Core Counts, and many of the principles of WeBS apply equally to these additional methods. Details of these methods are provided in subsequent sections. For clarity, these sections identify only the differences from the general WeBS approach, and thus should be used in conjunction with the section on WeBS Core Counts.

If you see any waterfowl with colour-rings, please note details of the species, date, time, location (with OS grid reference), ring colour(s) and inscription (if any), which leg the ring was on, the presence of any other rings, the number of birds in the flock and, if possible, whether the bird was paired or had young. This information should be sent to the following coordinators:

Cormorant Robin Sellers, Rose Cottage, Ragnall Lane, Walkley Wood, Nailsworth, Gloucestershire GL6 0RU.

Wildfowl Richard Hearn, The Wildfowl & Wetlands Trust, Slimbridge, Gloucestershire GL2 7BT.

Waders Stephen Browne (Wader Study Group Colour-ring Register), c/o British Trust for Ornithology, The Nunnery, Thetford, Norfolk IP24 2PU.

Large gulls Peter Rock, 32 Kersteman Road, Redlands, Bristol BS6 7BX.

In return, details of the bird's history, including previous sightings, will be sent to you.

Wetland Bird Survey (WeBS) Core Counts

National, co-ordinated counts of non-breeding waterfowl on wetlands have been undertaken in the UK since 1947 (Cranswick et al 1996). The initial scheme was restricted to wildfowl, with waders and other waterfowl (grebes, cormorant, gulls) counted later as part of a complementary scheme. These schemes were merged in 1993 as the Wetland Bird Survey (WeBS), ensuring consistent methods for monitoring all waterfowl species in all habitats at wetland sites throughout the UK.

WeBS is a partner scheme of BTO, WWT, RSPB and JNCC and relies on the efforts of a network of over 3,000 largely volunteer counters. National Organisers co-ordinate the survey through information provided in regular newsletters and are assisted by a network of volunteer Local Organisers who co-ordinate counts on a county or site basis.

Most UK estuaries and the most important inland still waters (eg reservoirs, gravel pits) are covered regularly for WeBS Core Counts, and most are counted once per month during the winter. Other habitats, eg rivers and the open coast, and small sites or those furthest from centres of population are relatively poorly covered.

Counts are compatible with the International Waterfowl Census (IWC), and January data are submitted to Wetlands International for inclusion in the IWC database.

Always contact the WeBS Secretariat before beginning counts at a site. They will be able to check whether the site is already covered and, if so, supply you with the necessary details, including a recording form, a map of the count area and details of the Local Organiser. This will ensure that your counts are co-ordinated with those on adjacent count areas.

Note that the specialised methods given below for particular species or groups of species should only be used for those surveys. When undertaking a WeBS Core Count, only WeBS Core Count methods should be used. Thus, for example, counts of swans and geese submitted on WeBS Core Count forms should be daytime counts of birds on wetlands made at the same time as counts of other species, not counts of birds in fields or of roosting birds. The systematic errors that arise from WeBS are known and can be accounted for, providing that the data submitted are consistent with WeBS, eg it is expected that only a very small number of pink-footed geese would be recorded during a daytime count at a major roost site.

Information required

- total number of individuals of each species present
- additional data elements listed on the WeBS recording form
- map of the count area if not previously covered.

Number and timing of visits

One count per month on predetermined dates (see below), September–March. The January count is particularly important for inclusion in the

IWC. Dates are provided for counts during summer months also, but these are of lower priority. Any number of additional counts can be undertaken to assess site importance.

Time of day/state of tide

During the day; at estuarine sites or at those close to the coast, within two hours (three at most) either side of high tide. At inland sites, counts should be completed within four hours, preferably in the morning.

Weather constraints

Sites should be visited on the allocated date if safe to do so. If poor weather or other factors prevent a proper count, make a repeat count as close to the recommended date as possible, although consideration should be given to co-ordination with adjacent sites. If monthly counts are made throughout the winter period, the absence of one count is unlikely to have a significant effect on the objectives of the survey.

Sites/areas to visit

All wetland sites. Most of the important sites for waterfowl, eg estuaries and large still waters, are already covered.

Equipment

- telescope (preferably 30 × magnification; a wide-angle lens may be an advantage)
- 1:25,000 map showing boundary of the area to be covered
- field notebook and two(!) pencils
- WeBS Core Count recording form
- tally counter (optional)
- dictaphone (optional).

Safety reminders

Ensure that someone reliable knows of your whereabouts and when you will return. Ensure that she/he knows what to do if you are late. Check that you are not liable to be cut off by the tide. Beware of slipping into saltmarsh creeks, which can be hidden by vegetation, and very soft sediments from which it may be difficult to extricate oneself.

Disturbance

Avoid undue disturbance to birds near shorelines, as it may confuse the count if they change location. This is especially true at tidal sites, where there may be few or no alternative safe roost locations, and during cold conditions, when birds may waste energy moving location or be denied access to feeding areas.

Methods

To ensure that the data collected can be used to estimate national populations and trends, synchronisation of counts between different sites is important. Given the large proportion of volunteer counters, 'priority dates' are recommended for the once-monthly WeBS Core Counts. This enables counts across the whole country to be synchronised, thus reducing the likelihood of birds being double-counted or missed. Such synchronisation is imperative where a number of counters are required to cover large sites, due to the likelihood of local movements affecting count totals. Local Organisers ensure co-ordination in these cases.

WeBS priority dates are set to coincide with spring high tides occurring during daylight hours on most estuaries. If it is not possible to count on these dates, eg because of differences in the tidal regime in some parts of the UK, the site should be counted in a systematic manner that reduces the possibility of double- or under-counting. Similarly, if synchronisation is not possible between adjacent sites, coverage should be organised in such a way as to reduce the possibility of bird movements going undetected. Exceptionally, at some sites counts may best be made at different times of the tidal cycle.

All sites counted in recent years are currently being mapped at 1:25,000. Counters should contact the WeBS National Organiser or relevant Local Organiser for a map and use the same count boundary if the site has been covered previously, ensuring comparable data are collected over time.

Large sites are divided into sectors, each of which can practicably be counted by a single person in a reasonable time (up to four hours). If the site has not been counted previously, or if sector boundaries are being determined for the first time, choose boundaries that are easily recognisable. Divisions should be based on permanent features that can be readily identified in the field. Only decide on the count boundary to use when you are sufficiently familiar with the site.

Changes to existing boundaries should only be necessary in exceptional circumstances, usually when the area of wetland habitat changes, eg when a new gravel pit is dug or an old one infilled. Counters should consult the WeBS Secretariat before any other proposed changes are made in order that the implications for the WeBS database and for other counters can be considered.

Always obtain permission for access to privately owned land. Often, an initial approach and explanation of the work being undertaken is sufficient, but a letter from the WeBS Secretariat can be provided if necessary. Ensure that you follow subsequent instructions from the site owner, eg many water companies require counters to comply with safety regulations when counting reservoirs.

Choose good vantage points from which to count. Use landscape or other features to ensure you do not re-count the same area when moving to a new position. Make a note of any birds that move during the count, in particular the position to which they have moved, so that they are not missed or double-counted.

All waterfowl species, as defined by Wetlands International (see Rose and Scott 1997), should be counted. In the UK, this includes divers, grebes, cormorants, herons, swans, geese, ducks, rails, waders, gulls and terns (including vagrants, introductions and escapes, eg flamingos). In addition, WeBS records numbers of kingfishers.

Ensure that all areas are searched, and pay particular attention to water edges and the edges of reedbeds, although birds hidden in channels and secretive or cryptic species, eg snipe, are likely to be systematically undercounted. Only counts of birds *seen* should be recorded on the form; estimates of birds thought to be hidden should not be recorded. If large numbers of birds are known to be out of sight, eg a flock of waders seen to fly into a creek that could not be viewed, the count of the species should be noted as an undercount (see below). Counts of naturally secretive or cryptic species should *not* be recorded as undercounts; this is taken into account as part of the methodology. Counts of gulls and terns for WeBS Core Counts are optional.

Any points peculiar to the coverage of the site (eg a particular route used to avoid disturbance) should be documented to enable future counts to be made in exactly the same way. The route should be designed to ensure that the whole count area is viewed (the reliable range of most binoculars is around 500 m, especially if species are in mixed flocks; with a telescope, if topography and weather conditions allow, experienced observers can identify birds up to 3-4 km away). Although winter roosts are often at traditional sites, they may move between years, so always search the survey area thoroughly.

It is important to record the accuracy of the count, especially if it was conducted in poor weather or visibility, or disturbance has caused birds to fly around. The overall accuracy of the count should be recorded as 'OK' or 'Low'; 'Low' indicates that the count was a gross underestimate and that caution should be used when using count data. If a count of an individual species is low, flag this count accordingly. Even experienced observers may only be within 10% of the true total when estimating large flocks (experiments have shown that smaller numbers are overestimated and very large flocks are underestimated); however, these counts should still be treated as 'OK' in terms of accuracy.

If, in exceptional circumstances, you are unable to make a count of a particular species, ensure that you note its presence on the form, otherwise it will be assumed that no birds were present. Also, if you visit a site and no birds are present, eg the site is covered in ice or there is much disturbance, always complete and return the form indicating a nil count, otherwise it will be assumed that either the site was not counted or that the birds had moved and were recorded by another counter at the site to which they moved.

Count birds present in relatively small numbers or dispersed widely individually. Estimate the number of birds in large flocks by initially counting five or ten individuals, dividing them mentally into groups of the same size and counting the number of groups. Estimate very large flocks by counting in groups of 50 or 100, or exceptionally even 1,000. In these cases, make allowances for varying densities of birds in the flock, making the block size larger or smaller as appropriate. Tally counters are particularly useful for this approach.

If large numbers of birds are moving, or are thought likely to leave (eg because of disturbance), the following should allow at least an approximate count:

1. Make a quick total count (don't separate species).
2. Make a quick assessment of proportions of species.
3. Start with the most common species; if all birds leave, you can probably make a reasonable guess at the others (eg recording that pochard are twice as common as tufted duck is better than nothing).
4. Re-scan slowly for less common species; slow scans also help to locate diving species in mixed flocks which may have been missed on the first scan.
5. Scan slowly through the whole flock (a dictaphone can be useful).

OFFICE USE | R | PB | I | C | IB

1 NAME AND ADDRESS - ☐ ✓ if change of address

Office use:

Home tel:

Work tel:

Postcode:

(a) GRID REFERENCE

2 SITE:

SUB-SITE:

(b) Complete only if counts below refer to part of a large site

SECTOR:

Office use:

3 DATE:

TIME: start / finish

TOTALS

4 WILDFOWL: please write NIL in the next row if no wildfowl were present

	CODE
Nil birds (wildfowl)	XW
Red-throated Diver	RH
Great Northern Diver	
Little Grebe	
Great Crested Grebe	
Slavonian Grebe	
Black-necked Grebe	
Cormorant	
Grey Heron	
Mute Swan	
Bewick's Swan	
Whooper Swan	
Bean Goose	
Pink-footed Goose	
European White-front	
Greenland White-fron	
Greylag Goose	
Canada Goose	
Barnacle Goose	
Dark-bellied Brent Go	
Light-bellied Brent Go	
Shelduck	
Wigeon	
Gadwall	
Teal	
Mallard	
Pintail	
Shoveler	
Pochard	
Tufted Duck	
Scaup	
Eider	
Common Scoter	
Goldeneye	
Smew	
Red-breasted Mergan	
Goosander	
Ruddy Duck	
Water Rail	
Moorhen	
Coot	

DATE or SECTOR:

TOTALS

WADERS: Please write NIL in the next row if no waders were present

	CODE
Nil birds (waders)	XS
Oystercatcher	OC
Avocet	
Little Ringed	
Ringed Plov	
Golden Plov	
Grey Plover	
Lapwing	
Knot	
Sanderling	
Little Stint	
Curlew San	
Purple Sand	
Dunlin	
Ruff	
Jack Snipe	
Snipe	
Woodcock	
Black-tailed	
Bar-tailed G	
Whimbrel	
Curlew	
Spotted Re	
Redshank	
Greenshank	
Green Sand	
Wood Sand	
Common Sa	
Turnstone	

GULLS: Ol

Nil birds (gu | Black-heade | Common Gu | Lesser Blacl | Herring Gul | Great Black

TERNS: Ol

Nil birds (te | Sandwich T | Common Te | Arctic Tern | Little Tern

Kingfisher

DATE or SECTOR:

5 ALTERNATIVE METHODS: this section has been removed. Please ensure that all species counts written on the form relate to the precise date and time stated in section 3. A separate form is available from your local Goose Organiser or WWT for roost counts of geese. Roost counts of other species should be written in a separate column, with specific details of date, time and count accuracy etc. In section 9, record the dates and times of any roost counts.

6 COUNT CONDITIONS: please circle the appropriate number (refer to key below line)

State of tide (not inland sites)	1 2 3 4	1 2 3 4	1 2 3 4	1 2 3 4	1 2 3 4	1 2 3 4

1 Rising 2 High 3 Falling 4 Low

% ice cover

7 COVERAGE: please circle the most appropriate choice below

Were you able to cover "all" or only "part" of your count area?

Area covered	All/Part	All/Part	All/Part	All/Part	All/Part	All/Part

Were you able to complete your count within about 3-4 hours?

Within about 3-4 hours?	Yes/No	Yes/No	Yes/No	Yes/No	Yes/No	Yes/No

Visibility: please indicate range 1 Excellent (more than 2km) 2 Good (1-2km) 3 Moderate (250m-1km) 4 Poor (less than 250m)

Visibility	1 2 3 4	1 2 3 4	1 2 3 4	1 2 3 4	1 2 3 4	1 2 3 4

Disturbance: please indicate the overall level of disturbance 1 None 2 Moderate 3 High 4 Very high

Disturbance level	1 2 3 4	1 2 3 4	1 2 3 4	1 2 3 4	1 2 3 4	1 2 3 4

Do you feel your counts of wildfowl and waders were reasonably accurate (circle "OK") or did factors (e.g. weather, disturbance) prevent you from recording a significant number of wildfowl or waders present (circle "Low")?

Count accuracy	OK/Low	OK/Low	OK/Low	OK/Low	OK/Low	OK/Low

8 ACTIVITY: OPTIONAL SECTION. ✓ if completed ☐ ☐ ☐ ☐ ☐ ☐

Activity type: please circle the numbers corresponding to those activities occurring at the site, e.g. ②, and cross those affecting the birds, e.g. ⊗

1 2 3 4 / 5 6 7 8 / 9 10 11 12 / 13 14 15 16 (per column)

1 Walkers 2 Dogs 3 Horse riders 4 Anglers 5 Shooters 6 Bait-diggers 7 Shellfishers 8 Unpowered boats 9 Powered boats 10 Vehicles 11 Micro-lights 12 Wind-surfers 13 Jet skis 14 Aircraft 15 Others (please specify) 16

Birds of prey: circle species codes for the birds of prey present at the site, indicating which were disturbing waterfowl with a crossed circle

MR HH SH / K. ML PE / BZ SE O. (per column)

MR Marsh Harrier HH Hen Harrier SH Sparrowhawk K. Kestrel ML Merlin PE Peregrine BZ Buzzard SE Short-eared Owl O. Other species (please specify)

9 ADDITIONAL INFORMATION: please continue on a separate sheet if necessary

Figure 1
An example recording form for WeBS Core Counts.

If it is not possible to positively identify the species, record birds as 'unidentified' or within a given category, eg unidentified *Aythya* spp, unidentified scoter spp, unidentified grey goose spp. If you cannot be accurate, be honest!

Only record birds using the site, eg do not record geese simply overflying the area. Do not separate males and females; simply provide one count per species. If making counts during the breeding season, only young that are at least three-quarters grown should be included in count totals; small ducklings should be excluded.

Use a notebook in the field; fill out the WeBS recording form (Figure 1) as soon after the count as possible.

Completed WeBS recording forms should be returned to the WeBS Secretariat via the relevant WeBS Local Organiser directly after the March count.

Additional information

During particularly cold periods, numbers of birds may be swollen by influxes from the continent or other areas. An additional count under these conditions may be useful to identify potential cold-weather refuges. Extra care should be taken to avoid disturbance to birds at this time as they are likely to be stressed.

Some sites may be particularly important as roost sites for birds which feed in different habitats during the day (eg geese return to wetland roosts after feeding on agricultural land during the day, goosanders feeding on rivers often use still waters as roosts); count methods for species which regularly exhibit such behaviour are provided in the following sections. Nevertheless, some sites are also used as roost sites by species such as mallard and teal. Additional counts late in the evening may thus be important in assessing site importance. Such counts should be made using the same methodology as above. It is particularly important to record the start and end times of these counts.

Analytical methods have been developed specifically for the WeBS dataset to allow estimation of low or missing counts for inclusion in calculating national population estimates and trends (Underhill and Prys-Jones 1994). Please contact the WeBS Secretariat for details.

For further details, contact: Mark Pollitt, WeBS National Organiser (Core Counts), WeBS Secretariat, The Wildfowl & Wetlands Trust, Slimbridge, Gloucestershire GL2 7BT; tel 01453 890333 ext 255/280.

Contributed by Peter Cranswick

References

Cranswick, P A, Kirby, J S, Salmon, D G, Atkinson-Willes, G L, Pollitt, M S and Owen, M (1996) A history of wildfowl counts by The Wildfowl and Wetlands Trust. *Wildfowl* 47: 216–229.

Rose, P M and Scott, D A (1997) *Waterfowl Population Estimates* (second edition). Wetlands International Publ. 44. Wageningen.

Underhill, L G and Prys-Jones, R P (1994) Index numbers for waterbird populations I. Review and methodology. *J. Applied Ecology* 31: 463–480.

National wintering swan census

All swan species in the UK may spend a significant proportion of the day in areas that are poorly covered by WeBS, eg non-wetland habitat or areas in west Scotland. These are covered every five years as part of a Europe-wide census, currently scheduled for winters 2000–01, 2005–06, 2010–11, etc. Additionally, age counts and brood sizes may be made to measure productivity, and habitat use is also noted.

Differences and additions to the methods and requirements of WeBS Core Counts are given below.

Information required

- total number of individuals of all swan species
- the proportion of young, brood sizes and the habitat (all optional).

See *WeBS Core Counts* (above) for additional information.

Number and timing of visits

One on a predetermined date in mid-January; in the UK, usually the WeBS January date. A repeat count is recommended if the first was considered to be inaccurate, eg due to poor weather.

Time of day

See *WeBS Core Counts*. At important roost sites, it may be preferable to count birds at first light before they depart to non-wetland areas.

Sites/areas to visit

See *WeBS Core Counts*. Non-wetland areas known or thought likely to support migrant swans should also be searched.

Equipment

See *WeBS Core Counts*.

- swan census recording form.

Safety reminders, Weather constraints

See *WeBS Core Counts*.

Disturbance

See *WeBS Core Counts*. Swans feeding in fields can often be approached more readily using a car than on foot.

Methods

See *WeBS Core Counts*.

Check all areas known or thought likely to hold birds, including non-wetland habitat, eg arable fields.

Where dawn counts are made at roost sites, ensure the same birds are not counted again during the day on adjacent non-wetland habitat.

If the international census date differs from the chosen WeBS date, counts at the most important Bewick's swan roosts should ideally be undertaken on both dates in case there is a major influx of birds from the continent, eg as a result of cold weather.

Also see *Productivity of swans and geese* (below) for assessing the proportion of young and brood size.

Completed recording forms should be returned to the WeBS Secretariat via the relevant WeBS Local Organiser immediately after the March count.

For further details, contact: WeBS Secretariat, The Wildfowl & Wetlands Trust, Slimbridge, Gloucestershire GL2 7BT; tel 01453 890333 ext 255/280.

Contributed by Peter Cranswick

National censuses of geese

Most goose species in the UK may spend a significant proportion of the day widely distributed over non-wetland habitat, using wetlands only as night-time roosts. These species are best counted as they fly from or to these roosts at dawn or dusk. Most species are monitored annually, and dates are set for each species for co-ordination of national censuses. Other species are best counted using WeBS Core Count or similar methodology. Additionally, daytime age counts and brood sizes may be made to measure productivity.

Methods and requirements for generalised goose counts (where these differ from those for WeBS Core Counts) are given below. Following this, the national census methodology for each species/population is given.

Information required

- total number of individuals of each goose species.

See *WeBS Core Counts* (above) for additional information.

Number and timing of visits

Between one and three times per year, depending on the individual species (see list of species, below), on predetermined dates. Additional counts are necessary to assess site usage through the winter (see below).

Time of day

Preferably dawn, but can be dusk.

Weather constraints

Cloud, heavy rain or fog may make accurate counts impossible. However, since these conditions are often localised, especially in upland areas where many roosts occur, sites should always be visited on the national census date if safe to do so. The presence or absence of birds may still be noted by their calls.

Sites/areas to visit

See *WeBS Core Counts*.

Equipment

- goose roost recording form.

See *WeBS Core Counts* for additional information. NB a telescope is usually not very useful for this type of count.

Safety reminders

See *WeBS Core Counts*. Shooting occurs at many goose roosts, so take extra care if the count position is away from roads or other areas generally accessible to the public. It is advisable to contact the site owner and/or gamekeeper in advance at sites with regular shoots. If counting from a car, park in appropriate and safe places off main roads to view fields. In more remote or upland areas, ensure someone knows where you are and when you are due back. Always carry a compass. If possible, surveyors should not work alone in remote upland areas, and spare warm clothing, a plastic survival bag, whistle, first-aid kit and food supplies should be carried.

Disturbance

See *WeBS Core Counts*. At sites where the count position is close to the favoured roosting location, try not to 'spook' the birds when approaching or leaving the site, particularly during the shooting season when birds are more wary. Geese feeding in fields can often be approached more readily using a car than on foot.

Methods

The predetermined count dates are timed to record the peak population size in the UK. However, this may give a misleading picture of individual site importance during the winter. Thus, counts at other times of year are needed to determine how numbers change during the season. Also, if the peak is very short-lived, a series of counts, each just a few days apart, is necessary to record the maximum number using the site.

Dawn counts are recommended, since birds may continue to arrive at the site in the dark during dusk counts, particularly during mid-winter, when days are shorter, or at roosts where feeding grounds are distant. The departure of birds in the morning is generally thought to be more predictable, and only rarely do birds depart in complete darkness. National census dates for species counted at roosts are timed to coincide with new moons, since birds may remain on fields during bright, moonlit nights rather than return to the roost site. Nevertheless, the choice of dawn or dusk counts may differ for different sites, according to usage by the birds, topography and time of year.

Birds are best counted in flight as they arrive at or depart from the roost. At many sites, these flightlines are well-defined, as birds travel to suitable feeding areas. Even in half-light, birds can generally be counted with relative ease against the sky. Depending on the topography of the site and the direction of flightlines, certain positions around the site may enable birds to be counted more easily. At particularly large sites, where flightlines are well spaced, or where very large numbers of birds arrive on a number of different flightlines, two or more counters may be necessary. Alternatively, use more than one tally counter. Note that a telescope is usually not very useful for roost counts.

For dawn counts, aim to be on site before first light, especially during mid-winter when the shorter daylength may mean that birds depart the roost earlier relative to sunrise. Remain until all birds have left or at least until it is sufficiently light to count any birds still remaining at the site.

Similarly, start dusk counts either before birds begin to arrive, or while it is sufficiently light to count any birds already present. Continue counting until it is too dark to count birds against the sky. If birds continue to arrive in the dark, record the fact that birds were heard arriving and record the count as an undercount.

In certain parts of Scotland in particular, mixed roosts of greylag and pink-footed geese occur. Separation of these species is possible under reasonable light conditions or if birds are calling. However, if specific identification is not possible, the total number of unidentified geese should also be recorded.

It should be noted that identifying and counting geese on flightlines requires a measure of practice even for counters experienced at counting species during daytime. Knowledge of the use of the site and surrounding areas for feeding will also help in obtaining an accurate count.

Exceptionally, even for those species best counted at roosts, some birds may best be counted during the day, particularly if the roosts are difficult to access or topography makes counting difficult. Daytime counts require a good knowledge of bird distribution and care is needed to avoid counting birds in feeding flocks that have been counted at other roosts.

Note that fewer birds may use roosts and instead remain on fields during moonlit nights or if there is extensive flooding.

Although counts are made of birds on flightlines, and not of birds actually on the wetland site, it is useful to produce a map which clearly marks any areas of shoreline or land used by the birds, identifies which waterbodies are used in a complex of wetlands or shows the approximate area of mudflat used at estuarine sites.

Provide the location and central grid reference for birds on agricultural land or other non-wetland habitat. Birds may range widely in this habitat, so site definition and demarcation of boundaries is not normally practical.

See also *Productivity of swans and geese* (below) for assessing the proportion of young and brood size.

Where relevant, completed recording forms should be returned to WWT via the relevant Goose Local Organiser, immediately after the census date or after the last winter count is made. All data sent to WWT will be incorporated into WWT databases or forwarded to the relevant organisation.

National census methodologies

Pink-footed goose and Icelandic greylag goose

Roost counts, on predetermined dates in October and November. Co-ordinated by Goose Local Organisers for WWT. Additional roost counts can be made at any time during the winter and are incorporated into WWT databases.

North Scottish greylag goose

Outer Hebrides: daylight counts of birds in fields, August and February. National census: late summer 1997; next scheduled 2005–06. Co-ordinated by WWT.

Bean goose

No national census. Majority of population occurs at two sites, which are already monitored by RSPB and the Scottish Bean Goose Working Group.

Greenland white-fronted goose

Roost or field counts, depending on the site.
National census: late autumn (November/December) and spring (March/April), co-ordinated by the Greenland White-fronted Goose Study (GWGS).

Greenland barnacle goose

Daytime field counts.
Islay: regular all-island counts throughout the winter by Scottish Natural Heritage (SNH).
International census: spring, every five years, including aerial census, co-ordinated by WWT.

Svalbard barnacle goose

Daytime field counts.
Solway-wide monthly counts (since 1997–98) throughout winter; November, January and mid-March are priorities. Co-ordinated by WWT.

Dark-bellied brent goose

WeBS Core Count methods.
Extra effort made in January and February, when peak numbers usually occur in the UK, co-ordinated by WWT.

Canadian/Irish light-bellied brent goose

WeBS Core Count methods.
Annual international census: October, when all birds have usually arrived on the wintering grounds, co-ordinated by the Irish Brent Study Group. WeBS counts in Northern Ireland and counts by the Irish Wetland Bird Survey (I-WeBS) in the Republic of Ireland provide a total count of the population.

Svalbard light-bellied brent goose

WeBS Core Count methods. Nearly all birds in the UK are found at Lindisfarne.
Fornightly counts have been made in recent winters by local counters.

Naturalised/introduced geese

WeBS Core Count methods.
Periodic national censuses of Canada geese (recently extended to include all other naturalised/introduced species): late June to mid-July, co-ordinated by WWT.

Useful addresses

WWT co-ordinates many surveys, and is in close contact with SNH, GWGS, I-WeBS and other organisations and individuals undertaking goose counts. WWT can provide information and contact addresses for relevant organisations or individuals appropriate for the species, region or census concerned.

General. Goose Surveys, The Wildfowl & Wetlands Trust, Slimbridge, Gloucestershire GL2 7BT; tel 01453 890333.

Greenland white-fronted goose. Greenland White-fronted Goose Study, c/o Tony Fox, Department of Wildlife Ecology, National Environmental Research Institute, Kalø, Grenåvej 12, DK-8410 Rønde, Denmark.

Irish Brent Goose Study Group, c/o Kendrew Colhoun, BirdWatch Ireland, Ruttledge House, 8 Longford Place, Monkstown, Co Dublin, Ireland.

Contributed by Peter Cranswick

Low Tide Counts

WeBS Core Counts on estuaries are usually made at high tide when most birds concentrate at traditional roosts and, as a consequence, are generally easier to count. While such counts identify accurately the total number of birds using the site, and are thus used in the production of national population estimates and trends, they present only a snapshot of bird distribution within dynamic sites. Low Tide Counts assess the spatial distribution of non-breeding waterfowl and the relative importance of the various parts of the estuary for feeding birds. WeBS Low Tide Counts are conducted once-monthly from November to February on UK estuaries that hold more than 5,000 waterfowl. Each estuary in the scheme is counted at least once every seven years.

Information required

- total numbers of feeding and non-feeding birds for each species.

See *WeBS Core Counts* (above) for additional information.

Number and timing of visits

Four visits, once a month, November–February. The exact dates will depend on the tides.

Time of day

Within two hours either side of low tide, during the day. Avoid counting earlier than one hour after sunrise or later than one hour before sunset to avoid dawn and dusk flighting.

Sites/areas to visit

Any estuary.

Equipment

- WeBS Low Tide Counts recording form (Figure 2).

See *WeBS Core Counts* for additional information.

Safety reminders

Ensure that someone reliable knows of your whereabouts, when you will return and what to do if you are late. Check that you are not liable to be cut off by the tide. Beware of slipping into saltmarsh creeks, which can be hidden by vegetation, and very soft sediments that may be difficult to extricate yourself from.

Methods

Visit and get to know the estuary before counting. Contact your local WeBS organiser or the BTO to find out whether it is already counted and, if so, how it has been subdivided for counts. If it is already counted, liaise with the local Low Tide Count organiser over where and when to count. If you are tackling an estuary for the first time, unless it is very small you will need to divide it into a number of counting sections, each of which can be covered by a single observer in four hours or, preferably, less. Mark the boundaries of each of these sections clearly on a map. For each count section, establish and mark on the map the best vantage points from which to carry out counts. These are probably already used by the local WeBS counters. Details of these observation positions (location and compass direction of observation) should be kept and used each year.

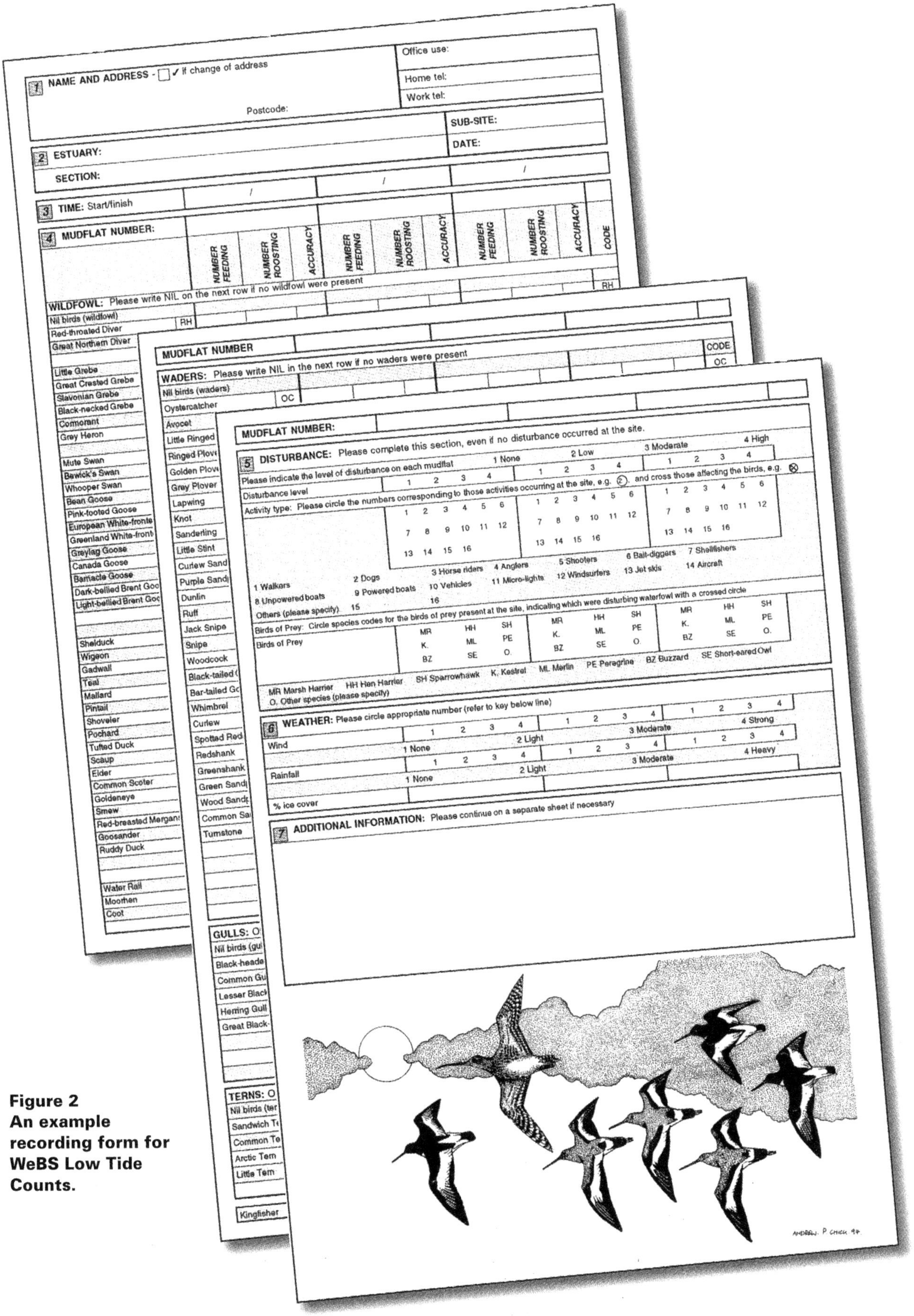

1 NAME AND ADDRESS - ☐ ✓ if change of address

Postcode:

Office use:

Home tel:

Work tel:

2 ESTUARY:

SECTION:

SUB-SITE:

DATE:

3 TIME: Start/finish

4 MUDFLAT NUMBER:

NUMBER FEEDING | NUMBER ROOSTING | ACCURACY | NUMBER FEEDING | NUMBER ROOSTING | ACCURACY | NUMBER FEEDING | NUMBER ROOSTING | ACCURACY | CODE

WILDFOWL: Please write NIL on the next row if no wildfowl were present

Nil birds (wildfowl)
Red-throated Diver RH
Great Northern Diver
Little Grebe
Great Crested Grebe
Slavonian Grebe
Black-necked Grebe
Cormorant
Grey Heron
Mute Swan
Bewick's Swan
Whooper Swan
Bean Goose
Pink-footed Goose
European White-fronte
Greenland White-front
Greylag Goose
Canada Goose
Barnacle Goose
Dark-bellied Brent Goo
Light-bellied Brent Goo
Shelduck
Wigeon
Gadwall
Teal
Mallard
Pintail
Shoveler
Pochard
Tufted Duck
Scaup
Eider
Common Scoter
Goldeneye
Smew
Red-breasted Mergan
Goosander
Ruddy Duck
Water Rail
Moorhen
Coot

MUDFLAT NUMBER | CODE

WADERS: Please write NIL in the next row if no waders were present

Nil birds (waders)
Oystercatcher OC
Avocet
Little Ringed
Ringed Plove
Golden Plove
Grey Plover
Lapwing
Knot
Sanderling
Little Stint
Curlew Sand
Purple Sand
Dunlin
Ruff
Jack Snipe
Snipe
Woodcock
Black-tailed
Bar-tailed Go
Whimbrel
Curlew
Spotted Red
Redshank
Greenshank
Green Sand
Wood Sand
Common Sa
Turnstone

GULLS: O
Nil birds (gu
Black-heade
Common Gu
Lesser Black
Herring Gull
Great Black-

TERNS: O
Nil birds (ter
Sandwich T
Common Te
Arctic Tern
Little Tern

Kingfisher

MUDFLAT NUMBER:

5 DISTURBANCE: Please complete this section, even if no disturbance occurred at the site.

Please indicate the level of disturbance on each mudflat: 1 None, 2 Low, 3 Moderate, 4 High

Disturbance level: 1 2 3 4 | 1 2 3 4 | 1 2 3 4

Activity type: Please circle the numbers corresponding to those activities occurring at the site, e.g. ②, and cross those affecting the birds, e.g. ⊗

1 2 3 4 5 6 / 7 8 9 10 11 12 / 13 14 15 16 (repeated for each mudflat)

1 Walkers, 2 Dogs, 3 Horse riders, 4 Anglers, 5 Shooters, 6 Bait-diggers, 7 Shellfishers, 8 Unpowered boats, 9 Powered boats, 10 Vehicles, 11 Micro-lights, 12 Windsurfers, 13 Jet skis, 14 Aircraft

Others (please specify) 15 16

Birds of Prey: Circle species codes for the birds of prey present at the site, indicating which were disturbing waterfowl with a crossed circle

Birds of Prey: MR HH SH / K. ML PE / BZ SE O. (repeated for each mudflat)

MR Marsh Harrier, HH Hen Harrier, SH Sparrowhawk, K. Kestrel, ML Merlin, PE Peregrine, BZ Buzzard, SE Short-eared Owl, O. Other species (please specify)

6 WEATHER: Please circle appropriate number (refer to key below line)

Wind: 1 2 3 4 | 1 2 3 4 | 1 2 3 4 — 1 None, 2 Light, 3 Moderate, 4 Strong

Rainfall: 1 2 3 4 | 1 2 3 4 | 1 2 3 4 — 1 None, 2 Light, 3 Moderate, 4 Heavy

% ice cover

7 ADDITIONAL INFORMATION: Please continue on a separate sheet if necessary

ANDREW. P CHICK '94

Figure 2 An example recording form for WeBS Low Tide Counts.

Unless you are using the mapping technique, it is usually helpful to subdivide each count section, because it makes counting more systematic and less daunting. If doing so for the first time, try to divide it up according to sediment type, eg sand, mud, etc. If the section is fairly uniform, however, subdivisions can be made using permanent features such as channel markers and groynes in the estuary itself or on sightlines with features on the opposite shore. Ideally, the whole estuary should be counted simultaneously, although this will require several people. It is, however, acceptable to count over more than one day, and therefore a single observer may be sufficient at small sites. Mark any subdivisions clearly on a map and show clearly how they were made, so that future counters are able to use them.

If the data are to be included in the national WeBS database, send copies of any maps showing the subdivisions, observation points, etc, to the BTO.

When carrying out a count, scan slowly across the mudflat, counting birds as you scan. It is often easiest to do this species by species, starting with the most numerous. Count the whole of each subdivision in turn, recording the number of individuals of each species. Record feeding and non-feeding birds separately. Be as systematic as possible. While you are counting a particular subdivision, ignore birds which arrive in or leave areas or subdivisions, other than the one you are counting at the time but, if possible, record the times of flying and landing birds for comparison with other counters' results later.

When time and conditions permit, it is sensible to repeat the counts of groups, particularly if there are several species present or if count conditions are less than ideal. In this case, the average of the two counts should be recorded.

If more a detailed representation of bird distribution is required, mark the location of all birds on a map. This can be done by marking either the number of individuals or the location of any flocks of each species. Use BTO two-letter species codes and protocol (see Appendix 1) when mapping, taking care to record the landing positions of any flying birds already counted. If there is any likelihood of counting birds twice when moving between vantage points, or if you require accurately fixed group positions for a detailed study, then you can plot flocks on the map with a compass.

When the survey is complete, transfer all information to a recording form (Figure 2). The visit totals for each species in each subdivision should be recorded on the form. The count totals should include birds resting, feeding, bathing, etc, within the site. Only include birds flushed off the site and birds moving around within the site if you are sure these have not been picked up by another observer. Do not include in the count total any birds seen flying overhead from one unknown point to another.

Include on the recording form the timing of the visits and the count conditions. For each winter, quote the mean density of individuals recorded in each subdivision during the four visits.

For further details, contact: Andy Musgrove, WeBS National Organiser (Low Tide Counts), British Trust for Ornithology, The Nunnery, Thetford, Norfolk IP25 2PU; tel 01842 750050.

Waterfowl on non-estuarine coastlines

In winter, large proportions of the populations of several wader species – such as purple sandpiper, turnstone, ringed plover and sanderling – are present on non-estuarine coasts, particularly rocky shores (Browne et al 1995a, b, 1996). When surveying waterfowl on non-estuarine coasts, use the methods developed for the Non-estuarine Coastal Waterfowl Survey (NEWS) which were based on the methods used for the Winter Shorebird Count (Summers et al 1975, Moser and Summers 1986). It is planned to conduct national censuses of waders on the UK open coastline every nine years.

Information required

- total number of all species.

See *WeBS Core Counts* (above) for additional information.

Number and timing of visits

A single visit any time between 1 December and 31 January.

Time of day

Within 3½ hours either side of low tide (half-ebb to half-flood).

Weather constraints

Avoid counting on days when visibility is affected by fog or heavy rain, or when safety is jeopardised by high winds (greater than Beaufort force 5).

Sites/areas to visit

All safely accessible non-estuarine shores including cliff-bound shores, and the inland areas visible from or near the high-water mark.

Equipment

- recording form
- mobile phone (if available)
- compass (optional).

See *WeBS Core Counts* for additional information.

Safety reminders

Ensure that someone reliable knows where you are, when you will return and what to do if you are late. Check that you are not liable to be cut off by the tide. Rocky shores in particular are always hazardous because they are uneven, slippery and weed-covered. Beware of slippery vegetation at cliff edges, and of undercut or loose strata.

Methods

Visit and get to know the site before counting. Mark the survey area on a map. Establish and mark on the map the best vantage points from which to count. Details of these observation positions (location and compass direction of observation) should be kept and used each year.

The shoreline should be split into sections defined by major changes in habitat (substrate, slope, seaweed cover). Each section should be no longer than 4 km (this allows the data to be incorporated into the WeBS database). For new, previously uncounted sections, this splitting can be done during the survey. Bear in mind that only 3–10 km of rocky shores can be surveyed for waders in a single day, since only about 0.5 km can be counted per hour on the most productive stretches.

Use a notebook or dictaphone and transcribe counts onto recording forms as soon as possible afterwards. Count the number of birds in each section separately. If there is any likelihood of counting birds twice when moving between vantage points, or if you require accurately fixed group positions for a detailed study, use a compass and plot flocks on a map. When carrying out counts from vantage points, use a telescope and be systematic. Where appropriate, and when time and conditions permit, repeat each count (particularly if there are several species present or if count conditions are less

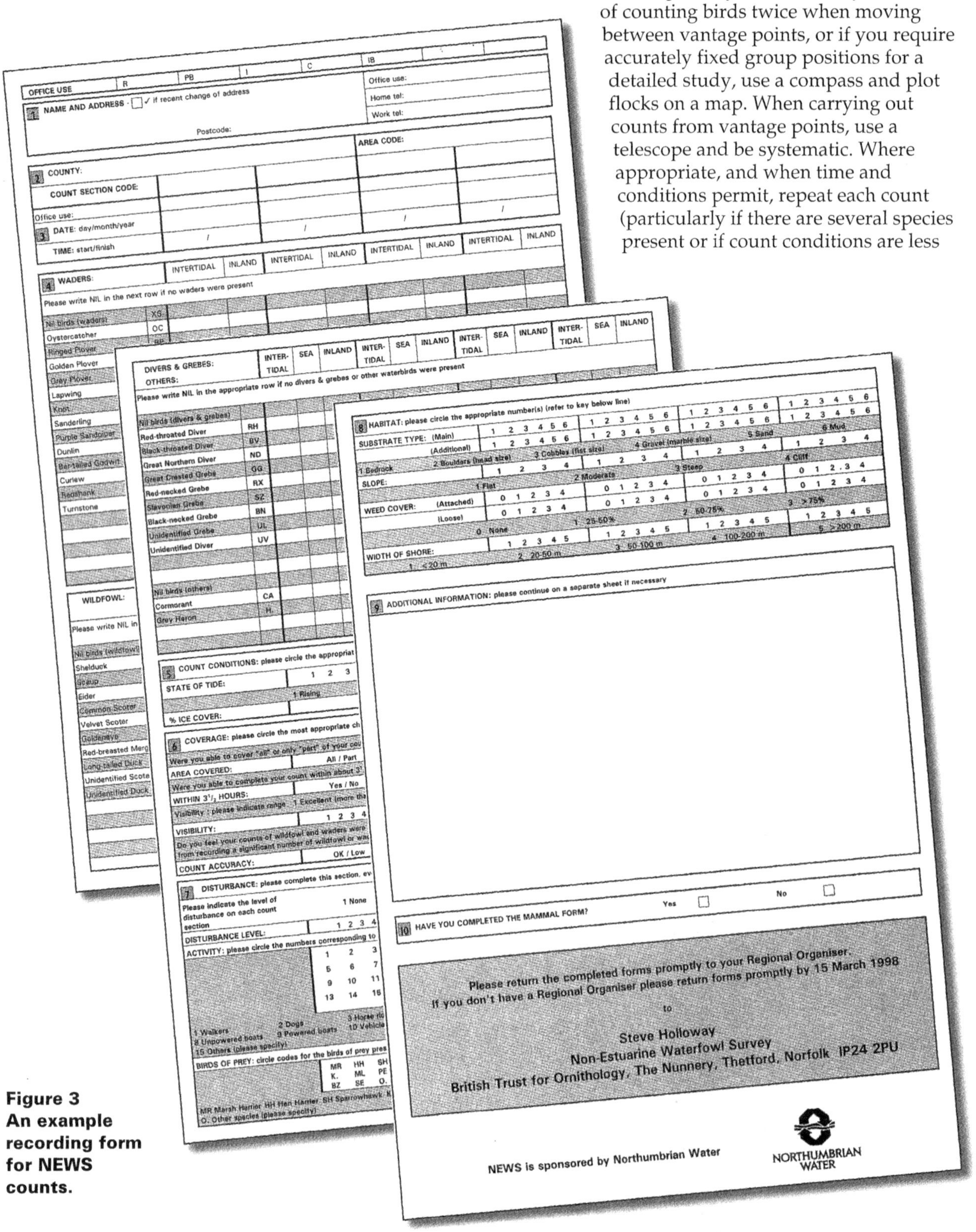

Figure 3
An example recording form for NEWS counts.

than ideal) and record the average of the two counts. If poor weather or other factors prevent a proper count, repeat it as soon as possible. Try to ensure that your count is co-ordinated with others being carried out on adjacent stretches of coast.

To count waders, walk through each section, keeping well down the shore to avoid missing or double-counting birds. Ideally, commence the counts on a falling tide. You must always be aware of the possibility of the tide cutting you off from the landward side. Once the littoral zone has been counted, walk back through the section and count the landward strip. Flock sizes should rarely have to be estimated, because they are usually sufficiently small to allow individual birds to be counted. Where cliffs prevent access to the littoral zone, count from the cliff top.

Make a note of the flight paths of the birds that are disturbed. Count those that fly inland, out to sea or behind you, and those that you walk past, but exclude those that fly forward. Do not count those that fly forward but note their landing position so that they can be counted on the return trip. Note the landing position of those that fly past from behind but *do not* count them on the return journey as they have probably already been counted.

For each species, add together the numbers of individuals recorded at each observation point within each section and transfer the information to a NEWS recording form (Figure 3). The count totals should include birds resting, feeding, bathing, etc, within the site, birds flushed off the site and birds moving around within the site. Do not include birds flying overhead from one unknown point to another. Note the timing of visits and the count conditions on the form. If more than one visit is made, quote the peak number of individuals recorded during any one visit.

For further details contact: Mark Rehfisch, British Trust for Ornithology, The Nunnery, Thetford, Norfolk IP24 2PU; tel 01842 750050.

References

Browne, S J, Austin, G E and Rehfisch, M M (1995a) *A Comparison of Data Collected for the Winter Shorebird Count and the Non-estuarine Coastal Waterfowl Survey, Pilot Count.* BTO Research Rep. to WeBS partners.

Browne, S J, Austin, G E and Rehfisch, M M (1995b) *Evaluation of Bird Monitoring Requirements for the United Kingdom's Non-estuarine Coastline.* BTO Research Rep. to WeBS partners.

Browne, S J, Austin, G E and Rehfisch, M M (1996) Evidence of decline in the United Kingdom's non-estuarine coastal waders. *Wader Study Group Bull.* 80: 25–27.

Moser, M E and Summers, R W (1986) Wader populations on the non-estuarine coasts of Britain and Northern Ireland: results of the 1984–85 Winter Shorebird Count. *Bird Study* 34: 71–81.

Summers, R W, Atkinson, N K and Nicoll, M (1975) Wintering wader populations on the rocky shores of eastern Scotland. *Scottish Birds* 8: 299–308.

Contributed by M M Rehfisch and S J Holloway

Inshore marine waterfowl (divers, grebes and sea-ducks)

With the exception of little and great crested grebes, this group of species is most numerous in coastal waters during winter. Counts of birds on the sea can be very difficult, particularly during bad weather and when the birds are a long way offshore. A number of methods for counting waterfowl in coastal areas have been developed using aeroplanes and boats (see *Waterfowl and Seabirds at Sea,* below). However, since these are costly and require considerable training, they are normally undertaken by professionals. Shore-based methods for counting inshore marine waterfowl are detailed below. All birds on inland waters can be monitored using WeBS Core Count methodology.

Information required

- total number of all species.

See *WeBS Core Counts* (above) for additional information.

Number and timing of visits

Once a month, October–February. The January count is highest priority. Additional counts can be made in late summer (August–September) at known moult sites and in spring (April–May) at known staging sites.

Time of day

Around high tide. Avoid times of day which would mean looking directly into the sun. Do not start earlier than one hour after sunrise or finish later than one hour before sunset to avoid dawn and dusk flighting (although specific roost counts may be the most suitable at some sites for some species).

Weather constraints

Avoid counting on days when visibility is affected by fog, heavy rain or high winds (greater than Beaufort force 4). Avoid days with heavy swell, which may persist for several days after gales.

Sites/areas to visit

All estuarine and non-estuarine shores. Most sea-ducks favour areas of sea less than 10–15 m deep; bays with large areas of shallow water are likely to be most productive. Mussel scars and sewage or cooling-water outflows may attract sea-duck flocks.

Equipment

- recording form
- compass (optional).

See *WeBS Core Counts* (above) for additional information.

Safety reminders, Disturbance

See *WeBS Core Counts.*

Methods

Sea-duck, grebes and divers are counted as part of local or regional initiatives in many places. WWT is in close contact with many of the organisations and individuals undertaking such counts, and can provide the information and contact addresses appropriate for the species, region or census concerned.

Specific notes on individual species are provided below. These may aid survey planning where particular species predominate:

> Divers are normally spread very thinly over large and often remote areas. **Red-** and **black-throated divers** often use sandy embayments; **great northern divers** may be found off rocky headlands and considerable distances offshore.
>
> **Red-necked**, **Slavonian** and **black-necked grebes** occur in small numbers at traditional sites, usually sheltered sandy bays. **Great crested grebes** may occur in large concentrations, particularly on some estuaries, with generally small numbers off open coasts. **Little grebes** are rare in coastal waters.
>
> **Scaup** normally congregate in large flocks in estuaries and sheltered bays, much less often on open coasts. Some notable concentrations occur around outfalls. **Eiders** are often found relatively close to shore, and observers may need to look over cliff tops or into small inlets to ensure none are missed; note also that tidal and diurnal changes in numbers at sites are well documented for this species and it is important to time the count appropriately. **Long-tailed ducks** are often found much further offshore than other species, spread over wide areas in small groups; they are frequently seen flying in small groups during the day, presenting problems when counting, and in the Moray Firth counts of birds flying to or from roosts have proved the most effective means of obtaining accurate counts. **Scoters** generally favour sandy substrates, and may occur in large, dense flocks. Birds may be mobile, and move frequently between adjacent sites during the course of a winter. Flocks may be a considerable distance from shore, although velvet scoter occur relatively close to the shore at some sites. Like scaup, **goldeneye** favour shallow waters, but have a relatively localised distribution in UK waters This includes some notable concentrations, particularly near outfalls. **Red-breasted mergansers** generally occur very close inshore, generally in dispersed groups on sandy, sheltered estuaries and bays, but are also found along open coasts.

Unless there is good reason to do otherwise, count areas should match established boundaries used for WeBS Core Counts. Alternatively, provide a map of the count boundaries used. The seaward boundary of the count area should extend as far as it is possible to see, even though this may differ on different visits; the range of visibility should be recorded on the form.

Establish suitable (normally high) vantage points from which to count; these should be used on each count. If relevant, record details of these observation positions (location and compass direction of observation). Due to the difficulty of determining the area already counted when moving to a new vantage point, use reference points such as land features on the far side of large bays or estuaries, and buoys or boats on open coasts. Alternatively, obvious gaps in the sea-duck flocks may be used or, if the flock is spread over a large distance, distinctive groups of birds, eg a tight group of rarer grebes or divers. A compass may also be useful in determining the position of flocks and whether or not they have been counted previously.

Spend sufficient time counting from each position to ensure that thorough checks have been made for the different species (see below). Often, it may be necessary to make at least one scan far from shore to record large sea-duck flocks and divers, and one close to shore for grebes and mergansers. A dictaphone can be particularly useful when making a slow scan of mixed-species flocks. Try to ensure that your count is co-ordinated with others being carried out on adjacent stretches of coast.

Only those wildfowl actually offshore from a section should be included within the counts for that section. Do not record birds simply flying through the count area. Under ideal conditions, many species can be identified at considerable distances. Nevertheless, it is probable that a proportion of birds will remain unidentified. These should be recorded to species group as far as possible, eg scoter spp, small grebe spp, though the most distant birds may simply have to be recorded as sea-duck spp. Beware of the presence of auks at some sites. Secondary identification features may help with distant birds, eg diving behaviour.

Counts of these species sent to WWT will be included in the annual WeBS report on waterfowl monitoring in the UK.

For further details, contact: WeBS Secretariat, The Wildfowl & Wetlands Trust, Slimbridge, Gloucestershire GL2 7BT; tel 01453 890333.

Contributed by Peter Cranswick

Waterfowl and seabirds at sea

Brief instructions on how to conduct aeroplane and ship surveys of waterfowl and seabirds offshore are give here. Before undertaking such surveys, you are strongly advised to read Komdeur et al (1992) which is a much more detailed guide.

1. Aeroplane surveys

Advice on species identification from the air can be found in Komdeur et al (1992).

Information required

- total number of all species.

See *WeBS Core Counts* (above) for additional information.

Number and timing of visits

Cover the whole area to be surveyed once.

Time of day

Any time of day.

Weather constraints

Ideal conditions are calm winds and sea, a complete but light cloud cover at high altitude, and visibility of more than 10 km.

Sites/areas to visit

Any wetland habitat, but particularly areas of sea not visible from the shore.

Equipment

- recording form
- GPS (optional)
- map
- dictaphone.

See *WeBS Core Counts* (above) for additional information.

Safety reminders, Disturbance

See Komdeur et al (1992).

Methods

The count boundaries should be determined using appropriate features or a GPS and recorded on a map enabling the count to be repeated exactly in the future. The quality of estimates of flock size from an aeroplane varies according to weather conditions, the bird species, flock density, distance of the flock from the aeroplane and the mood and ability of the observer. Aerial surveys allow quick coverage of large waterbodies and it is easy to move between lakes, meadows, coastal stretches and offshore waters. However, the high speed of the aircraft leaves the observer with only a very short time for each observation and no time to record additional information on sex or age ratios and behaviour. The short time makes accurate counting of flocks difficult, and some flocks remain unidentified. The chance of spotting rarer species is also reduced, especially when they are mixed with common ones.

Two methods have been developed for surveying waterfowl and seabirds by aeroplane, as follows.

Total survey

Split the survey area into manageable, easy-to-recognise sections (for narrow waters, 1–30 km^2; open coast, 20–80 km^2; offshore areas, 40–200 km^2). Fly along the coast at a distance of about 300–400 m, counting all birds. Repeat this at increasing distances from the coast with about 2–3 km between the plane's lines of flight; this distance is probably large enough to avoid counting the same flocks twice. Record flocks directly onto a map and record all other information directly into a dictaphone. Summarise the results as the total number of individuals of each species in each area.

Transect survey

This method requires navigational equipment (preferably for satellite navigation) on the aeroplane and relies on all birds within a given transect band being counted. Decide on the overall area of the study site. Plan the route in detail in advance. Decide on the number of transects that will be flown within the study site, and the transect width to be adopted. This should be about 180 m in areas with a great variety of species or 280 m where sea-ducks are the main focus. Check the transect width by flying over a training area which has the width marked out on the ground. Maintain the same height above the ground that will be adopted during the survey and mark the aircraft windows in such a way that the width of the transect can be correctly judged from the air. Calibrate the navigational equipment and, at the start of each transect, record the time, starting position and direction in which the transect is to be flown. It is generally preferable to fly from north to south to avoid sun glare. Allow the dictaphone to record without any breaks and record all chosen species within the transect area as well as their numbers and the side of the aircraft on which they were observed. As the number of each species in the transect band is known, the density of each can be readily calculated and the number in the total survey area estimated. The more transects that are flown, the better will be the estimates of density and thus of overall population sizes.

2. Ship surveys

A small boat with an outboard motor has been used successfully to count the large eider flock in the mouth of the Tay. However, the shape and size of other sites may render this method inappropriate. Ship surveys are also useful for wintering divers, which are often several kilometres offshore.

Information required

In order of priority:

- number of individuals present 'in transect' or 'not in transect'
- species or taxon
- behaviour (on sea or flying)
- distance from the ship
- plumage moult, age and sex
- approximate flight direction.

Number and timing of visits

Cover the whole area to be surveyed once.

Time of day, Weather constraints

As above.

Sites/areas to visit

Any wetland habitat, but particularly areas of sea not visible from the shore.

Equipment

- recording form.

See *WeBS Core Counts* (above) for additional information.

Safety reminders, Disturbance

See Komdeur et al (1992).

Methods

Three methods have been used for counting birds from ships, as follows.

(a) Scan

Count all birds seen ahead of the ship, on the water or in flight, within a 90° or 180° scan. Information from the count produces an index of abundance of birds seen per 10 minutes of observation time. Convert results, according to the ship's speed, to birds recorded per km travelled.

(b) Scan with band transect

As (a) but, additionally, note those birds which occur on the water within a band transect (usually 300 m wide) on one side of the ship, as being 'in transect'. This technique attempts to quantify the numbers of birds seen per km^2 surveyed. However, it deals ineffectively with birds in flight. With this survey, all birds on the water which are seen within the band are recorded as being 'in transect', whereas birds in flight are not.

(c) Scan with band transect using 'snapshot technique'

As (b) but additionally include all birds seen in flight within the band transect as being 'in transect'. This method is the recommended standard for use in European waters.

Reference

Komdeur, J, Bertelsen, J, Cracknell, G (eds) (1992) *Manual for Aeroplane and Ship Surveys of Waterfowl and Seabirds*. IWRB Spec. Publ. 19. Slimbridge.

Productivity of swans and geese

Information required

- total number of birds in the flock
- total number of birds aged
- total number of young
- brood sizes
- date and time
- survey boundary
- flock locations (including OS grid reference)
- habitat
- name of observer.

Number and timing of visits

No set number. Ideally, at least three visits to areas supporting large numbers of birds. Count period varies between species, as follows:

Whooper swan	Early November to late January
Bewick's swan	Early December to late January
Pink-footed goose	Mid-October to mid-November
Bean goose	Early November to late December
European white-fronted goose	Mid-October to late December
Greenland white-fronted goose	December
Icelandic greylag goose	Late October to mid-December
North Scottish greylag goose	August
Greenland barnacle goose	Late October to late December
Svalbard barnacle goose	Mid-November to early December
Dark-bellied brent goose	Mid-October to mid-December
Canadian/Irish light-bellied brent goose	Mid-October to late November
Svalbard light-bellied brent goose	Late October to late December
Naturalised/introduced geese	Generally only undertaken in conjunction with national surveys (see *WeBS Core Counts* and *National censuses of geese*, above)

Time of day

Counts of birds in fields or on estuaries should be made during the day, ideally at least one hour after dawn and one hour before dusk.

Sites/areas to visit

Any open wetland, agricultural or non-wetland habitat, including temporary grasslands and temporarily flooded areas.

Equipment

- window-clamp or bean-bag if using a telescope from a car.

See *WeBS Core Counts* (above) for additional information.

Safety reminders

See *WeBS Core Counts* and *National censuses of geese*.

Disturbance

See *WeBS Core Counts*. Swans and geese feeding in fields can often be approached more readily using a car than on foot.

Methods

Ideally, counts should commence only when the majority of the population has arrived, to take into account differences in arrival dates for adults with and without young. Generally, agricultural land supports a higher proportion of young. Ideally, counts should be representative of the distribution of birds or, alternatively, delayed until fields have been ploughed and flocks are using just one or relatively few habitat types. As winter progresses, the proportion of young for many species will fall due to differential mortality between juveniles and adults, particularly for quarry species; in addition, young birds become increasingly difficult to differentiate from adults beyond mid-winter. For other species, the proportion of young may rise due to the arrival of birds from other parts of the wintering range. Thus, the count period should allow some assessment of these factors.

Most of the species concerned breed in Greenland/Iceland or in arctic Russia/Europe. A longer period of assessment should be allowed for those breeding in Russia/Europe, since a large proportion of the population may stage on the continent, especially in the Netherlands, before moving to Britain in mid-winter. In contrast, birds breeding in Greenland/Iceland tend to arrive in the UK *en masse* in early autumn.

Map the boundary of the area searched, and record the date and time of the survey and the route taken – which should allow coverage of the whole area. It is not usually possible to assess the proportion of young and brood sizes at the same time; separate scans of the flock are required to obtain these data. Assessment of the overall proportion of young should take priority over brood-size assessment.

The proportion of young varies according to date, habitat and flock size, eg more young are found on the edge of feeding flocks in fields, so all of these variables should be recorded for the results to be properly interpreted. Try to assess a range of flock sizes on a range of habitats. Ideally, the whole of each flock should be aged. However, if this is not possible, record the flock size and ensure that you spread your observations throughout the flock, so that they are representative.

To assess the proportion of young, scan slowly through the flock using a telescope to identify juvenile birds as follows:

> Juvenile **whooper** and **Bewick's swans** have grey plumage, some of which will be retained well into the following calendar year. Juvenile Bewick's and whooper swans have pink bills (not yellow as in adults), turning quickly to chalk-white in whooper swans.
>
> Juvenile **grey geese** have the wing coverts narrower and more rounded/pointed; on adults, these feathers are larger and square-ended. In addition, juvenile **pink-footed geese** show mottling on the breast, neck and belly in good light. Juvenile **greylag geese** have a dark nail on the bill (not pink as in adults). Juvenile **white-fronted geese** lack the white blaze on the forehead characteristic of adults, at least until mid-winter; beyond this time, a bird without a white forehead is a juvenile, but birds with white may be adults or juveniles. Similarly, juveniles lack black belly-bars until at least mid-winter.
>
> Juvenile **barnacle geese** lack a distinct white trailing edge to the greater coverts and show a contrast between a black neck and grey mantle (both neck and mantle are black in adults).
>
> Juvenile **brent geese** are easily separated by extensive white barring on the folded wings, formed by white tips to the coverts; adults lack the white tips and have all-dark wings.

Juvenile **Canada geese** quickly moult body feathers so that they resemble adults very closely after September, making age assessments in the field impractical.

NB Unlike the skills needed in the monitoring of many bird species, the ageing of geese, and the assessing of brood size, is not straightforward. A degree of experience is necessary before assessments can be made with confidence, and observers should practise these techniques before contributing data to national schemes.

Count the number aged (eg on a tally counter) and the number of young (eg in your head) as you scan through the flock. Do not spend time struggling with birds which, for example, are partly obscured or show some intermediate characters; these are best left unaged.

To assess brood size, watch birds until you are confident that particular young and adult birds are associating. Record brood sizes only for families with two adults; otherwise, you cannot be certain that the whole family is present. Again, spread your observations throughout the flock.

Location and central grid reference should be provided for birds on agricultural land or other non-wetland habitat, although, since birds may range widely in this habitat, demarcation of flock boundaries may be difficult.

Completed records should be returned to WWT or the appropriate organisation after the last count is made.

Useful addresses

As for *National censuses of geese* (above).

Pinewood bird survey

This is a detailed transect method which has been used by the RSPB over a number of winters and was developed as part of the Pinewoods Birds Project (RSPB/ITE/Game Conservancy; Buckland and Summers 1992).

The method can be used to estimate the density of woodland birds in winter, in particular birds which are difficult to survey accurately in the breeding season. For example, it has been used to survey capercaillie, crossbills (Summers and Ellis 1994) and woodcock (Summers and Buckland 1996).

Densities and population sizes of each species are estimated using the distance sampling methods of Buckland et al (1993). Even though the specially written software package DISTANCE (Laake et al 1993) is available to help undertake these analyses, they can become complex, and specialist help may be required.

Information required

- total number and density of birds of both sexes, by habitat type.

Number and timing of visits

One or more visits, between November and March. Visits should be at least a month apart.

Time of day

0900–1600 BST.

Weather constraints

None.

Sites/areas to visit

Areas of mature open pinewood.

Equipment

- 1:25,000 OS map of the area to be visited
- compass
- prepared recording form
- tape measure.

Safety reminders

Nothing specific. See *Health and safety guidelines* in the *Introduction.*

Disturbance

Keep disturbance to a minimum.

Methods

This method can be used for a variety of pinewood species which are difficult to survey during the breeding season, eg crested tit, crossbill, capercaillie and woodcock. A single survey can encompass all these species.

If necessary, stratify sampling according to habitat type, eg as native pinewood, non-native pine woodland, and other woodland. The location of transects in a particular stratum/habitat, eg native pinewood, should be selected systematically. For example, every 1-km square that has the grid reference xx8yy5 and lies within native pinewood can be chosen, or (for a larger area) alternate 1-km squares with that grid reference. A sufficient number of transects need to be walked to obtain about 50 registrations of (ideally) each species, though this may not always be practical.

Transects should be triangular in shape. After choosing the six-figure grid reference points, locate them on the map. These points should lie halfway along the side of an equilateral triangle, the length of whose sides total 2 km (the transect length). Adopt a standard method for locating the transect around each point, eg always have the triangle pointing north, or choose a random angle from the point for the first leg.

The entire transect may not always fall within woodland. A transect which runs out of woodland should have extensions along the length of one of its arms so that a full 2 km of woodland is still surveyed for all transects. However, non-woodland areas within 50 m of identified woodland boundaries should still be included in the wooded category. Split the 2-km length of the transect into 20 numbered 100-m segments. This makes it easier to record the positions of any birds seen, and there is a column for this on the recording form (Figure 4). If possible, place permanent markers at the corners of each transect and at the divisions between the different sections.

Prepare recording forms (Figure 4) and maps for each transect.

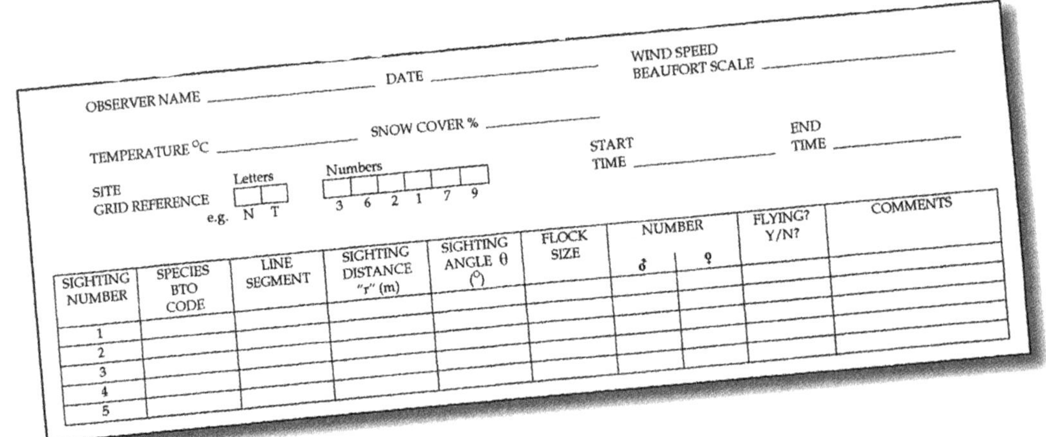

OBSERVER NAME ______ DATE ______ WIND SPEED BEAUFORT SCALE ______

TEMPERATURE °C ______ SNOW COVER % ______ START TIME ______ END TIME ______

SITE GRID REFERENCE Letters [ ][ ] e.g. N T Numbers [ ][ ][ ][ ][ ][ ] 3 6 2 1 7 9

SIGHTING NUMBER	SPECIES BTO CODE	LINE SEGMENT	SIGHTING DISTANCE "r" (m)	SIGHTING ANGLE θ (°)	FLOCK SIZE	NUMBER ♂	NUMBER ♀	FLYING? Y/N?	COMMENTS
1									
2									
3									
4									
5									

Figure 4
An example of the type of recording form for bird recording during winter transects (method 1) for surveys of capercaillie, Scottish crossbill and crested tit. Use the standard BTO codes for species names (see Appendix 1).

On the map, record the habitat and transect route and, if necessary, adjust the transect to the different types of habitat that fall along it. Record enough information to allow any future fieldworker to identify section end-points. Take the map on each visit to the transect and record any birds seen and heard relative to the numbered segments of the transect. This part of the recording form is the same for each species. Use standard BTO species codes (Appendix 1).

On the form, record each sighting, the segment number the observer was in, the number of birds seen, their sex (where possible), the angle q of the bird(s) from the transect and the sighting distance *r* between the bird and the observer (see Figure 5). Sighting numbers should be cross-referenced between the recording form and a mapped transect. Measure the angle q accurately (to 1°) using a compass, and pace the distance *r*, then return to your original position on the transect line.

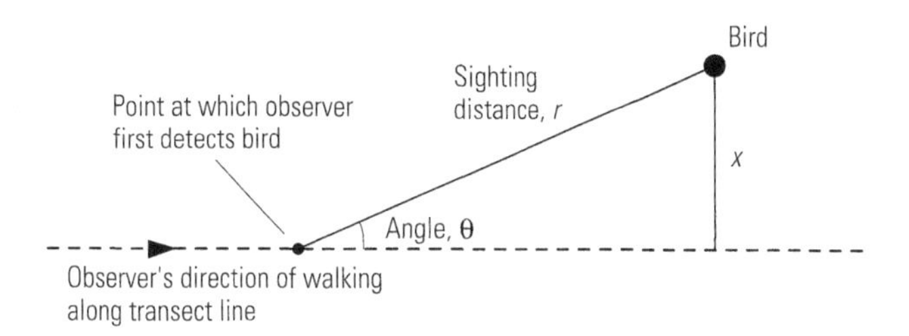

Figure 5
A plan view of the relationship between angles and distances that should be recorded during transects. q and *r* are recorded in the field, *x* is not.

Record whether any birds were flying and their distance from the line when they passed 'abeam' (in a line perpendicular to the observer). If they are out of detection range at the time they pass abeam, record the perpendicular distance to the transect as 'U'. If they land before passing abeam, record where they land. If they land behind, record them as 'B'.

Follow the course of the transect closely, even in dense conifer plantations. If it proves impossible to penetrate dense stands and at the same time to detect most birds close by (say within 10 m), skirt the densest portions and mark any deviations on the map. Enter the raw data (number of birds, distance *r* and angle q) into the Laake et al (1993) DISTANCE program to give bird densities. Multiply the total area of the study site by the density to give the population size. If stratified by habitat, do this separately for each stratum and sum the individual totals. If you do not have the DISTANCE program, the density can be calculated using the methods outlined in Buckland et al (1993).

References

Buckland, S T and Summers, R W (1992) *Pinewoods Birds Project: Protocol for a Sample Survey of Conifer Woods.* Unpubl.

Buckland, S T, Anderson, D R, Burnham, K P and Laake, J L (1993) *Distance Sampling: Estimating Abundance of Biological Populations.* Chapman and Hall, London.

Laake, J L, Buckland, S T, Anderson, D R and Burnham, K P (1993) *Distance User's Guide Version 2.0.*

Summers, R W and Ellis, J A (1994) *An Assessment of the Scottish Crossbill, Capercaillie and Crested Tit Populations and Woodland Habitat in Kinveachy Forest SSSI.* RSPB unpubl.

Summers, R W and Buckland, S T (1996) Numbers of wintering woodcock *Scolopax rusticola* in woodlands in the Highlands of Scotland. *Wader Study Group Bull.* 80: 50–52.

Appendix 1: BTO recording codes

BTO standard species recording codes

	BTO code
Red-throated diver	RH
Black-throated diver	BV
Great northern diver	ND
Little grebe	LG
Great crested grebe	GG
Red-necked grebe	RX
Slavonian grebe	SZ
Black-necked grebe	BN
Fulmar	F.
Cory's shearwater	CQ
Great shearwater	GQ
Sooty shearwater	OT
Manx shearwater	MX
Storm petrel	TM
Leach's petrel	TL
Gannet	GX
Cormorant	CA
Shag	SA
Bittern	BI
Little bittern	LL
Grey heron	H.
Purple heron	UR
White stork	OR
Spoonbill	NB
Mute swan	MS
Bewick's swan	BS
Whooper swan	WS
Bean goose	BE
Pink-footed goose	PG
White-fronted goose	WG
Greenland white-fronted goose	NW
European white-fronted goose	EW
Greylag goose	GJ
Snow goose	SJ
Canada goose	CG
Barnacle goose	BY
Brent goose	BG
Dark-bellied brent goose	DB
Pale-bellied brent goose	PB
Egyptian goose	EG
Shelduck	SU
Mandarin duck	MN
Wigeon	WN
Gadwall	GA
Teal	T.
Mallard	MA
Pintail	PT
Garganey	GY
Shoveler	SV
Red-crested pochard	RQ

	BTO code
Pochard	PO
Ferruginous duck	FD
Tufted duck	TU
Scaup	SP
Eider	E.
Long-tailed duck	LN
Common scoter	CX
Velvet scoter	VS
Goldeneye	GN
Smew	SY
Red-breasted merganser	RM
Goosander	GD
Ruddy duck	RY
Honey buzzard	HZ
Red kite	KT
White-tailed eagle	WE
Marsh harrier	MR
Hen harrier	HH
Montagu's harrier	MO
Goshawk	GI
Sparrowhawk	SH
Buzzard	BZ
Rough-legged buzzard	RF
Golden eagle	EA
Osprey	OP
Kestrel	K.
Merlin	ML
Hobby	HY
Gyr falcon	YF
Peregrine	PE
Red grouse	RG
Ptarmigan	PM
Black grouse	BK
Capercaillie	CP
Red-legged partridge	RL
Grey partridge	P.
Quail	Q.
Pheasant	PH
Golden pheasant	GF
Lady Amherst's pheasant	LM
Water rail	WA
Spotted crake	AK
Corncrake	CE
Moorhen	MH
Coot	CO
Crane	AN
Oystercatcher	OC
Black-winged stilt	IT
Avocet	AV
Stone-curlew	TN
Little ringed plover	LP

cont.

	BTO code
Ringed plover	RP
Kentish plover	KP
Dotterel	DO
Golden plover	GP
Grey plover	GV
Lapwing	L.
Knot	KN
Sanderling	SS
Little stint	LX
Temminck's stint	TK
Pectoral sandpiper	PP
Curlew sandpiper	CV
Purple sandpiper	PS
Dunlin	DN
Buff-breasted sandpiper	BQ
Ruff	RU
Jack snipe	JS
Snipe	SN
Woodcock	WK
Black-tailed godwit	BW
Bar-tailed godwit	BA
Whimbrel	WM
Curlew	CU
Spotted redshank	DR
Redshank	RK
Greenshank	GK
Green sandpiper	GE
Wood sandpiper	OD
Common sandpiper	CS
Turnstone	TT
Red-necked phalarope	NK
Grey phalarope	PL
Pomarine skua	PK
Arctic skua	AC
Long-tailed skua	OG
Great skua	NX
Mediterranean gull	MU
Little gull	LU
Sabine's gull	AB
Black-headed gull	BH
Ring-billed gull	IN
Common gull	CM
Lesser black-backed gull	LB
Herring gull	HG
Iceland gull	IG
Glaucous gull	GZ
Great black-backed gull	GB
Kittiwake	KI
Sandwich tern	TE
Roseate tern	RS
Common tern	CN
Arctic tern	AE
Little tern	AF
Black tern	BJ
Guillemot	GU
Razorbill	RA
Black guillemot	TY

	BTO code
Little auk	LK
Puffin	PU
Rock dove	DV
Feral pigeon	FP
Stock dove	SD
Woodpigeon	WP
Collared dove	CD
Turtle dove	TD
Ring-necked parakeet	RI
Cuckoo	CK
Barn owl	BO
Snowy owl	SO
Little owl	LO
Tawny owl	TO
Long-eared owl	LE
Short-eared owl	SE
Nightjar	NJ
Swift	SI
Kingfisher	KF
Hoopoe	HP
Wryneck	WY
Green woodpecker	G.
Great spotted woodpecker	GS
Lesser spotted woodpecker	LS
Woodlark	WL
Skylark	S.
Shore lark	SX
Sand martin	SM
Swallow	SL
House martin	HM
Richard's pipit	PR
Tawny pipit	TI
Tree pipit	TP
Meadow pipit	MP
Rock pipit	RC
Water pipit	WI
Yellow wagtail	YW
Grey wagtail	GL
Pied wagtail	PW
Waxwing	WX
Dipper	DI
Wren	WR
Dunnock	D.
Robin	R.
Nightingale	N.
Bluethroat	BU
Black redstart	BX
Redstart	RT
Whinchat	WC
Stonechat	SC
Wheatear	W.
Ring ouzel	RZ
Blackbird	B.
Fieldfare	FF
Song thrush	ST

cont.

	BTO code
Redwing	RE
Mistle thrush	M.
Cetti's warbler	CW
Grasshopper warbler	GH
Savi's warbler	VI
Aquatic warbler	AQ
Sedge warbler	SW
Marsh warbler	MW
Reed warbler	RW
Icterine warbler	IC
Melodious warbler	ME
Dartford warbler	DW
Barred warbler	RR
Lesser whitethroat	LW
Whitethroat	WH
Garden warbler	GW
Blackcap	BC
Yellow-browed warbler	YB
Wood warbler	WO
Chiffchaff	CC
Willow warbler	WW
Goldcrest	GC
Firecrest	FC
Spotted flycatcher	SF
Red-breasted flycatcher	FY
Pied flycatcher	PF
Bearded tit	BR
Long-tailed tit	LT
Marsh tit	MT
Willow tit	WT
Crested tit	CI
Coal tit	CT
Blue tit	BT
Great tit	GT
Nuthatch	NH
Treecreeper	TC
Short-toed treecreeper	TH

	BTO code
Golden oriole	OL
Red-backed shrike	ED
Great grey shrike	SR
Jay	J.
Magpie	MG
Chough	CF
Jackdaw	JD
Rook	RO
Carrion crow	C.
Hooded crow	HC
Raven	RN
Starling	SG
House sparrow	HS
Tree sparrow	TS
Chaffinch	CH
Brambling	BL
Serin	NS
Greenfinch	GR
Goldfinch	GO
Siskin	SK
Linnet	LI
Twite	TW
Redpoll	LR
Common crossbill	CR
Scottish crossbill	CY
Parrot crossbill	PC
Scarlet rosefinch	SQ
Bullfinch	BF
Hawfinch	HF
Lapland bunting	LA
Snow bunting	SB
Yellowhammer	Y.
Cirl bunting	CL
Ortolan bunting	OB
Reed bunting	RB
Corn bunting	CB

BTO standard activity recording codes

The standard BTO list of conventions is designed to help you make your field notes clear and unambiguous, and the following are examples of their use. Symbols can be combined where necessary (see *Bird Census Techniques*). Additional activities of territorial significance, such as display or mating, should be noted using an appropriate clear abbreviation. In all cases the standard BTO species recording codes (see above) should be used.

CH juv CH♂ CH♀	Chaffinch sight records, with age, sex or number of birds if appropriate. Use CH⚥ to indicate one pair of chaffinches, so that 2CH⚥ means two pairs together.
R. fam	Juvenile robins with parent(s) in attendance.

R.	A calling robin.
R.	A robin repeatedly giving alarm-calls or other vocalisations (not song) thought to have strong territorial significance.
(R.)	A robin in song.
R. R.	An aggressive encounter between two robins.
*R.	An *occupied* nest of robins; do not mark unoccupied nests, which are of no territorial significance by themselves.
[*] BT	Blue tits nesting in a specially provided site (eg nest-box). Please remember to use this special symbol for a nest in a tit box.
*PW on	Pied wagtail nest with an adult sitting.
PW mat	Pied wagtail carrying nest material.
PW food	Pied wagtail carrying food.

Movements of birds can be indicated by an arrow using the following conventions:

—GR→	A calling greenfinch flying over (seen only in flight).
(D.) →	A singing dunnock perched, then flying away (not seen to land).
→B.♂	A male blackbird flying in and landing (first seen in flight).
WR——→WR	A wren moving between two perches. The solid line indicates it was definitely the same bird.

The following registrations indicate when registrations relate to different birds, and when to the same bird. Their proper use is essential for the accurate assessment of territorial clusters.

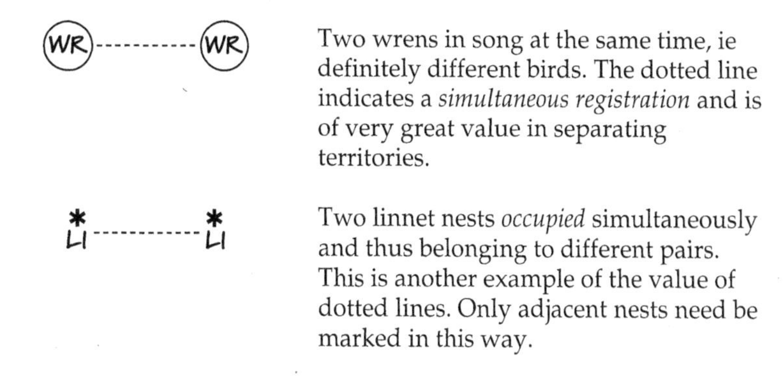

The solid line indicates that the registrations definitely refer to the same cuckoo.

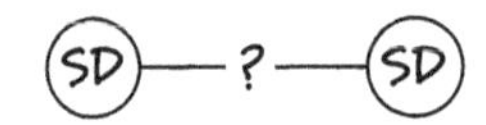

A question-marked solid line indicates that both registrations of stock dove *probably* relate to the same bird. This convention is of particular use when the census route returns to an area already covered – it is possible to mark new positions of (probably the same) birds recorded before, without the risk of double-recording. If birds are recorded without using the question-marked solid line, overestimation of territories will result.

WR   WR mat

No line joining the registrations indicates that these wrens are *probably* different, but depending on the pattern of other registrations they may be treated as if only one bird was involved (you may if you wish use a question-marked dotted line, indicating that the registrations were almost certainly of different birds).

C.*   C.*

Where adjacent nests, such as these of carrion crow, are marked without a line, it will often be assumed that they were first and second broods, or a replacement nest following an earlier failure.

Lightning Source UK Ltd.
Milton Keynes UK
UKHW050949270819
348658UK00004B/6/P